Drones and Unmanned Aerial Systems

Aleš Završnik
Editor

Drones and Unmanned Aerial Systems

Legal and Social Implications for Security and Surveillance

Springer

Editor
Aleš Završnik
Institute of Criminology at the Faculty of Law
Ljubljana
Slovenia

ISBN 978-3-319-23759-6 ISBN 978-3-319-23760-2 (eBook)
DOI 10.1007/978-3-319-23760-2

Library of Congress Control Number: 2015951633

Springer Cham Heidelberg New York Dordrecht London

Printed on acid-free paper

Springer International Publishing AG Switzerland is part of Springer Science+Business Media (www.springer.com)

Acknowledgments

The editor wishes to express his gratitude to those individuals and institutions that have made this book possible.

I have greatly profited from the constructive atmosphere with the coauthors of this book at the international conference "Spy in the Sky: Regulatory Issues of Drones and Unmanned Aerial Systems" organised at the Faculty of Law of the University of Ljubljana in May 2013 in Ljubljana (Slovenia). This was one of the first, if not the first, conferences worldwide dedicated to regulatory aspects of drones. This was the period of the initial political decisions of governments around the world that drones should be integrated into the civilian space and the time many of us realised that drones are here to stay. I also greatly appreciate what I gained from my coauthors, who were in many ways much more versed in this area, and also from my colleagues at the Institute of Criminology at the Faculty of Law—I owe many ideas to our inspiring and sometimes necessarily turbulent conversations.

My greatest thanks go to all my coauthors. I met some of them for the first time at the conference. They have been courageous enough to trust the editing of the book to me. I have known others for a while, and I am happy to be able to work with them on a common book as I admire and appreciate their work and thoughts.

I would like to thank the Slovenian Research Agency for awarding a postdoctoral grant that enabled me to research the topic and organise a conference in Ljubljana. My thanks go also to the Institute of Criminology in Ljubljana and its Director, Professor Matjaž Jager, who showed full understanding of the book.

Finally, I would like to thank the team at the publisher for being very supportive and encouraging the whole process of preparing this book throughout.

Ljubljana, 4 June 2015 Aleš Završnik, Editor

Contents

Contributors

Mark Andrejevic Department of Media Studies, Pomona College, Claremont, USA

Vasja Badalič Institute of Criminology at the Faculty of Law, Ljubljana, Slovenia

David Goldberg Glasgow, Scotland

Primož Gorkič Faculty of Law, University of Ljubljana, Ljubljana, Slovenia

Mélanie De Groof The Group for Research and Information on Peace and Security (GRIP), Brussels, Belgium

Kamila Krajčíková Heinrich-Heine-Universität Düsseldorf, Düsseldorf, Germany

Luisa Marin Department of Law, School of Law, Economics and Governance, Utrecht University, Utrecht, The Netherlands

Institute for Governance Studies, University of Twente, Enschede, The Netherlands

Pablo Mendes de Leon International Institute of Air and Space Law, Leiden University, Leiden, Netherlands

Sanja Milivojevic School of Social Sciences, University of New South Wales, Sydney, Australia

Kristin Bergtora Sandvik Norwegian Centre for Humanitarian Studies, Peace Research Institute Oslo (PRIO), Oslo, Norway

Benjamyn Ian Scott International Institute of Air and Space Law, Leiden University, Leiden, Netherlands

Aleš Završnik Institute of Criminology at the Faculty of Law, Ljubljana, Slovenia

About the Editor

Asst. Prof. Aleš Završnik Doctor of Law (LL.D.) is a researcher at the Institute of Criminology at the Faculty of Law in Ljubljana. He was a postdoctoral Yggdrasil fellow at the Department of Criminology and Sociology of Law at University of Oslo (2012) and a postdoctoral fellow at the Max-Planck-Institut für Ausländisches und Internationales Strafrecht, Freiburg im Breisgau (2009) and a fellow of the World Federation of Scientists, Geneva (2008). He was collaborating with the European Cooperation in Science and Technology (COST) Action "Living in Surveillance Societies" and the COST Action "Cyberbullying". Aleš Završnik edited a book *Crime and Technology: How Computers Transform Surveillance and Privacy, Crime and Crime Control?* (2010), published a book *Homo Criminalis: Images of a Criminal in High-Tech Risk Society* and *Cybercrime* (2015). He has researched and published on theoretical criminology, criminal law and "law and technology", i.e. cybercrime, cyberlaw and surveillance. He is a member of the editorial board of the *Masaryk University Journal of Law and Technology*.

Chapter 1
Introduction: Situating Drones in Surveillance Societies

Aleš Završnik

The object of this book is to shed light on the growing regulatory framework for "drones", on the one hand, and on the accompanying moral and political landscape and knowledge production process related to drones on the other. The regulatory framework relevant for drones is multifaceted, and the book sheds light on the relevant aspects of privacy and personal data protection law, aviation law, public international law, media law and migration law—in several countries. The charted moral and political landscape of drones employed in various contexts (e.g. in war and in peace, by commercial and noncommercial users, in national and international contexts) offers insight into the wider picture of "security theatre" (Schneier 2003) and "surveillance games" in the contemporary "big data" society. In more general terms, drones are also used as a point of departure for mapping the knowledge production process in contemporary societies: Should we ground political decisions on social, psychological and anthropological considerations ("deep ontology"), or should we ground such decisions on the stockpile of data ("on the surface") gathered from various sensing and recording devices, including drones? While the book dissects the complex and geographically and substantially diverse regulatory frameworks, it sheds light on regulatory processes permeated with the contrasting interests and values of multiple stakeholders. In such a manner, the book sheds lights on both the legal and social implications of "drones" for security and surveillance.

As definitions are crucial for any regulation, discussions about "drones" in the book are, more precisely, discussions about "unmanned aircrafts", defined as aerodynamic flying systems that can be piloted remotely via a joystick or digital interface supported by different levels of automatic control (Houses of Parliament 2014). As such aircrafts are supported by infrastructure, such as control stations and data links, the book likewise uses the term "unmanned aircraft systems" (UAS),

A. Završnik (✉)
Institute of Criminology at the Faculty of Law, Poljanski nasip 2,
SI-1000 Ljubljana, Slovenia
e-mail: ales.zavrsnik@pf.uni-lj.si

A. Završnik (ed.), *Drones and Unmanned Aerial Systems*,
DOI 10.1007/978-3-319-23760-2_1

and, moreover, "aircraft systems" as in most cases aircrafts are not yet completely unmanned, but with operators on the ground instead of on board; thus, policymakers have also termed the technology "remotely piloted aircraft systems—RPAS" (European Commission (EC) 2012).

Drones come in all sorts of sizes, ranging from robotic moths and hummingbirds, to Boeing's Phantom Eye with its 150-ft wingspan that can cruise at 65,000 ft for up to 4 days. They come in all sorts of shapes (helicopter-like with rotary blades, airplane-like with fixed wings, zeppelin- or balloon-like and insect- and bird-mimicking devices—for the latter see, e.g., the Nano unmanned aerial vehicle (UAV) "Black Hornet", BBC 2013) and are used for a variety of purposes: from killing people on the basis of their abstract behaviour ("signature strikes") in Yemen, to filming wedding ceremonies in churchyards—and everything in between, for example, peering into hurricanes (e.g. NASA Global Hawk, Gannon 2014), supporting construction site planning and mapping cemeteries (Crowe 2015). They are used for commercial and noncommercial purposes. Ordinary citizens, hobbyists and do-it-yourself (DIY) enthusiasts use them; governments use them to execute "adversaries" in remote areas of the planet and for search and rescue operations; and corporations employ them to deliver pizzas or parcels (DHL and Amazon being at the forefront of this game, see Hern 2014 Amazon 2013). Drones are used for "registering" data (by looking, measuring, smelling or otherwise sensing airborne chemicals, pathogens and radioactive materials, or for reconnaissance) and for carrying payloads (letters, medicine, blankets for refugees or bombs). They can fly alone or in swarms (D'Andrea 2013). They can keep track of anything or anyone equipped with a radio frequency identification tag (wildlife, livestock, a suspect's vehicle) and, increasingly, also anything or anyone without a tag, based solely on "smart" vision. They can be visible or "hidden in plain sight", such as drones that mimic the size and behaviour of bugs and birds that are indistinguishable from the real thing (Spaeth 2013). With the advent of new applications, it would seem, at least for the time being, that the list is endless.

What makes drones an interesting object of inquiry is that they have triggered diametrically opposing reactions—enthusiastic acceptance on the one hand and relentless opposition on the other. From the first perspective, drones are perceived as technology that will transform the way we live and work. In her keynote address at a conference on remotely piloted aircraft systems that adopted the "Riga Declaration on Civil Drones" (6 March 2015), the European Union (EU) Commissioner for Transport, Violeta Bulc, stated that drones will change our lives and businesses (Bulc 2015). Drones are here to stay, she stated. They are supposed to boost creativity and create jobs. The EC (2012), for instance, claims drones will be "creating new markets of aerial services, in the same way that the iPad created an entirely new and unpredicted market for mobile data services" (EC 2012). Similarly, Internet giants, such as Google (Etherington 2014) and Facebook with its *Connectivity Lab* (Lapowsky 2014) are mapping out a very bright future for drones, and they have acquired companies developing drones as proxy satellites for delivering Internet connectivity to remote places.

Others perceive drones in a dystopian fashion. One commentator in the *New York Daily* stated, "this is worse than communism [...] America wake up, do not let government do this to you [...] You will be watched 24/7, from sun up to sun down" (Kuruvilla 2013). A study on the public perception of drones in the USA showed that 63 % of respondents feel it would be a change for the worse if the USA were to open its skies to personal drones, while 22 % think it would be a change for the better (Smith 2014). In a Reuters/Ipsos poll (Scott 2015), 42 % of more than 2000 respondents said that they oppose private ownership of drones and expressed concerns about safety and privacy. Despite the fact that our privacy is far more vulnerable in the face of surreptitious phone photography/video, closed-circuit television (CCTV) and ubiquitous Internet tracking than it is due to a conspicuous noisy device hovering in plain sight (Scott 2015), concerns over "killing robots" and privacy violations dominate public discussions about drones. "It seems we fear more fanciful tragedies than the regular ones—those that come at us in ways we have not yet accepted as commonplace" (Cobb 2015).

On the spectrum from techno-dystopian to techno-enthusiast views, one can find techno-libertarian views—concerned with fundamental liberties and minimal interventions by the state in the private lives of citizens—as well as techno-sceptical views—expressing doubts that technical paraphernalia can improve our "lives and business".

The huge emotional charge felt in the face of drones has also led to "instrumental" uses of drones. The fear associated with drones is a mechanism that breeds attraction—the more bizarre the invention, the more attention it gets. For instance, a company in Singapore wants to introduce "drone waiters" in restaurants that could lift a load of 2 kg—the argument being that "in Singapore food is a national obsession" and the only way to address the shortage of waiters is to introduce drones to help satisfy the insatiable clients (Wong 2015). The image of drones whizzing above clients' heads with napkins flying all around is just a gimmick, but it nevertheless attracted the attention of "respectable" global media (see BBC report in Wong 2015).

From flying "drone waiters" to flying "drone pets" (e.g. by turning dead cat into a drone, Price 2015), the intense approvals or repulsions show how drones raise important human values—the more important the value, the more mental energy is attached to the technology.

Situating Drones

Drone technology forms a part of several legal and social developments that are at work in contemporary Western societies. The development of the "Internet of Things"—an infrastructure in which billions of sensors are embedded in common, everyday devices in order to record, process, store and transfer data—carries an incentive to digitise all sorts of information from our immediate environment, such as from smart grids and even refrigerators (McOwan and McCallum 2014).

It normalises the idea of extensive data collection and processing of data, and fuels the development of drones as all-sensing machines collecting all sorts of data. This is closely linked to the notions of "pervasive" and "ubiquitous" computing, as stated by Article 29 Data Protection Working Party in the Opinion on the Internet of Things (Article 29 Data Protection Working Party 2014). "Wearable computing"—the concept of wearing recording devices and even putting them into apparel (Kaplan 2015)—similarly reflects the contemporary obsession with data, which bears strong moral consequences. The functioning of our bodies is monitored in order to disclose "weakness" and anticipate "health issues". The subject thus has to refashion behaviour—"act now or bear the consequences". This is done in a supposedly neutral manner in a "direct" translation of nature (bodies) via "wearable gadgetry" to the desired type of behaviour. These technologies can help, for instance, the visually impaired sense objects around them. However, the technologies may increase surveillance of our daily lives ("dataveillance", Clarke) to the levels of the darkest dystopian science fiction novels. Additionally, such gadgetry orders how to (or not to) act and behave as the "numbers do not lie".

Drones represent a part of these "things" that collect, store and process large quantities of data that are then here to be acted upon. They abolish narrative and subjectivity and introduce data doubles or *"dividuums"* (Deleuze 1995) that no longer need to speak—as the "data speaks for itself". Bodies become "coded" and function as "passwords" as they do not lie (Franko Aas 2006). The recent developments show the proximity of both trends—wearable computing and drones—when drones themselves become wearable (e.g. Nixie 2014).

Related to the above-mentioned trends is the development of the "big data" paradigm: the idea of being able to mine a large quantity of diverse data—either machine-readable or user-produced—in real (or near-real time) to make "informed" decisions about the future. The central idea of "big data", usually defined as the use of a greater volume of a great variety of data with greater speed, is the production of "actionable" data, or more precisely, to predict the future. Algorithmic predictions have entered various domains of our lives, including security—IBM's *Blue Crush* produces probability reports for crime yet to be committed (IBM 2015)—and the criminal justice system (Labi 2014).[1]

Drones used for intelligence purposes, similarly, produce what has been dubbed "big drone data" (Ackerman 2013). Big drone data is used to create collision detection systems, where algorithms can turn the imagery collected by cameras into something the drones can use. Besides the human problem of managing the enormous amount of data that Predators, Reapers, Global Hawks and Sentinels are generating (e.g. see the ARGUS-IS system, i.e. the Autonomous Real-Time Ground Ubiquitous Surveillance Imaging System, Limer 2013), there is a risk of *fishing expeditions* similar to the one stemming from "big Internet data". Like server farms

[1] The author of algorithm for predicting which of the 50,000 parolees would commit a serious crime within 2 years claims: "I use tens of thousands of cases to build the system, [as well as] asymmetric costs of false positives and false negatives, real tests of forecasting accuracy, the discovery of new forecasting relationships, and yes, machine learning" (Labi 2014).

in the cyber surveillance domain, "big drone data" will enable ex post facto processing of data (e.g. looking back into the past for traces of the suspect). Technologies such as Wide Area Airborne Surveillance (WAAS) and the ARGUS-IS developed by the Defense Advanced Research Projects Agency (DARPA) enable the clock to be rewound.

In answering the central challenge of how to mine mountains of surveillance data, the drone industry even looks to reality TV production and CCTV surveillance to learn how to improve "extensive processing, exploitation, and dissemination" power, which is essential for converting the raw data collected into usable intelligence (RAND Corporation 2012).[2]

Law enforcement drones disclose another two shifts in the security domain—a shift towards private or "for-profit" policing and a shift towards "preventive" justice. As Beck (2002) puts it when addressing the risks, we are talking about calculating the incalculable, colonising the future. Drones fit perfectly into this "pre-emptive" paradigm—a "pre-emptive" criminal justice model (Zedner 2007) or "preventive" law enforcement (Cole and Lobel 2007). As tools for collecting large quantities of data, such as the above-mentioned ARGUS-IS system, drones furthermore facilitate "intelligence-led" policing that has become a template for all police work (Lemieux 2008). Drones can be used as platforms for other police tactics and methods. The police could, for instance, use a *stingray*—which simulates a cell phone tower in order to trick nearby mobile devices into connecting to it and revealing their location—and attach it to a drone. It is only a matter of degree and economic rationale to nudge the idea into reality (e.g. Cessna planes are already used for such purposes).

In the immigration domain, border policing drones are entering a highly controversial area that is extensively debated in this book. Border policing has significantly changed in the past 30 years due to increasing global inequality in wealth. The "Western" countries ("Global North" countries) have employed pre-emptive and fortification strategies of border control, for example, by building "Digital Fortress Europe" as critics of the EU's immigration policy have claimed, by protecting the shores with exterritorial detention camps (e.g. Australia), or with new walls (e.g. between the USA and Mexico and between Bulgaria and Turkey—this time to keep people out of the country). In such an immigration policy landscape, border policing drones are used to create a "pre-frontier intelligence picture"—with the Eurosur intelligence system developed by the European agency Frontex (i.e. European Agency for the Management of Operational Cooperation at the External Borders of the Member States of the EU) being but a typical example of such immigration policy development.

Drones enter the domain of a new "spatial Web 3.0". By providing location data on the items or people recorded, they complement the ubiquitous monitoring devices that are forming a new geolocation services market. The estimations are that there will be 1 billion devices with built-in satellite receivers in 2015 and projec-

[2] Drone operations and Reality Television Production exploit multiple streams of motion imagery in near-real time; both of them operate 24/7 for extended periods, and both of them dynamically re-task and cross-cue sensors.

tions are that this will increase up to 3 billion in 2020 (Radovan 2015). Geolocation services are already used for automobile navigation and online cartography, and 50 % of Internet users already use online maps (Radovan 2015).

Drones are used as platforms for carrying specific sensors applied in the development of new visualisation techniques. It is not only that they are becoming fully autonomous, with auto piloting systems and drone collision detection systems (Finley 2015), they also raise a variety of normative questions related to responsibility and liability: If a drone "decides" to crash into a building in order to avoid a collision with a plane, who is responsible for such a decision built into the drone computer algorithm? They also are components of the testing of detection systems within the robotics industry.

The toy industry and drone hobbyists are beating the military–industrial complex at its own game. DIY enthusiasts want to make drones cheap, available and ubiquitous. Their open source and community-based endeavours resemble to a great extent the early development of the Internet with start-ups in garages—a particular digital version of the "American dream" where everyone can achieve success with "honest hard work". Like the early personal computer movement, the main use of drones at this point is experimentation—simple, geeky fun. As *Wired* forecasts, just as the 1970s saw the birth and rise of the personal computer, this decade will see the ascendance of the personal drone (Anderson 2012). We are entering the "Drone Age" (Anderson 2012) and a "gold rush on drones" (Rettman 2012).

Why do we need this book then? We are witnessing the above-mentioned rapid advancements in technology and are facing fundamental shifts in security and surveillance, especially related but not limited to crime and migration, and regulators need to recognise the wider social, cultural and political impacts of drones. They have to recognise the necessity of privacy and personal data protection regulation evolution and aviation law adaptation, and they need to balance security efforts against fundamental liberties in crime and migration policy. As drones are here to fly and inevitably also to crash, they need to design third-party liability and insurance (European RPAS Steering Group 2013).

Similar to the observations regarding "Janus-faced" surveillance (Lyon 2007), drones come with such a double face as well. They can be seen as geeky toys of DIY enthusiasts or as weapons of dragnet surveillance and the killing machines of the "industrial-surveillance complex" (Ball and Snider 2013). What the book attempts to address is this need to situate drones in the cultural, socioeconomic and existing legal framework. Technology, including drones, has become perceived as a means to solve social problems, particularly those of crime, migration and terrorism. These new "techno-fixes" purport to be neutral, but it is necessary to analyse them from ethical, legal, human rights and criminological perspectives, for technology is inescapably political in its uses. By protecting, for example, environments characterised by social, political and economic inequalities it can reinforce the conditions already at work there, to the detriment of social equality, justice and social cohesion.

The Structure of the Book

The first part of the book, *Political Technology of Drones,* is dedicated to the forms of knowledge produced by drones and the moral and political economy of drones. In the chapter *Theorizing Drones and Droning Theory,* Andrejevic pushes the boundaries of understanding drones beyond drones themselves. He describes the logic of "droning" in the context of our always-on society that creates an "avalanche of numbers" to be acted upon. He describes the manifold aspects of "drone logic". For instance, "de-differentiation" blurs the lines between the battlefield and other spaces as the logic of drones populate the civilian space with the space of the battlefield and blur the crucial division—for regulation at least—between civilian and combatant space and time. He then locates "drone logic" in several "spaces and places", for example in media, commerce, security, and pedagogy. In such a manner, the author unfolds "drone logic" to create a "psychoanalytically informed dialectical approach against strands of a so-called new materialism that ends up replicating and reinforcing drone logics".

By observing networked surveillance and a quasi-automated response through the figure of the drone, the chapter shows the ideology supporting drones and the power of the "digital redoubling of space and time" that is used to reconstruct the past and to predict the future. In the field of security, the chapter discloses the underlying assumption of what is understood as a threat and as a crime. "Drone logic" rests on the hidden presumption that crimes are primarily visible recordable events. However, (critical) criminology theory has often resisted such definitions and demonstrated that the real harm is produced by the silent operations of the economic system and the corrupt practices of corporations and governments, which are most often hidden from plain sight and not "recorded" by any surveillance technology.

Drone logic fulfils the fantasy of pre-emption by other means, the author claims, and shows how drone fantasy goes a step further and envisions forms of confrontation without threat. Drones are pushing the horizon of pre-emption: the "war on terror" version of drone warfare relies on the identification of external enemies at a distance, while the notion of drone security envisions a domestic space already permeated by threat. In such a manner, the author unfolds the ideology of drones: the idea of 24/7 coverage, of sleepless sensing and sense-making—the idea of gaplessness and wholeness that Lacanian theorists would name the Big Other. What drones attempt to do is thus to hide the fact that the Big Other, the symbolic order itself, is barré, crossed out by a fundamental impossibility, structured around a central lack (Žižek 2008).

In the third chapter, *The Political and Moral Economies of Dual Technology Transfers,* Bergtora Sandvik deconstructs the rhetoric of the "good", "public order drone"—defined as a police or firefighting drone—and demonstrates the mobilisation of the moral economy underlying the "good" drones that is at work in the attempts of the drone industry to transfer the technology from the military to the civilian domain. The author critically presents the literature on the militarisation of policing by claiming it fails to properly describe how military drones are moving into the civilian sphere. She elucidates the "normalisation" process at work, where drones are becoming normal tools for the police and subsequently normal tools for carrying weapons—a process that should be resisted for several reasons discussed in the chapter in great detail.

The chapter continues by dissecting an example of the mobilisation of the moral economy of the "Switchblade" microdrone—also known as a "surgically precise strike solution". The author shows how the microdrone is in fact an example of a deceptive and failed policy for maintaining public order. The "Switchblade" lowers the threshold for its use and substantially changes for the worse the decision-making process that is in place for the employment of drones. It shows how the "Switchblade" is a tool for keeping boots on the ground and not vice versa, as advocates of armed drones claim.

The central part of the chapter shows the mechanisms used by the drone industry to change the moral landscape of civilian uses of drones, for example, by engaging in educating the public about the benefits of drones in the civil airspace, by contesting the semantics of drones, by showing the shared moral values of the army and those of the public (e.g. saving time, money and lives). The author deconstructs the central arguments—pre-emption, precision and price—used for justifying police drones and the underlying assumption that the open-ended possibilities for the use of drones should be regarded as inherently valuable. The author is also critical about the "innovation boost argument" for supporting the drone industry and how the industry attempts to benefit from the good public image of start-ups. The innovative DIY scene is not that disconnected from the military surveillance complex as supposedly utterly civilian uses of drones also regularly spark interest from the military. The DIY enthusiasts should thus be more auto-reflective of their accomplishments as there is no value-free or value-neutral technology.

The second part, *Drones Between Privacy and Security,* tackles the legal aspects of drones and the employment of drones for (migration) security purposes.

The fourth chapter on *The (F)utility of Privacy Laws* tests the established legal mechanisms of privacy protection from the perspective of aerial drone surveillance. By analysing the jurisprudence of the Supreme Court of the United States, the European Court of Human Rights and the European Court of Justice, Gorkič demonstrates the need for the evolution of privacy law, for instance, how the test of the reasonable expectation of privacy cannot adequately protect an individual's privacy from drone surveillance.

The chapter shows the history of the relevant case law of aerial surveillance and the development of the reasonable expectation of privacy test in the USA and the EU. It shows the test's limited scope for the protection of privacy in the USA as there can be no reasonable expectation of privacy in so-called open fields and beyond the limits of an individual's home. The development of ubiquitous surveillance technologies changes both elements of the test of the reasonable expectation of privacy, that is, whether a person will exhibit an actual subjective expectation of privacy and whether such expectation is recognised by the society as reasonable.

The recent attempts in US jurisprudence, such as *Kyllo v. United States* (2001) and *United States v. Jones* (2012), show that the courts are well aware of such limitations, but are still not willing to go "all the way" in protecting privacy: *Kyllo* limits the scope of the protection to technology that is not in general public use, while *Jones* turns the focus to proprietary elements of the breach of privacy. In contrast, the author demonstrates, the European courts neglect the trespassory nature of the conduct of law enforcement agencies. European courts recognise that the processing of data gathered from the public space may lead to new insights into the personal life of an individual and that this fact alone is "particularly serious" with "chilling effect"—as the European Court of Justice noted in the Digital Rights judgement (C-293/12, C-594/12, 2014) on preventive indiscriminate data retention regulation. The European courts have focused on the dignity of the individual and not the property of the individual. Regardless of where data have been collected, from one's home or in the public space, the European courts would most probably refer to the dangers of a chilling effect due to "feelings of constant surveillance" in the case of indiscriminate drone surveillance.

The fifth chapter, *Re-bordering the Peripheral Global North and Global South,* scrutinises the deployment of drones in the domain of border policing and border management in the "Western" or "Global North" states, and contrasts these policies with those in place in the "Global South". Its author, Milivojević, sheds light on the predominant narratives supporting the use of surveillance technologies in border control in order to unpack the military rhetoric that underpins the usages of drones in border policing. She demonstrates how finding a technological "solution" to social problems is a common strategy taken by politicians who are unable—or more often unwilling—to tackle the real reasons for, for example, mass migrations from impoverished to more developed countries that are resulting in mass graves in the seas around the world (e.g. in the Mediterranean for those embarking in Maghreb, in the Andaman Sea for those fleeing from Myanmar and Bangladesh or in the Timor Sea and Indian Ocean for those trying to reach Australia from South Asia; see International Organization for Migration 2014). Unwanted migration to the Global North countries leads to the fortification of, for instance the EU/Schengen states—which has already been dubbed "Cyber-Fortress Europe" (Guild et al. 2008)—in the context of its digitised immigration policy. The author shows how drones fit in an immigration policy that is grounded in the heavy use of sophisticated technologies and led by an intelligence-driven approach (e.g. the European Border Surveillance System Eurosur with its "situational awareness pictures" being a typical example).

The chapter situates the discussion of border patrol drones in the "re-bordering" process enhanced by surveillance technologies and "crimmigration"—defined as the convergence of immigration and criminal law and a "tough on migrants" border policing strategy. It depicts technology advancements in border policing and presents the central dilemma of utilising drones in the border policing domain as they can carry weapons or help to rescue ships abandoned by crews and full of migrants, but the author remains critical of domestic surveillance and border-policing drones for several reasons. For instance, the analysis of the language used and semantic shifts show the aspirations of the protagonists of the drone surveillance systems (from Lord Darth Vader, for the VADER border surveillance systems in the USA, to mythological hunting names of machines such as Trion and Predator).

The chapter then tackles drone-led border policing with two in-depth case studies in order to illuminate the perspective from both global sides of the borderland: a case from Australia and one from Serbia, interesting for its political situation regarding the integration process in the EU. The case from Serbia discloses several insights identified elsewhere around the globe—the process of "offshoring" and "contracting out" border policing to non-EU states, a shift from policing to militaristic types of activity and how drones simply fail to deter mobile bodies. Analysis of media coverage of drones in Australia shows significant differences between military and border police drones; however, the author claims, both domains of drones are underpinned by military rhetoric, especially the notion of a hunt. The drones' gaze "reduces mobile bodies to targets, numbers, collateral damage, or transferees, law-breakers or illegals that need to be detected, immobilised and removed, with the help and under the watchful gaze of new, non-human 'eyes in the sky'".

In this way, the chapter critically—with massive human rights violations of migrants in mind—situates the use of drones in border policing in the wider processes of the regulation of migration: (1) the stratification of global mobility regarding differentiating the super-fast kinetic elites from the immobilised underclass, (2) offshoring border control and (3) pre-emptive strategies of border control.

The sixth chapter, *Deploying Drones in Policing Southern European Borders,* in Part 2, combines the theme of the previous two chapters by tackling the policing of the southern borders of Europe from the personal data protection law perspective. Its authors, Marin and Krajčiková, delve deep into the Schengen process that led to the strengthening of the external border controls and introduced an integrated management system for the external borders of the Schengen country members. It shows how centralisation of the powers in European integration is developing border policing into an increasingly autonomous policy field, with the agency Frontex at the forefront of the European border policing agenda. The chapter illuminates several trends in European migration policy, such as framing migrants as security threats, developing large information technology (IT) databases with biometric data and deploying and developing drones.

The chapter maps the actors in EU border policing, such as the EU member states that already use drones in the border policing context (e.g. Italy), the security industry, national bodies with exceptional powers (e.g. the Italian Navy) and the European agency Frontex. The authors critically present Frontex's activities and policies, for example, its lack of transparency, "push-back" operations and silent compliance with hidden bilateral agreements of the EU member states with third countries (e.g. such as the one between former prime minister of Italy Berlusconi and Libya's Gaddafi). Frontex, the authors claim, inter alia, operates as a coordinating body for the EU member states merely at the surface level, but in fact it is strongly immersed in support of the techno-fixes on the southern European maritime borders that also clearly breach international migration law (e.g. the non-refoulment principle or the prohibition of collective expulsion). By supporting drone-related research, hosting workshops on drone technologies and demonstrations of drone performance, Frontex is in fact one of the most active actors in the movement of drones into the border policing domain.

The central part of the chapter tackles the legal framework: the basic elements of aviation law and, in more detail, personal data protection law and the Schengen *acquis*. It presents relevant personal data protection regulation in the EU and its applicability to border policing drones. It analyses respect for the principle of proportionality in cases of indiscriminate surveillance of the Mediterranean. It meticulously presents one of the weakest points of the supposedly humanitarian employment of drones in the Mediterranean—the collaboration of Frontex with neighbouring third countries with poor human rights safeguards for the purpose of personal data exchange that puts migrants' lives at risk. In this way, the chapter demonstrates how the process of "surveillance off-shoring", that is, where surveillance is conducted in the territorial waters of third countries, is quite deadly for the fleeing "kinetic underclass". The authors clearly show how pre-emptive surveillance in North Africa does not improve search and rescue capacities, and is not proportionate to the risk these people represent to the security of the member states of the EU.

In the third part of the book, *Drones, the "War on Terror", and Public International Law,* two chapters tackle military uses of drones in an international context.

In the seventh chapter, *Death from the Sky,* De Groof shows the provisions of public international law governing whether a drone strike is considered legal. These are, the author shows, extremely stringent and thus the practice of existing attacks is "regularly…carried out in flagrant violation of international humanitarian law". The author arrives at this conclusion—amongst others—after mapping the relevant legal framework for the use of armed drones with scrupulous attention to details, that is, rules pertaining to: (1) *Jus ad Bellum* (international law of the use of force)—a corpus of rules regulating the initiation of war—which refers to the question of whether a drone strike respects central principles, such as proportionality and the last resort principle; (2) *Jus in Bello* (international humanitarian law)—a corpus of rules regulating wartime conduct—which refers to the question, inter alia, of whether a drone attack reaches the threshold of an "armed conflict"; and (3) international human rights law which refers to cases when armed violence does not reach the threshold of an "armed conflict" and the only justification of a drone strike may be found under these sets of rules.

If a state cannot justify a drone attack under these three branches of public international law, then the state is internationally responsible for the attack, and mechanisms can be triggered to remedy damages. The chapter thus also tackles the implications and legal consequences of the illegal use of armed drones for not only countries that use armed drones but also countries that willingly support or facilitate drone attacks by other means (e.g. by producing drones).

Before reviewing the central legal regulation of armed drones, the chapter tackles several definitional challenges, such as the types of drones, the types of targets of drone attacks and the contexts of drone attacks. With regard to the type of drones, the author explains that drones are legally not weapons, but systems of weapon transmission, and regulations on illegal weapons do not apply. In general, claims the author, drones can satisfy the general principles of international humanitarian law (e.g. necessity, proportionality); however, the "advantages of the use of drones are not always exploited." With regard to targets, she points to the so-called "signature strikes", where the presumed link of a target to a terrorist group is calculated from behavioural patterns and/or visual characteristics.

The legal section on *Jus ad Bellum* explains under which conditions extraterritorial drone strikes can be considered legal, that is, by consent of the country involved (intervention upon request), self-defence and with the authorisation of the Security Council of the United Nations. By offering examples of real cases of attacks, the chapter reveals substantial gaps in the legal justification of drone attacks with the argument of self-defence. The *Jus in Bello* section lays out the conditions for the legality of targeted attacks, such as the existence of an armed conflict. The chapter concludes with recommendations for enhancing respect for international law and promoting the transparency of, and accountability for, illegal drone strikes.

In the eighth chapter *The Predators' Rule of Terror,* Badalič tackles the USA's "war on terror" justification for extraterritorial drone killings from the Foucauldian disciplinary perspective. He scrutinises the "rapid dominance doctrine" for imposing "shock and awe" on adversaries in order to show that the goal of drone warfare is not solely to decapitate "terrorists", but rather to "normalise" the population living in areas where drones are employed.

The chapter begins with the argument that "the legitimacy of the use of violence at any particular moment in history depends on power relations that permit the dominant class to grant itself the right to decide whose force is legitimate". Drone policy is no different in this respect. In the USA, this power is vested in the president as the executioner-in-chief, who is vested with "the right of the sword" (Foucault)—the authority to make the final decision whether to launch a strike or not.

The chapter then proceeds with an analysis of the "surgically precise weaponry" argument—and dismisses it. On the basis of analyses of several drone strikes in which those killed or injured were either exclusively or predominantly civilians—strikes were executed in places such as a bazaar, an Islamic school, a funeral and a wedding procession—the author demonstrates that strikes are in fact random. He presents five reasons for this randomness such as the "structural" failure of the idea of identifying an individual's terroristic "pattern of life" in "signature strikes", creative interpretation of images by drone operators, faulty intelligence provided by local informants motivated by monetary incentives and unethical "double tap" attacks consisting of a sequence of two drone strikes in which the second strike hits rescuers of the victims of the first strike.

It is exactly the inaccuracy of such "surgically precise weapons", continues the author, which introduces a new power regime in areas under drone surveillance: A power regime in which anyone, "terrorist" or civilian, can be targeted and that seeks to terrorise the entire population living in drone-surveilled areas. The drone's panoptic gaze then also serves as a mechanism that directly, on its own, inflicts terror on the surveilled population. In this manner, the drone's gaze generates effects of power through a specific distribution of visibility and invisibility typical of a panopticon: (1) supervisors (drone operators) are constantly invisible and the surveilled subjects constantly visible, and this (2) creates the awareness in the surveilled populace that the "supervisors" might always be present—this is induced by the drones' constant visibility as well as their sound, as the local population reports.

The chapter then shows how drones' panoptic gaze spreads from closed institutions into open areas inhabited by targeted populations and how the power effect of drones' panoptic gaze is supported by the constant threat of the use of the "right of the sword". The chapter concludes by showing differences between "the political technology of drones" and Foucauldian disciplinary power. While the latter's aim was to create productive individuals who will carry out specific tasks in accordance with the dominant class's interests, the power regime established by drones is intended to create a passive and frightened population—the "killing drone" programme is producing docile bodies and disintegrated communities. However, because indiscriminate strikes have become the norm, the only solution for the subjected population is to do whatever it takes to prevent the launch of such strikes: "Therefore, the seeds of armed rebellion are already planted by drone warfare itself".

In the fourth part of the book, *Drones and International Air Law,* Mendes de Leon and Scott elaborate—in the chapter *An Analysis of Unmanned Aircraft Systems Under Air Law*—the current legal regulation of civil UAS by devoting particular attention to the set of rules of the International Civil Aviation Organization (ICAO) and the EU. They map the often unclear legal framework governing UAS (unmanned aircraft systems, the authors' preferred term, as explained in the chapter) and underline how important it is that regulations be drafted swiftly with regard to the safety, security, market access, privacy and liability issues of UAS.

The chapter addresses international civil aviation regulation, for example, the central Chicago Convention (1944) with annexes, produced by the ICAO. The authors analyse the legal implications of the differences between "remotely piloted" and "unmanned" aerial vehicles, "model aircraft" and "toy aircraft". They briefly show the relevant history of civil aviation rules from the Paris Convention (1919) and Chicago Convention (1944) to the Beijing Convention and Protocol (2010). The older conventions are relevant as the state-centred division of the world grants each state the "unilateral and absolute right [...] to permit or deny entry into its territory and to control all movements therein".

International law includes several areas relevant for UAS, and the authors of the chapter show the relevance of the distinction between "state" and "civil" aircraft—depending on the use—and the difference between commercial and noncommercial "civil" uses of UAS.

The body of European air law relevant for UAS is furthermore discussed, for example, sections on operating licenses, the right to operate intra-community "air services" and aviation insurance. The chapter also presents the "Single European Sky" initiative and the applicability of air service agreements between the EU/member states and third countries permitting foreign aircraft and UAS to enter the sovereign airspace of other states (such as the US–EU Open Skies Agreement).

The chapter continues by presenting the European Commission's proposal with six actions to promote the growth of the civil UAS market, with actions such as the development of common certification processes and standards for UAS, supporting research and development (R&D) and addressing the security aspects of UAS (e.g. protection against cyber attacks and the protection of personal data).

The chapter then focuses on three legal domains and shows how these pertain to UAS: safety rules adopted by the European Aviation Safety Agency (EASA) and national aviation authorities, rules setting the liability for damage caused by UAS (second party liability—passenger and cargo liability—and third-party liability—damage that the aircraft does to third-party property) and criminal offences that may be committed in relation to UAS (e.g. a criminal act against an aircraft as criminalised in the Montreal Convention of 1971).

The fifth and last part of the book, *Domain-Specific Uses of Drones,* presents arguments in favour of "good drones" used for civil purposes and the use of drones to counter the dominant power. In the tenth chapter, *Droning on about Journalism,* Goldberg focuses on drones—"aerial camera platforms"—used for newsgathering by either media or citizens as journalists. The key claim of the chapter is that such application exercises the fundamental human right to freedom of expression and, specifically, the right to receive ideas and information and the right to access communication technologies. The author acknowledges the competing interests regarding these rights, but he adopts the stance that only if, *in casu,* there exist extremely compelling and overriding considerations defending and promoting another protected interest, for example, someone's purported right to respect for their private and family life or home, would it be right to limit the use of drones in that specific context.

The message the chapter conveys is that besides several laws limiting the freedom of expression and, specifically, the right to receive ideas and information (e.g. laws protecting the reputation of subjects of media reporting or national security, advertising rules, broadcasting regulation, hate speech laws), aviation regulation should also be taken into account. The chapter seeks, inter alia, to equip journalists and citizens to act as one with a legal defence based on the exercise and protection of fundamental rights. The author situates drone journalism in twenty-first-century photojournalism as a distinctive form of photography and proposes that "responsible journalism"—"journalistic activities [offering] significant public benefit"—be the benchmark for newsgathering drones.

The chapter presents several pro-drone arguments focused on three "scenarios": (1) A scenario where the deployment of drones is generally prohibited on the basis of the argument that drone journalism represents a commercial or "business" use. The author counters the argument by connecting drones with the (human) right to freedom of expression: The central point of newsgathering is not "business" but "securing the free flow of information that inevitably feeds the 'free trade of ideas'". The chapter supports this view by citing European Court of Human Rights case law on protecting the identity of journalists' sources and protection against unjustified search and seizure of journalistic material. (2) A scenario where anyone operating newsgathering drones faces accusations of serious infringement of civil liberties, especially the right to privacy, regarding which the author claims that the deployment of drones cannot be inherently problematic, but their use can raise conflicting legitimate interests that trigger the need for a complex weighing of these interests. (3) A scenario where the opposition to newsgathering drones rests on the claim that the subjects of investigation might not realise that they are being surveilled. The author counters this claim with the distinction between snooping and legitimate subterfuge and offers examples from UK journalistic practices. In doing so, the chapter confronts what the author calls "anti-drone *lawfare*".

In the chapter *Drones, Resistance, and Countersurveillance,* Završnik examines whether drones can be used for resistance, denial, subversion and distortion of contemporary surveillance systems. It takes as a starting point Foucault's insight that supervisory power also generates resistance to the (same) power. The "new surveillance" (Marx 2002) is forming "complex networks of power relations and resistances" (Green 1999), and thus the increasing power of the "new surveillance" not only increases the capacity of the authorities (watchers) to monitor the watched majority but also carries emancipatory potential. The chapter analyses the objects of surveillance as empowered agents, not as passive and powerless parties (Ball 2003). They can be active entities reflecting on surveillance practices and trying to escape and avoid them, or at least draw public attention to such practices as illegitimate, discriminatory or otherwise unfair. Such resistance practices have already been tackled from different theoretical perspectives; for example, Mann (2002) talks about *"sousveillance"* (as viewing "from below"), others about "inverse surveillance", "co-veillance", where surveillance and countersurveillance coexist or "countersurveillance" that Monahan (2006) defines as a form of tactical interference, often leading surveillance technologies to work against themselves in order to redress institutional power asymmetries.

Acknowledgements The author wishes to express his thanks to Ajda Gorjanc (www.helivideo.si) and Dr. Danijel Skočaj (Computer Vision Laboratory, Faculty of Computer and Information Science, University of Ljubljana) for technological insights, and Laura Matjašec and Sara Železnik who assisted with research. Special thanks are due to Dean J. DeVos, an English language editor, for his patience and diligence during our collaboration.

References

Ackerman, S. (25 April 2013). Welcome to the age of big drone data. *Wired*. http://www.wired.com/2013/04/drone-sensors-big-data/. Accessed 2 June 2015.

Amazon. (2013). Amazon Prime Air. http://www.amazon.com/b?node=8037720011. Accessed 2 June 2015.

Anderson, C. (22 June 2012). How i accidentally kickstarted the domestic drone boom. *Wired*. http://www.wired.com/2012/06/ff_drones/all/. Accessed 2 June 2015.

Article 29 Data Protection Working Party. (2014). Opinion 8/2014 on the on recent developments on the Internet of things. Adopted on 16 September 2014.

Ball, K. S., & Webster, F. (Eds.). (2003). *The Intensification of Surveillance*. London: Pluto Press.

Ball, K., & Snider, L. (Eds.). (2013). *The surveillance-industrial complex: A political economy of surveillance*. London: Routledge.

BBC. (2013). Black Hornet spycam is a 'lifesaver' for British troops. http://www.bbc.com/news/uk-21450456. Accessed 5 April 2015.

Beck, U. (2002). The terrorist threat: World risk society revisited. *Theory, Culture & Society, 19*(4), 39–55.

Bulc, V. (2015). The future of flying. Conference on remotely piloted aircraft systems, Riga, 6 March 2015. http://ec.europa.eu/commission/2014–2019/bulc/announcements/future-flying-conference-remotely-piloted-aircraft-systems-riga-6-march-2015_en. Accessed 2 June 2015.

Cobb, J. (12 February 2015). What our Paranoia about drones says about us. *The New York Times Magazine*. http://www.nytimes.com/2015/02/15/magazine/what-our-paranoia-about-drones-says-about-us.html?_r=0. Accessed 2 June 2015.

Cole, D., & Lobel, J. (2007). *Less safe, less free: Why are we losing the war on terror*. New York: New Press.

Conference on remotely piloted aircraft systems. (6 March 2015). The future of flying. http://ec.europa.eu/commission/2014–2019/bulc/announcements/future-flying-conference-remotely-piloted-aircraft-systems-riga-6-march-2015_en. Accessed 2 June 2015.

Crowe, S. (5 February 2015). Meet plotbox the cemetery drone that maps your grave. *Robotics Trends*. http://www.roboticstrends.com/article/meet_plotbox_the_cemetery_drone_that_maps_your_grave. Accessed 14 March 2015.

D'Andrea, R. (11 June 2013). The astounding athletic power of quadcopters. https://www.youtube.com/watch?v=w2itwFJCgFQ. Accessed 2 June 2015.

Deleuze, G. (1995). Postscript to control societies. In G. Deleuze (Ed.), *Negotiations: 1972–1990*. New York: Columbia University Press.

Etherington, D. (14 April 2014). Google acquires Titan aerospace, the drone company pursued by Facebook. *TechCrunch*. http://techcrunch.com/2014/04/14/google-acquires-titan-aerospace-the-drone-company-pursued-by-facebook/. Accessed 2 June 2015.

European Commission. (2012). Working document. Towards a European strategy for the development of civil applications of Remotely Piloted Aircraft Systems (RPAS), Brussels. 4.9.2012 SWD(2012) 259 final.

European RPAS Steering Group. (2013). Final report. *Roadmap for the integration of civil Remotely-Piloted Aircraft Systems into the European Aviation System*. http://ec.europa.eu/enterprise/sectors/aerospace/files/rpas-roadmap_en.pdf. Accessed 2 June 2015.

Finley, K. (9 January 2015). The autopilot tech that could be a total game-changer for drones. *Wired*. http://www.wired.com/2015/01/autopilot-tech-total-game-changer-drones. Accessed 2 June 2015.

Franko Aas, K. (2006). The body does not lie: Identity, risk and trust in technoculture. *Crime, media, culture, 2*(2), 143–158.

Gannon, M. (11 October 2014). NASA drone captures amazing footage of Hurricane's Eye. *Livescienece*. http://www.livescience.com/48088-drone-flies-over-hurricane-video.html. Accessed 24 April 2015.

Green, S. (1999). A plague on the Panopticon: Surveillance and power in the global information society. *Information, Communication and Society, 2*, 26–44.

Guild, E., Carrera, S., & Geyer, F. (2008). The Commission's New Border Package. Does It Take Is One Step Closer to a "Cyber-Fortress Europe"? CEPS Policy Brief, No. 154/2008.

Hern, A. (25 September 2014). DHL launches first commercial drone 'parcelcopter' delivery service. *The Guardian*. http://www.theguardian.com/technology/2014/sep/25/german-dhl-launches-first-commercial-drone-delivery-service. Accessed 24 Feb 2015.

Houses of Parliament. (October 2014). The parliamentary office of science and technology, civilian drones, Number 479.

IBM. (2015). Memphis PD: Using analytics to fight crime before it happens. http://www.ibm.com/smarterplanet/us/en/leadership/memphispd/. Accessed 4 June 2015.

International Organization for Migration. (2014). Fatal journeys: Tracking lives lost during migration. http://publications.iom.int/bookstore/free/FatalJourneys_CountingtheUncounted.pdf. Accessed 4 June 2015.

Judgment of the Court Digital Rights Ireland and Seitlinger and Others. (2014). Joined cases C-293/12 in C-594/12, 8 April 2014.

Kaplan, B. (6 January 2015). Robotic spider dress powered by Intel smart wearable technology. *Intel iQ*. http://iq.intel.com/smart-spider-dress-by-dutch-designer-anouk-wipprecht/. Accessed 4 June 2015.

Kuruvilla, C. (21 February 2013). Police officers, archaeologists, and farmers are using high-tech drones for civilian purposes. *New York Daily News*. http://www.nydailynews.com/news/national/photos-high-tech-drones-find-non-war-civilian-article-1. 1270332. Accessed 4 June 2015.

Kyllo v. United States. (2001). (533 U.S. 27 (2001)).

Labi, N. (2014). Misfortune Teller. *The Atlantic, January/February issue*. http://www.theatlantic.com/magazine/archive/2012/01/misfortune-teller/308846/. Accessed 12 Nov2014.

Lapowsky, I. (23 September 2014). Facebook lays out its roadmap for creating Internet-connected drones. *Wired*. http://www.wired.com/2014/09/facebook-drones-2/. Accessed 4 June 2015.
Lemieux, F. (2008). Information technology and criminal intelligence: A comparative perspective. In S. Leman-Langlois (Ed.), *Technocrime. Technology, crime and social control* (pp. 139–168). Portland: Willan Publishing. (Cullompton [England]).
Limer, E. (27 January 2013). Watch the world's highest resolution drone-mounted camera in action. *Gizmodo*. http://gizmodo.com/5979372/watch-the-worlds-highest-resolution-drone-mounted-camera-in-action. Accessed 4 June 2015.
Lyon, D. (2007). *Surveillance studies: An overview*. Cambridge: Polity Press.
Mann, S. (2002). "Reflectionism" and "diffusionism": New tactics for deconstructing the video surveillance superhighway. In T. Y. Levin, U. Frohne, & P. Wiebel (Eds.), Rhetorics of Surveillance from Bentham to Big Brother (pp. 531–543). Karlsruhe: ZKM Centre for Art and Media.
Marx, G. T. (2002). What is new about "new surveillance"? Classifying for change and continuity. *Surveillance & Society, 1*(1), 9–29.
McOwan, P., & McCallum, L. (8 September 2014). When fridges attack: The new ethics of the Internet of things. *The Guardian*. http://www.theguardian.com/science/alexs-adventures-in-numberland/2014/sep/08/when-fridges-attack-the-new-ethics-of-the-internet-of-things. 4June 2015.
Monahan, T. (2006). Counter-surveillance as political intervention? *Social Semiotics, 16*(4), 515–534.
Nixie drone. (25 September 2014). Make It wearable finalists: Meet team Nixie. https://www.youtube.com/watch?v=_VFsdPAoI1g. Accessed 4 June 2015.
Price, R. (26 January 2015). The Dutch artist who turned his dead cat into a drone is keeping a Badger in his freezer to build a submarine. *Business Insider*. http://www.businessinsider.my/badger-submarine-bart-jansen-2015-1/#Xsq7fGzpy0bUig6L. 97. Accessed 4 June 2015.
Radovan, D. (2015). *Geostoritve z odprtimi prostorskimi podatki. In 1. dnevi prava zasebnosti in svobode izražanja, Zbornik 2015*. Ljubljana: IusSoftware.
RAND Corporation. (2012). The future of air force motion imagery exploitation: Lessons from the commercial world. http://www.rand.org/content/dam/rand/pubs/technical_reports/2012/RAND_TR1133.pdf. Accessed 4 June 2015.
Rettman, A. (17 February 2012). EU firms join gold rush on drones. *EuObserver*. https://euobserver.com/defence/115283. Accessed 4 June 2015.
Schneier, B. (2003). *Beyond fear: Thinking sensibly about security in an uncertain world*. Copernicus Books.
Scott, A. (5 February 2015). Americans OK with police drones—private ownership, not so much: Poll. *Reuters*. http://www.reuters.com/article/2015/02/05/us-usa-drones-poll-idUSKBN0L91-EE20150205. Accessed 4 June 2015.
Smith, A. (17 April 2014). U.S. views of technology and the future. *Pewreasearch Center*. http://www.pewinternet.org/2014/04/17/us-views-of-technology-and-the-future/. Accessed 4 June 2015.
Spaeth, R. (19 February 2013). WATCH: The deadly, insect-sized drones of the future. *The Week*. http://theweek.com/articles/467538/watch-deadly-insectsized-drones-future. Accessed 4 June 2015.
United States v. Jones. (2012). (132 S. Ct. 945 (2012)).
Wong, T. (8 February 2015). Drone waiters to plug Singapore's service staff gap. *BBC*. http://www.bbc.com/news/world-asia-31148450. Accessed 4 June 2015.
Zedner, L. (2007). Pre-crime and post-criminology? *Theoretical Criminology, 11*(2), 261–281.
Žižek, S. (2008). *The sublime object of ideology*. London: Verso.

Aleš Završnik Doctor of Law (LLD), assistant professor, is a researcher at the Institute of Criminology at the Faculty of Law in Ljubljana. He was a postdoctoral Yggdrasil fellow at the Department of Criminology and Sociology of Law at the University of Oslo (2012) and was a postdoctoral fellow at the Max-Planck-Institute für ausländisches und internationales Strafrecht, Freiburg i. Br. (2009) and a fellow of the World Federation of Scientists, Geneva (2008). He collaborated in the European Cooperation in Science and Technology (COST) Action "Living in Surveillance Societies" and the COST Action "Cyberbullying". Aleš Završnik edited a book, Crime and Technology: How Computers Transform Surveillance and Privacy, Crime and Crime Control? (2010) and published a book, Homo Criminalis: Images of a Criminal in High- Tech Risk Society (2009). He has researched and published works on theoretical criminology, criminal law and "law and technology", i.e.that is, cybercrime, cyber- law and surveillance. He is a member of the editorial board of the Masaryk University Journal of Law and Technology

Part I
Political Technology of Drones

Chapter 2
Theorizing Drones and Droning Theory

Mark Andrejevic

Overview

This chapter uses the figure of the drone to unpack the emerging logics of ubiquitous, always-on sensor-based monitoring. The following argument is not about drones, per se, but about what might be described as drone logic: the deployment of ubiquitous, always-on networked sensors for the purposes of automated data collection, processing, and response. It treats the drone as an avatar of the emerging logic of increasingly passive interactivity. Thus, the intended contribution of the chapter is to reframe theoretical approaches to interactivity in the digital era. The goal of this reframing is to take into account the ways in which interactivity is becoming passive, distributed, mobile, and reliant upon quasi-centralized infrastructures for data sharing, processing, and deployment. The figure of the drone is useful because it offers a highly visible and controversial example of the deployment of networked surveillance and quasi-automated response and suggests the ways in which the implementation of drone logic across disparate spheres of social practice partakes of the military-inflected rationalization of everyday life. The chapter is called drone *theory* because it outlines a theoretical approach to the social implications of the "droning" of daily interactions and because it simultaneously critiques recent developments in theory that, wittingly or not, align themselves with the forms of knowledge anticipated by automated forms of information collection, processing, and response. Thus, the chapter offers a social-theoretical critique of the embrace of drone logic in both the practical and theoretical realms. The goal is to intervene in the formation in which theoretical and technological developments align themselves in ways that have significant and troubling implications for questions of power, control, and democracy.

M. Andrejevic (✉)
Department of Media Studies, Pomona College, Claremont USA
e-mail: markbandrejevic@gmail.com

A. Završnik (ed.), *Drones and Unmanned Aerial Systems*,
DOI 10.1007/978-3-319-23760-2_2

Abstracting away from the figure of the "unmanned" flying device, the notion of drone logic, broadly construed, unfolds the shared characteristics of distributed forms of networked data collection and response. The promise of the drone as hyperefficient information technology is fourfold—it extends and multiplies the reach of the senses, it saturates the times and spaces in which sensing takes place (entire cities can be photographed 24-h a day), it automates the sensemaking process, and it tends toward the automation of response (see, for example, Sharkey 2011). In this regard, the figure of the drone comes increasingly to characterize the operation of interactivity in an era of proliferating, distributed, digital devices. We might describe it as contributing to the increasing automation of both interaction and response. If interaction implies some kind of conscious contribution to a relationship, encounter, or exchange, the "droning" of interactivity anticipates a world in which feedback is passively collected, analyzed, and responded to. It also anticipates a world in which the database aspires to capture all of reality: drone logic implies the ongoing unfolding of new distributed, always-on sensors to capture new data dimensions. In this regard, the mapping of the territory and its construction go hand in hand. The term "drone" is suggestive in this context, since it connotes a continuous background presence—as distinct from other forms of more targeted and discrete forms of interactivity. The forms of interactivity described here have typically been associated with so-called "ubiquitous computing" and "the internet of things"—but neither of these formulations invoke the increasingly instrumental and infrastructure-intensive character of automated data collection and response. Such formulations typically invoke the notion of a world of "smart" objects that collect information from their interactions and are able to respond accordingly. The figure of the drone, by contrast, necessarily invokes the underlying and quasi-centralized "data-link"—the costly information network behind the interface, and the entities that can access and operate it. Here again, the history of the term is suggestive, insofar as the "drone" nickname traces one of its roots back to US Admiral William Stanley's name for remote-control aircraft used for target practice. He developed the planes after seeing a demonstration of the British Navy's DH 82B Queen Bee—another automated craft—and, in homage to this innovation, decided to call the remote-control planes he developed for the US Navy "drones"—not least because, as one account put it, "a drone could only function when controlled by an operator on the ground or in a 'mother' plane" (Zimmer 2013). Thus, the notion of drone logic invokes the infrastructures associated with drone deployment—and the political and economic implications of control over these infrastructures.

The following sections offer a rethinking of the operation of interactivity against the background of the figure of the drone. In this regard, the chapter intervenes in ongoing theorizations of digital media and their current state of development. To do so, it explores a series of case studies that consider the deployment of drone logic in a range of spheres of social practice. Each case study adds to a consideration of the dimensions and implications of this logic. (The issues raised, for example, by the generation of huge amounts of data that must be processed automatically, the attempt to make humans more "efficient" by finding ways of automating their own sensory and data-analysis processes, and so on.) Along the way, the chapter develops the beginnings of a critical, theoretical approach to the shared logics of "droning." In particular, it pits what might be described as a psychoanalytically informed dialectical approach against strands of a so-called new materialism that ends up replicating and reinforcing drone logics. With this in mind, the chapter outlines an alternative critical theorization of the assumptions and practices that underwrite the drone model of interactivity.

Enter the Dronosphere

> As billions and even trillions of sensors are placed around the globe and in our atmosphere, we will gain the ability to literally hear our world's 'heartbeat.'
> (Cisco Systems 2014)

> What would you say if I told you that in fifteen years, the FBI or local police might be able to identify you from a distance—perhaps from a drone hovering far out of sight of the human eye—using a technique called Eulerian Video Magnification and your biometric heart rate information? In other words, by reading your heart rate—unique to you—from a distance?
> (Sosadmin 2012)

What might it mean to think of non-drone apparatuses through the lens of drone logic? First of all, it means considering the ways in which distributed probe networks automate interactivity and result in large amounts of data that are centrally processed. For example, Microsoft researchers are developing an application that monitors smartphone use in order to infer the users' "mood" (and thus to more appropriately target advertising; LiKimWa 2012). Apparently, patterns of phone use, messaging, and Internet usage correlate unintentionally with the users expressed mood. Smartphones can serve, in other words, as mobile distributed mood sensors to collect data about users that they do not even realize is being generated. In a similar development, the US Homeland Security Advanced Research Projects Administration is developing an early warning system for chemical attack that would rely on sensors embedded in all smartphones: These sensors could "sniff" the surrounding air and automatically relay data about toxic substances to the program's headquarters—allowing for perpetual monitoring and rapid response (Department of Homeland Security 2013). We might describe both of these examples as partaking of the logic of "droning" insofar as they:

1. Extend the reach, scale, and scope of the senses.
2. Saturate the times and spaces being monitored. (The sensors are always on, and are distributed wherever members of the phone-equipped populace goes.)
3. Automate data collection.
4. Allow for automated response: Alarms are triggered, and response protocols are thereby initiated. This tendency toward the automation of response is one of the defining aspects of drone logic, insofar as the subtraction of the human element from the decision loop follows naturally from the automation of data processing. In the military arena, the goal of rapid response leads to the automation of targeting and strike. In the case of early warning chemical sensors, the goal is

similar to increase response speed. In the example of Microsoft's "MoodScope," targeted strikes take the form of customized, custom-timed advertising.

With these and a range of other examples in mind, we might start to make a case for the fruitfulness of using the figure of the drone as a means of interpreting and critiquing the emerging "sensor society"—in which the so-called "internet of things" comes to function as a distributed surveillant assemblage (Haggerty and Ericson 2000). Smart "things," from cars to toasters come to serve as probes, providing detailed information passively collected about their users and environments. This chapter abstracts away from the figure of the military drone to consider the imperatives that it addresses. It also situates the figure of the drone within the broader context of the goals of risk assessment, data management, and information processing associated with developments in the realms of commerce, sociality, and national security. The seeds of automation were sowed in what Ian Hacking calls the "avalanche of numbers" that accompanied the development of biopolitical strategies for managing populations in the nineteenth century (282). Indeed Foucault and Ewald's (2003) early formulation of biopolitics outlined the familiar connection between logics of securitization, the collection of data, and the management of "aleatory" forms of risk. As this avalanche grows, strategies for managing and interpreting the numbers come to rely increasingly on machine forms of data storage and processing and, inevitably, decision-making.

Drone Media

If we are to think of the drone as a medium—and I would argue that it is—then we might also consider the implications of the "droning" of the media. Here, I am invoking the ways in which media platforms not customarily associated with the figure of the drone qua "unmanned" aircraft come to serve as automated and distributed probes. For example, in a talk about digital media and monitoring, Eben Moglen succinctly described the monitoring model upon which the social media platform Twitter is based: "Every time you tweet a URL, Twitter is … arranging that anybody who clicks on that URL will be monitored by Twitter as they read. You are not only helping people know what's on the Web, but also helping Twitter to watch everybody you helped read, so they read over everybody's shoulder everything you recommend" (2014). Twitter is a communication medium that doubles as monitoring mechanism and a centralized probe. Twitter manifests its interface in multiple, mobile-distributed access points: smartphones, laptops, and electronic tablets. But each of these access points collects information about users and automatically captures, sorts, and stores them. Twitter is also what might be described as a metadata probe—it tracks information about which stories are receiving the most circulation. All forms of social media that allow for the sharing of stories, tweets, and other blip-memes operate similarly; they serve as both distribution platforms and distributed monitoring systems.

In this regard, we might describe Twitter as one model for the droning of communication media. As a communication medium it enables unique forms of one-to-many communication; as a probe it partakes of the logics of always-on data collection, analysis, and response. Of course, is it not alone—smartphones double as probe platforms, upon which new forms of sensors and new sensing techniques can be grafted. Drawing on the model of drone swarms (Bürkle et al. 2011), we might describe such devices and their supporting infrastructures as constituting probe networks. Interactive video games, for example, now incorporate technologies for viewing players and reading their facial expressions and biometric indicators. E-readers collect increasingly detailed information about user behavior: How long is spent on each page, which words are looked up or highlighted, whether and when a book is finished, which pages are revisited, and so on. Digital video recorders collect a growing range of metadata about user behavior and relay these back to providers who mine them for exploitable patterns.

Gmail is a form of probe that illustrates the logic associated with the droning of the media: in particular, the metadatification of everything insofar as all information comes to double as information about itself. Our emails are not read for their content (in the sense that the computer does not try to "understand" our thoughts) but for actionable patterns: Does the occurrence of a particular word in a particular sequence correlate with an increased likelihood of clicking on a particular keyword in an advertisement? Does one's pattern of email usage provide insight into other aspect of life? Does a particular viewing pattern indicate a greater likelihood of responding positively to a particular show (or to the advertisements embedded in it)? Such questions are not ancillary or incidental to the "primary" communicative function of online platforms like Google and Facebook—rather they lie at the heart of the emerging data-driven business model of the online economy.

Facebook and Google become valuable precisely because of the way in which they function as "always-on" probes able to gather and organize new forms of data that promise to lend themselves to a growing array of applications. These are probes that, like drones, are also apparatuses of "dedifferentiation": One of the hallmarks of the drone is its overcoming of distinctions between the battlefield and other (times and) spaces—or, more properly speaking, the drone enfolds all spaces within that of combat (Sharkey 2011). The drone patrols entire cities and regions and it documents not just exceptional moments, but the rhythms of daily life. In expanding the battlefield indefinitely (Packer and Reeves 2013), it proposes always-on monitoring and perpetual threat as a response to the mobilization of ubiquitous risk. The military figure of the drone has, in recent years, come to signify one side of so-called asymmetric warfare. The expansion of the battlefield it envisions corresponds to the multiplication of vectors of threat associated with the so-called war on terror. If the threat of attack is no longer limited to military personnel or targets, then all times and spaces fall within the reach of defense operations. If non-state actors engage in hostilities then the globe becomes the battlefield. The centralized drone network and the distributed terrorist network from two sides of the process of dedifferentiation between battlefield and home space; civilian and combatant. Drones are, in this context, dedifferentiating machines.

Our networked communications similarly fail to recognize or respect conventional divisions in the time and space of daily life: people text and surf, call and update at home and at work, in sites both public and private, in trains, buses, airports, and while walking down the street. Checking one's smartphone and email can be the last thing one does at night and the first upon rising. Even the pattern of "downtime" generates potentially valuable information about users. All times and spaces are enfolded into the competition: for information, speed, efficiency—combat by other means. The communication probes launched by applications like Facebook and Twitter weave themselves into the fabric of daily life. At the same time, the form and volume of data collection necessitates new ways of handling and using information—themes taken up in the following sections.

Drone Commerce

One of the hallmarks of the forms of dedifferentiation characteristic of "droning" is the ease with which military and commercial applications blend into one another and inform one another. There is nothing new about this connection—marketers have long availed themselves of various military technologies, and the military has played a crucial role in the development of media (c.f. Kittler 2010; Virilio 1989). However, the pace of the exchange and the conceptual blurring between targeting and target marketing, between surveillance and forecasting, and between various forms of risk management is unprecedented. The CIA bases its information processing on Google and then commandeers the data it generates (Hunt 2012; Greenwald 2014). With respect to drone logic, we might trace the connection between the data overload generated by distributed sensors and the proliferating forms of information to which consumers are subjected. Cutting through the clutter, as it were, becomes a key concern in both realms. Consider the example of a Defense Advanced Research Projects Agency (DARPA)-funded research project designed to maximize the ability of human viewers to respond to visual data. The so-called Cortically-Coupled Vision System uses electroencephalography (EEG) to determine when viewers' brains respond to particular images—prior to their conscious registration of this response (Gerson et al. 2006). That is, brains are treated as sensors that, when connected to a monitoring infrastructure, allow viewers "to find meaningful objects in mountains of images up to 10 times faster than they normally could" (Discover 2011). The problem faced by the military is that new drones generate imagery faster than humans can make sense of them (Discover 2011). The resulting search for more efficient ways of processing visual imagery underwrites attempts to, on the one hand, develop forms of automated processing of visual imagery and, on the other, to "drone" humans by turning their brains into automated sensors decoupled from conscious processes. The human brain becomes a more efficient probe than human viewers. Unsurprisingly, the shift from military to consumer applications is immediate: The project's director Paul Sajda emphasizes the potential benefits of his military research project for civilians:

> A miniaturized, wireless version of the device might be used to tag consumer items or even specialty shops that catch your fancy as you walk down a city street. Just a quick glance at a dress in a window, for instance, might elicit a neural firing pattern sufficient to register with the system. A program could then offer up nearby stores selling similar items or shops you might want to investigate. (Discover 2011)

The logic here is familiar: Your own behavior provides cues to your preferences that can be captured and put to use. In Sajda's example, the human brain is turned into a desire-detecting sensor. Suggestively, the implicit connection is made between the proliferation of surveillance data and that of commercial messaging: As one enthusiastic account puts it, "If you have ever felt overwhelmed by a multitude of choices—say, 10,000 items in an online catalog—this brain-boosting invention is for you" (Discover 2011). Consumers are analogized to intelligence workers—forced to make sense of the avalanche of data available to them, and thus to avail themselves (and subject themselves) to the logic of automated data processing. More generally, then, the use of distributed probes for commerce works not just to rationalize the production process, but also the consumption, promotion, and distribution network—indeed, it further dedifferentiates the distinctions between them: consumption contributes to the production process in new and significant ways, distribution networks are redoubled for the purposes of promotion; interactive promotional strategies feed into the production process, and so on. The end result is an ongoing process of data collection that embraces the twin logics of retrospection and prospection: the goal of both creating as complete an archive as possible and of using this as a means of projecting into the future.

Drone Security

In 2014, residents of Compton, California learned that their city had been the subject of a 2012 experiment in total video monitoring 2 years ago. The LA County Sheriff's Department contracted with a private surveillance company to test an airborne camera that monitored, "the entire city 24-hours a day using high resolution video of everything that happened inside the 10-square-mile municipality" (Friedersdorf 2014). For 48 h, everything in the city was tracked. As the city's privacy consultant put it, "We literally watched all of Compton during the times that we were flying, so we could zoom in anywhere within the city of Compton and follow cars and see people" (Friedersdorf 2014). The imagery was archived, so that if a crime had been reported during that period, the police could go back and retrieve the video and zoom in on the location of the crime. The demonstration envisioned the possibility of comprehensive real-time monitoring and archiving: the digital redoubling of space and time that would make it possible to reconstruct the past—and, through the power of data analytics—predict future threats. The surging interest in so-called predictive policing relies on this possibility—the more data that can be collected, the greater the reach and power of the algorithms that can be used to preemptively allocate policing power (c.f. McCue 2006). The realm of security is the native habitat of the drone, with its military provenance, its strategy of asymmetry, and its direct link to surveillance technology.

The Compton experiment illustrated the drive toward "diachronic omniscience" (Parks 2001) that characterizes drone logic. The fantasy of the drone is to cover all spaces all of the time: hence, the drive toward more high-resolution cameras with broader ranges of field carried by devices that can stay in the air as long as possible. Consider for example, the US Air Force's so-called Gorgon Stare project, which uses multiple cameras on a remote control aircraft "to transmit live video images of physical movement across an entire town" (Nakashima and Whitlock 2011). In keeping with drone logic more generally the shift is from individualized forms of targeting to treating the entire population/space as the target. As one military official put it in a news account, "With the new tool, analysts will no longer have to guess where to point the camera. …Gorgon Stare will be looking at a whole city, so there will be no way for the adversary to know what we're looking at, and we can see everything" (Nakashima and Whitlock 2011). The further, logically related goal, is that of drone autonomy, envisioned by The DARPA, "CODE" project (Collaborative Operations in Denied Environments), which anticipates the creation of smart swarms of drones. Or, as the agency describes it, the goal is "autonomy and inter-platform collaboration … Using collaboration algorithms, [drones] can provide services to each other, such as geo-locating targets with long-distance sensors and guiding less-capable systems within their sensor range, … [and] protecting each other by overwhelming defenses and other stratagems" (RT.com 2014).

In keeping with the generalization of the logic of prosthetic extension, DARPA has a host of related projects, including an expansion of its underwater drone project and the development of robot soldiers and silent vehicles, that is, it envisions the drone colonization of land, sea, and air—diachronic and synchronic omniscience (RT.com 2014).

From the perspective of "defense" and security, then, drone logic envisions the coupling of comprehensive surveillance with perpetual low-grade warfare enabled by structural asymmetry. The prosthetic distancing enabled by the drone promises to lower the cost of perpetual "low-intensity" conflict for the dominant party by removing fighters from the scene of battle. (However, the promise may well fall short of the reality, given the forms of anger, frustration, and desperation that ongoing long-distance attacks generate among the victims.) This structural asymmetry is characteristic of drone logic more generally. When it comes to surveillance, for example, there is the asymmetry between those who are perpetually and comprehensively monitored and those whose monitoring activities remain covert, nontransparent, and remote as they disappear into the weave of the fabric of daily life. It is perhaps a telling sign of the time that in publicity materials for the latest Superman movie it is revealed that Batman is semiretired, "controlling drones from the Batcave" (Caruso 2013).

Drone logic, in other words, fulfils the fantasy of preemption by other means (Bogard 1996). If one information age fantasy is the perfect anticipation of future events via forms of data modeling that allow threats to be identified and addressed before they materialize, the drone fantasy envisions forms of confrontation without threat. This is perhaps why there is an unfolding logic of internal versus external uses of drones for the purposes of security and policing. The "war on terror" version of drone warfare relies on the identification of external enemies at a distance: threats that can be defused "over there" before they arrive here, on "our" shores. In this version of drone warfare, the generalized ubiquitous threat that materializes at a distance is complemented by the image of a safe home space from which this threat can be handled: a sheltered shipping container in a Nevada suburb where life can proceed as normal.

By contrast, the notion of drone security envisions a domestic space already permeated by threat: one in which areas like Compton need to be photographed 24 h a day in order to record patterns of activity for the purposes of preemption or reconstruction. Perhaps this is why the figure of the drone attracts so much attention; it embodies the ways in which a militarized asymmetry comes to characterize domestic security operations. Hence, the convergence of critique from the left and the libertarian right around the issue of drone security. The high-profile libertarian US Senator Rand Paul has made a point of mobilizing the figure of the drone as avatar of totalitarian control, insisting that the US attorney general answer whether or not he thinks it would be legal for the president to "authorize … a drone strike, against a U.S. citizen on U.S. soil?" (Epps 2013). As recent (and less recent) events in the USA indicate, various policing entities already routinely exercise lethal power, sometimes against US citizens who do not pose any immediate threat—the invocation of the image of the drone adds the dimensions of remoteness and automation. These latter dimensions (coupled with ubiquity) represent the novel aspect of drone logic.

Drone Pedagogy

The range of examples explored in this and previous sections suggests the portability of "drone logic." In similar terms, one of the many productive insights of Foucault's work on the panopticon was his abstraction of its logic away from particular applications, buildings, or uses. He was interested in the panopticon as "the diagram of a mechanism of power reduced to its ideal form" (1977, p. 198). Thus, he argues that, "its functioning, abstracted from any obstacle, resistance or friction, must be represented as a pure architectural and optical system: it is in fact a figure of political technology that may and must be detached from any specific use" (1977, p. 198). This detachment of course, made it possible to rearticulate the technology to a wide range of disciplinary uses in a range of enclosures (the school, the workplace, the prison, etc.) and beyond. Drone logic invokes a similar diagram of power—one in which logics of remote sensing, networking, distributed ubiquity, mobility, and automation coalesce. As in the case of the panopticon, drone logic operates in the school as well as the workplace, and the more flexible and expansive spaces of digital interactive enclosures (smart cities or homes, mobile phone networks, Wi-Fi hotspots, and the internet). The tools for learning are fast becoming networked, sensorized, and automated. As educational materials become interactive, every student action will be redoubled in the database. Describing a Rupert-Murdoch-funded initiative to develop educational programming for electronic tablets, one press account captures the tone of the droning of the schools:

> Soon, games that know what a student has read … will be able to strategically sprinkle a particular word in his path based on how many times the research says you need to see a new word in order to learn it. In a few years … advances like "gaze tracking" and measurement of pupil dilation "will revolutionize" the gauging of cognitive response by making it possible to determine exactly what students are reacting to on the screen. (Rotella 2013)

The drone architecture in the schools then will be composed of networked portable, customizable, personalized tablets that serve as the converged platform for educational activities—reading, writing, viewing, listening, communicating, and so on. The result will be a constant flow of data upstream that feeds into automated forms of response—and, of course, the creation of comprehensive and increasingly complex student profiles:

> This growing stream of information, which can be analyzed down to individual keystrokes, yields a picture that will eventually progress in complexity from, say, a list of words a student looks up to a profile of metacognitive skills—like the ability to concentrate—and in time to a full-blown portrait of a developing mind. In theory, each student will generate the intellectual equivalent of a fantastically detailed medical chart. (Rotella 2013)

Such a chart will have a variety of potential purposes—useful not just for educators but for medical and mental health professionals, for college admissions, for employers, and, of course, for marketers. Murdoch's company, Amplify, insists (for the moment) that the data will remain solely in the hands of individual school districts (Rotella 2013). As it expands of course, and comes to be stored in "the cloud" it is sure to become available—at least in anonymized forms, to commercial data-warehousing companies. It could also become a very tempting asset for cash-strapped schools: Why not sell anonymized student data, if this can help recover some portion of the huge sums invested in "tabletizing" the schools?

The development of such technology is allegedly undertaken with an eye to providing guidance to teachers, but the logic of automation is irresistible: There is no way any individual teacher will be able to make sense of the huge amounts of data generated by the constant use of an interactive device by a class full of students. Of course the sorting, sensemaking, and response apparatus will be automated—just like the game that determine which words a student needs to work on and then shapes the reading curriculum to feature those words. There are surely some wonderful uses for tablets in the schools—but the droning of the schools is another matter. As pedagogy becomes automated, bypassing teachers themselves (realizing a recurring fantasy of the political right's ongoing campaign against public school teachers), it comes to be structured by the imperatives of those who craft the algorithms and mine the data. The operation of the algorithm becomes increasingly opaque in direct relation to the increasing amounts of data captured and processed.

If Amplify succeeds, Rupert Murdoch's private, commercial, media empire will control not only a vast swath of the news to which people are exposed but also the algorithms that shape the educational priorities and activities of a nation's schools. The paradox of drone logic is that individual specification and customization—the alleged push beyond the mass society model—coincides with increasingly powerful and efficient forms of centralization. Indeed, "total" individualization can only be provided by automated, networked devices: It would be impossible to have a teacher for each student, but not a tablet. And the tablet only operates in the mode of individuated instruction (as opposed to individuated use) to the extent that the process is remotely, centrally, and automatically managed. The interface is nothing without the infrastructure: This is the central lesson of drone logic. Or, as one Air Force general put it, "It's about the datalink, stupid" (Bowden 2013). A lone, disconnected tablet is like an untethered drone. It can perhaps record data, but it cannot share it, mine the combined data drove, and generate an automated response based on these aggregated findings. Droning, appropriately enough, is about the massification of individuation.

Drone Sociality

In keeping with the notion of "drone logic" as a portable model, we might ask the question regarding what happens to sociality itself when it is read through the lens of the drone. One possible answer lies in the example of an app called "NameTag" that received a lot of media attention in 2013 when it was marketed as the "first facial recognition app for Google Glass" (http://www.nametag.ws/). It portrayed itself as a kind of date-detecting drone: an application that would scan a room for faces that it could then link to social networking sites in order to assess compatibility with the user. As one breathless headline put it, "Could Google Glass Find Your Dream Date?" (ignoring the fact that Google, as of now, has not approved any app that uses facial recognition to identify anyone but the user; Prigg and Thornhill 2014). The app, in other words, transposes automated monitoring, detection, and profiling (associated with, say, the signature strike), into the realm of social life. Moreover, the app transforms Glass into a threat detection device: "Its makers claim it can even check the sex offenders register before deciding on a match" (Prigg and Thornhill 2014). The app caters to the lure of the missed opportunity. In a world in which we pass mysterious and perhaps eligible strangers every day on the street or at a party, "Why leave meeting amazing people up to chance? Don't miss out on the opportunity to connect with others who share your passions!" (http://www.nametag.ws/). The flip side of ubiquitous threat is ubiquitous opportunity. The technology offers to make the process of meeting compatible people much more efficient, continuing the logic of devices like LoveGety and exporting the search-and-sort function of online dating into the offline world (or, rather, folding the offline world into the online one, via Google Glass).

Drone sociality relies on automated opportunity identification and threat protection; it would take too long to interview everyone in a crowded room/street/party and make them fill out OKCupid style questionnaires. But if they have already done so online, NameTag can do the work of identifying them and looking them up for the user. The model is a familiar one in an era of information glut—when the (perceived) field of choices expands dramatically, automated forms of sorting and recommendation take over. The dependence of Web users on Google has become so taken-for-granted as to be almost overlooked—without a browser, the tremendous amounts of information available online would remain all but inaccessible. The same might be said about the expanding field of media content available to users—once upon a time a viewer could sort through the available channels without assistance. But with thousands of movies available on Netflix, users need some kind of automated or externally generated sorting and recommendation system. When it comes to control over information, organization is the new content. Whereas once upon a time, the attempt to limit access to particular kinds of information or knowledge depended on restricting or censoring content, in the era of infoglut, it depends on controlling the algorithms that search, sort, and recommend.

Drone sociality transposes this logic into the realm of social life. It is already deeply entrenched in social networking sites like Facebook which organize our exposure to messages from "friends" for us, deciding which are most likely to further the goals of the site (to get us to spend more time on Facebook, to post more information about ourselves, and to participate in online activities that contribute to the company's bottom line). The company's announcement of a smartphone app that would activate the user's microphone to sample the sonic environment and automatically detect (and share) information about what music or TV is playing in the immediate vicinity highlights the droning of Facebook (Hill 2014). The app would treat smartphones as distributed content probes, scanning the environment for information about users and relaying this not just to friends, but, of course, to the company's database of information about the listening and viewing habits of its users. More generally, sites like Facebook already partake of drone logic, insofar as they serve as ubiquitous, networked, automated forms of information capture and response. They do so in a double register: For users they serve as automated data collection devices about the sentiments and activities of "friends," whereas for their owners they operate as distributed probes gathering an increasingly comprehensive map of the social landscape: of patterns of interaction, communication, and affiliation (as well as of movement, media consumption habits, and an expanding array of information). Of course, these probes generate more data than would be humanly possible to make sense of, requiring automated forms of sensemaking and response. In this regard, such sites face the common problem of drone monitoring: how to manage the growing array of data made available by increasingly pervasive forms of information capture in a growing range of dimensions. Users, interestingly, face a similar problem—and the development of apps like NameTag indicate how social life comes to be viewed through the same lens as other forms of information retrieval and processing. NameTag promises to open up the field of opportunity for social relations in the way that the Web opened up the range of available information. With the help of the app, any face in the crowd can potentially be de-anonymized and vetted without the time and effort associated with introductions, conversation, and other forms of human contact that, paradoxically seem to gunk up the promise of friction-free sociality.

The invocation of logics of targeting and risk detection into social relationships is not particularly novel and is at least as old as their treatment in market terms. However, the droning of social life enacted by social networking sites and applications of various kinds (including dating and hookup apps like Grindr and Tinder) updates these logics for the digital era, in which the automation and digitization allows it to expand in scope and to permeate time and space in new ways. A person sitting on a park bench can use Tinder to search for potential dates up to 99 miles away. For the hypothetical user of NameTag, anyone within sight is a potential match. The increasing reliance upon the algorithm—and the automated forms of sorting and organization it enables—endows it with novel forms of power associated not so much with the content of the information being shared, but with the structure of sharing itself. As social life migrates onto online platforms, its dimensions become reliant on automated forms of sorting and response. Facebook notoriously uses algorithms to shape the information streams users receive from online contacts to meet its particular sets of priorities (fostering forms of engagement that contribute to its economic model). Along the way, of course, it engages in ongoing forms of experimentation to determine how tweaking the algorithm might influence user behavior. The 2014 revelation of an experiment designed to determine whether changing the news feed good influence users' moods generated some concern and controversy—but this was just a quick peek at the forms of ongoing experimentation that such sites do as a matter of course. As OkCupid's co-founder put it in a blog post after the Facebook revelations: "We experiment on human beings … But guess what, everybody: if you use the Internet, you're the subject of hundreds of experiments at any given time, on every site. That's how websites work" (Rudder 2014). And those with the power to craft the online environment and capture the information it generates stand to benefit from forms of monitoring and manipulation that are largely invisible to users. In this regard, every Web site and all networked communication devices operate in the drone dimension, all serving as probes that surreptitiously captured detail information that is collected, aggregated, and analyzed in order to provide automated forms of customized response. All the ingredients are there: distributed, always-on, ubiquitous monitoring devices, infrastructures for data collection and response, and automated forms of information processing and response.

Drone Politics

Of course, any site of potential power is also of interest in the political sphere. Thus, political consultants are jumping on the big data bandwagon in order to figure out how best to influence potential voters, contributors, or supporters in automated data-driven ways. As one somewhat retro scenario outlined in a political consulting trade magazine put it,

> The next frontier is embedding computer chips in direct mail pieces. No need to enter a web address or take a picture with your cell phone to access content … The computer chip will then match tested messaging to the user's profile and online history, generating a personalized message for the user. (Mack and Henry 2012)

In this scenario, the letter itself serves as part of a probe swarm, linking a particular recipient with other datasets, from online behavior to marketing history, to voting history: Whatever is available in the database. The goal, unsurprisingly, is to provide an automated customized appeal to a known target: "For example, if the user does web searches on education, the computer chip will calculate the dollar amount her neighborhood schools will cut if the wrong candidate gets elected in her state or district. If the user is a thirty-something female with two young children, the picture on the front of the website will be the same" (Mack and Henry 2012).

Viewed from the perspective of data miners, the broad sweep of interactive applications and devices serve as probes for inferring political preferences and proclivities even when these are not directly expressed. One group of researchers, for example, claims to be able to use Twitter and other social networking sites to infer a user's political leanings, which the researchers describe as "latent content": "even when limited or no self-authored data is available, language from friends, retweets and user mention communications provide sufficient evidence for prediction." (Volkova et al. 2014). Indeed, the researchers found that, using Twitter, "political preference can often be predicted using roughly 100 tweets" (Volkova et al. 2014). The music streaming company Pandora has developed algorithms to predict political preferences based on musical taste: "It can deconstruct users' song preferences to predict their political party of choice" (Singer 2014). The company says that it "does not analyze listeners' attitudes to individual political issues like abortion or fracking" (Singer 2014)—at least for the moment. However, a computer researcher, mulling the power of Pandora as a probe, speculated that, "looking at music choices, you could probably predict with high accuracy a person's worldview … You might be able to predict people's stance on issues like gun control or the environment" (Singer 2014). This possibility, of course, is what makes the automated data collection associated with networked media use, so interesting to political consultants and strategists seeking new ways to target and influence voters. Drone politics treats any and all information that can be collected about voters as potentially useful for inferential purposes, that is, for excavating information about preferences and possible points of influence. The looming disconnect between user expectations and the intentions of data miners is constituted by the difference between conscious forms of self-disclosure (I will let Pandora know which songs I like) and inferential data mining (We can figure out what political issues are important to you from your musical tastes). If the goal of drone surveillance is the redoubling of everyday life in the form of a data profile, every aspect of life is rendered productive for a growing array of uses: threat detection, marketing, and influence. Every activity that takes place within reach of some element in a drone assemblage finds itself at the business end of a sensor. The ambition of the network incorporates the fantasy of the drone: 24/7 coverage: sleepless sensing and sensemaking—a certain gaplessness. With this goal in mind, we might move beyond the notion of drone politics as yet another instantiation of the application of drone logic: A notion leading to the consideration of the ways in which political consultants, like marketers, enlist data collection and data mining to more effectively target and influence key voters and constituencies. The larger question here regards what the implicit politics of drone logic itself—that is, in what ways does this logic have political effects, construed in the sense of a broader concern with questions of control and governance.

A consideration of drone politics at this somewhat more general level suggests the following themes:

1. Drone logic augurs an era of the automation of influence, in the sense that customizing and targeting necessarily become the product of automated processes of data collection, analysis, and response. A politician can only tailor a message within the constraints imposed by the ability to absorb, understand, and personally implement knowledge about a voter base. Drone logic imagines the possibility of perfect specification: at the limit, a customized message for every inbox.
2. Drone logic only individualizes through the lens of the aggregate. Ultimately, it is not interested in individuals per se, but only in the patterns of influence that emerge from upstream data collection and downstream targeting of influence. The hallmark of the drone attack is the so-called "signature strike"—that is, the attack based on a data-generated profile, rather than on positive identification. As one news account puts it, "The term 'signature strike' is used to distinguish strikes conducted against individuals who 'match a pre-identified "signature" of behavior that the U.S. links to militant activity,' rather than targeting a specific person." (Greenfield 2013). Much the same might be said of the broader forms of drone targeting described in previous sections: The signature pattern is more important for the decision-making process than is personal identification.
3. In the drone era, sorting and mapping are the new forms of scarcity. Information is plentiful: It is control over the ability to make sense of this data that becomes the source of power. Thus, logistical media like search engines, algorithms, and other organizational platforms become central to the ability to make information useful. Without a good search engine, the huge amounts of information available online become useless. Much the same might be said of the troves of data accumulated by companies like Facebook, Google, and Twitter.
4. In drone logic, the process of automation tends toward the obliteration of what it relies upon. The goal is to surpass not only the human limitations on sensing and sensemaking, but on response time, and response specificity. The paradox is a familiar one: The whole system depends on the data generated by a category of raw material (humans) that it seeks to surpass. What makes the system possible at all is what "gums it up"—serving as the last remaining bit of resistance to the instantaneity of a feedback loop spiraling faster and faster.

Drone Knowledge

Perhaps the most important political aspect of drone logic is the form of knowledge it ushers in: What might be described as a post-referential, post-explanatory, and pragmatism. In the account I have been offering here, drone logic is inseparable from automated forms of data processing: that is, ubiquitous, networked, always-on sensors generate too much information to be handled by humans. Data processing at this scale in turn generates forms of knowledge that are, as one account puts it, "too big to know" (Weinberger 2011)—that is, when they are working as planned, they represent the complex interactions of large numbers of variables that are too complex to render in the form of written or spoken hypotheses or explanations. The patterns are emergent—they are a product of the process itself, and their reliability is simply a function of how robust they remain over time (as calculated using historical data and predictive analytics). Cybernetic control, in its drone manifestation evinces its roots in targeting and prediction.

Against this background we might reconsider the ways in which relatively recent developments in theory anticipate and reinforce drone logic. I am tempted to describe these forms of knowledge, disparate as they may seem, as partaking of drone logic, and hence as forms of "drone theory." By this term I mean to designate theoretical approaches that push in the post-explanatory, post-discursive, and, in some respects, post-human direction. Perhaps unsurprisingly these approaches tend to ally themselves with an allegedly resurgent materialism (albeit a decidedly non-dialectical one). That is, they push beyond the realm of "discourse" and language to study "things," evincing an interest in "material" histories, "material" cultures, and new "materialisms" of one kind or another. It is not difficult to discern in these approaches an echo of the high-profile data scientist Alex Pentland's admonition to pay attention not to what people say but to what they do (Eggers 2014)—and even more broadly, to what things do (as being not discontinuous with what people do)—or, what human-thing assemblages do.

In treating language as one more set of objects to be scanned, analyzed, and responded to in automated ways, drone logic participates in what might be described as the metadatification of language. When Google scans and analyzes messages, for example, it transforms the content into data *about* the content: A word in a message is not read for its communicative import to a receiver, but is registered as a fact about a particular communicative act—at a particular time and in a particular location, person A included the words "trip" and "Paris" in a message, so perhaps this person might be a good target for an advertisement for hotels or restaurants in Paris (for example). Certainly the data processing is becoming increasingly sophisticated, but again the goal is not to derive interpretations of the content, but rather to align data about messages with data about responses.

This process of "metadatification"—whereby a message is reconfigured into data about itself—comes to stand as the post-ideological or post-textual moment taken to its logical conclusion. That is, once the notion of a containable, interpretable, and transmissible "dominant" meaning (or meanings) is deconstructed beyond recognition—what we are left with is the circulation of affects, and eventually, from an instrumental point of view, their effects.

In the wake of what might be called the mistrust of signification—that is, the debunking of an ideological register over and above or otherwise distinct from an affective register, it is not particularly surprising that one of the recent tendencies embraced by post-discursive approaches to the media is toward the analysis of the circulation of affects rather than that of meanings, content, or representations. Such movements tend to fashion themselves as anti-discursive in their rejection of the linguistic turn's ostensibly surpassed focus on representation and cognition and toward bodies (and things) in their materiality (rather than their signification). Karen Barad, whose theory of "Agential Realism" has been influential in the development of "new materialist" approaches that have been taken up in media studies (and digital media studies in particular) argues, for example, that,

> Language has been granted too much power. The linguistic turn, the semiotic turn, the interpretive turn, the cultural turn: it seems that at every turn lately every 'thing'—even materiality—is turned into a matter of language or some other form of cultural representation. (2003, p. 804)

Unsurprisingly, in its antipathy to reading ("no humans read your email") Google might well agree with this anti-anthropocentric formulation (and drone-friendly logic).

In the same vein, one of the provocateurs of so-called object-oriented ontology expresses a disdain for the distractions of language as only a video game aficionado could: "When we spend all of our time reading and writing words … we miss opportunities to visit the great outdoors" (Bogost 2012, p. 90).

A different, but perhaps not unrelated tendency is also manifested in the growing recent interest in German-inflected medium theory, where the influence of Friedrich Kittler looms large. As Jeremy Packer puts it, Kittler's work represents a "cold-turkey cure for hermeneutic and ideological fixation" (2013, p. 295). There is a certain affinity between the post-human commitments that surface in "new materialist" approaches and the fact that, as John Durham Peters puts it (in part of his introduction to Kittler's *Optical Media* titled "This is Not Cultural Studies") the German theorist, "has no use for the category of 'the human' or 'experience'" (Kittler 2010, p. 5). For him the conditions of the latter are subordinated to core media processes in a particular historical period: the collection, storage, and processing of data. The virtue of such an approach, according to Packer, is that it addresses a shift in the way digital media operate—a shift that he describes in distinctly logistical terms: "digital media power is first and foremost epistemological, not ideological. It is computational" (2013, p. 297). We might say much the same of drone power: it is post-ideological and post-discursive. Drones do not monitor ideology, they track sequences of correlation as proxies for chains of affect and effect. Even a "drone" analysis of text bypasses the register of "meaning" proper to seek out those sequences of words that correlate with responsiveness to particular influences and behaviors. In this regard, drone theory amounts to a kind of post-theory that amounts to what Chris Anderson describes as "the end of theory" ushered in by the advent of big data: "Out with every theory of human behavior, from linguistics to sociology … Who knows why people do what they do? The point is they do it, and we can track and measure it with unprecedented fidelity. With enough data, the numbers speak for themselves" (2008). Like Anderson, Packer draws on Google's data mining prowess as a reference point for his analysis of what this chapter describes as the drone logic of automated monitoring and response:

> Google's computations are not content-oriented in the manner that advertising agencies or critical scholars are. Rather, the effect is the content. The only thing that matters are effects—did someone initiate financial data flows, spend time, consume, click, or conform? Further, the only measurable quantity is digital data. Google doesn't and couldn't measure ideology. (2013, p. 298)

After all "no humans read your email."

In this regard it is worth asking, with Alex Galloway, "Why, within the current renaissance of research in continental philosophy, is there a coincidence between the structure of ontological systems and the structure of the most highly evolved technologies of post-Fordist capitalism? … Why, in short, is there a coincidence between today's ontologies and the software of big business?" (2013, p. 347). Admittedly, the various strands of post-cultural theory referenced above (new materialism and affect theory, object oriented ontology, and German medium theory) have some very significant differences, but they partake to various degrees in countering the linguistic/discursive turn. It is not difficult to discern certain affinities between tendencies in these approaches and a "datalogical" turn more generally—that is, a growing interest in what automated forms of information processing can reveal about the object formerly known as "content." The attempt to get data to "speak for itself" parallels the appeal to a post-hermeneutic approach to the (vibrantly) "material"—and a precognitive approach to the effects mobilized by the circulation of affect.

In keeping with the post-human ethos, data miners are not particularly interested in interpretive narratives of human intentionality. In the great "democracy of things" represented by the database, data about air pressure, broken windows, and the number of daylight hours commingle with the more distinctly human statistics about poverty and abuse. They are all endowed with a role in the formation of potentially useful patterns from the data—and thus treated as contributors to the monitored outcomes, whether these be crime rates, purchase volumes, catastrophic events, the circulation of diseases, and so on. The multiplication of variables and the automaticity of the algorithm displace the roles of interpretation and comprehension—the data are not so much symbols that need to be understood as they are inputs that need to be sorted. The patterns generated by the database are not explanatory, but actionable. The goal is not to unearth the depths of significance, but to accumulate enough raw material to allow surface patterns to emerge. Humans take their place—not necessarily a central one, and not necessarily in the distinctive role as subjects—within the growing menagerie of data about inanimate objects, forces, interactions, flora, and fauna. We might describe this as the perspective of the drone: a description of drone experience—the experience anticipated and aped by drone theory.

While the resurgent focus on the significance of the material, the biological and technological has much to contribute to an understanding of the operation of emerging media forms and practices (perhaps not least because of the affinity noted by Galloway), their attempt to bypass or background the discursive dimension poses its own set of political concerns. Perhaps the defining one is the combination of an unfolding recognition of the complexity of networks of interaction with a correlative leveling tendency (all parties to the interaction become subsumed to its overall effects). Consider, for example, the substance of Jane Bennett's (2009) version of "vibrant" materialism, in which simple narratives of causality are displaced by the recognition of the constantly unfolding network of contributors to any event or outcome. These go far beyond the human "agents" and include the material properties and affordances of any of an expanding array of participants. A power blackout, for example, is not simply the outcome of human activity, but of a range of "very active and powerful nonhumans: electrons, trees, wind, electromagnetic fields" (2009, p. 47). The logic is both indisputable and indefinitely expandable (in ways that are familiar to older versions of materialism): The winds themselves are outcomes of a variety of "actors," (the rotating earth, the sun's activity, and so on) as are the material properties of the wood of which trees are composed, and on and on.

The result, suggestively, is that the "agent" is the *outcome* and not the cause of a particular "assemblage." (To transpose this into communicative terms, the message is the effect, as Packer suggests.) The corollary is that subjective notions of intention are subsumed by the assemblage: "an intention is like a pebble thrown into a pond or an electrical current sent through a wire or network: it vibrates and merges with other currents, to affect and be affected" (Bennett 2009, p. 32). Such a formulation does not simply disperse or distribute agency, it also reassembles it in the form of the "agentic capacity" of the assemblage—what Bennett describes as "the dynamic force emanating from a spatiotemporal configuration rather than from any particular element in it" (Bennett 2009, p. 34). The message is the effect, and the effect is, in a sense, de-narrativized (the only narrative that could adequately encompass the infinitely expanding horizon of contributing agents is that available to a ubiquitous, distributed, all sensing drone network). Indeed, the only understanding and accounting of such an assemblage would be that which is too big to know.

Taken to its post-human limit the drone perspective amounts to a kind of view from nowhere—and hence a resurgent idealism, despite the materialist posture. And the view from nowhere has a tendency to align itself a form of undifferentiated pluralism or imagined harmony that, wittingly or not, replicates the portrayal of the market as the most *natural* of cultural assemblages. Barad's ethical injunction is perhaps compelling, in general terms: "Intra-acting responsibly as part of the world means taking account of the entangled phenomena that are intrinsic to the world's vitality and being responsive to the possibilities that might help us and it flourish" (2007, p. 396). Whence the notion of a unified "it" or "us"—and the underlying assumption of the possibility that everything encompassed therein might flourish together (nuclear power plants and humans? Carbon combustion devices and oxygen breathing creatures? Flesh-eating bacteria and limbs? Capitalist and proletariat).

The fantasy of drone politics and the theory that underwrites it is that it might be either possible or desirable to permanently bracket the cultural, as in the complementary trajectories of, on the one hand, the forms of automated tracking and sensemaking that dispense with the need for comprehension and, on the other, the analysis of assemblages in their infinite complexity (a process that necessarily tends toward and will likely come to privilege automation—as in some strands of the digital humanities). It is perhaps not surprising that even in the humanities there has been a proliferation of grant requests to fund the hardware and software that allow for the automated analysis of large numbers of texts, ranging from historical bodies of literature to Twitter feeds. The goal here, in all of its post-representational glory, is to gain some kind of efficient "knowledge" without having to read, Google-style. The impulse resonates with that of automated data mining for security purposes: the generation of suspects that has nothing to do with the interpretation of individual narratives, with posited causal interpretations, but simply with patterns of interactions over time (the signature strike). Action can be taken without any interpretive understanding of why—without "reading," as it were.

There is an historical logic to the advent of what I am calling drone theory: on the one hand, it aligns itself with recent developments in other knowledge spheres—those which are coming to rely increasingly upon emerging automated processes. At the same time, it addresses the challenges posed by the interpretive impasses associated with discourse-oriented approaches. In this regard, it partakes of what might be described as the neo-pragmatism that seems to unite various strands of post-metanarratival theory while simultaneously eliding the realm of the subconscious, desire, and psychoanalytic subject. These are also, perhaps not surprisingly, the blind spots of the correlational logic of data mining and analytics, which do not seek to reconstruct subjects or parse their motivations, conscious or otherwise. When the data speaks for itself, the subject need not. The term "drone," with its connotations of asexual reproduction, is suggestive in this regard. For what both theoretical and practical versions of drone thinking elide is the relationship between language, the subconscious, and desire. Post-humanist thought of the materialist variety outlined above remains, in this regard, both post-discursive and, necessarily, postpsychoanalytic. Drone thought is asexual in the sense that it seeks to bypass the deadlock of the symbolic—whose defining feature is, as Slavoj Zizek (2011) argues, the nonexistence of the sexual relationship. The deadlock of the real, he argues, is "inscribed in the very core of human sexuality: 'There is no sexual relationship.' Human sexuality is marked by an irreducible failure, sexual difference is the antagonism of the two sexual positions between which there is no common denominator" (p. 65). Perhaps the clearest example of the resultant form of false pluralism—pluralism under duress (to invoke Adorno's notion of "reconciliation under duress") is the uncanny recurrence of harmony in the form of the assemblage—that is, the implicit disavowal of the type of structuring deadlock that interests Zizek (and that he associates, of course, with capitalism). In such analyses, the assemblage is the only remaining element of structure—and it is neither a causal nor an explanatory structure. Description can delve beneath the level of the assemblage—its elements can be accounted for, but the notion of a fundamental structuring deadlock within the assemblage remains foreign to the analysis. How else to account for Barad's imagery of "the world's vitality" and its shared flourishing with "us"? There may be interactions that cancel one another out, or result in one element overwhelming another—but no deadlocks or contradictions running like fault lines through these various assemblages. There may be split subjects, but there are no split assemblages. Perhaps the challenge posed by the disconcerting complementarity between these variants of post-humanist materialism and what I have been calling "drone theory" is to narrativize the post-narrative turn—to tell a story about why it might have come into being and what purposes it might serve; after all, try as it might, it has not (yet) wrested itself from the narrative logic it eschews.

References

Anderson, C. (2008). The end of theory. *Wired Magazine,* July, 16.7, 33–38.

Barad, K. (2003). Posthumanist performativity: Toward an understanding of how matter comes to matter. *Signs, 28*(3), 801–831.

Barad, K. (2007). *Meeting the universe halfway: Quantum physics and the entanglement of matter and meaning*. Durham: Duke University Press.

Bennett, J. (2009). *Vibrant matter: A political ecology of things*. Durham: Duke University Press.

Bogard, W. (1996). *The simulation of surveillance: Hypercontrol in telematic societies*. Cambridge: Cambridge University Press.

Bogost, I. (2012). *Alien phenomenology, or, what it's like to be a thing*. Minneapolis: University of Minnesota Press.

Bowden, M. (14 August 2013). The killing machines: How to think about drones. *The Atlantic,* http://www.theatlantic.com/magazine/archive/2013/09/the-killing-machineshow-to-think-about-drones/309434/. Accessed 15 May 2014.

Bürkle, A., Segor, F., & Kollmann, M. (2011). Towards autonomous micro UAV swarms. *Journal of Intelligent & Robotic Systems, 61*(1–4), 339–353.

Caruso, D. (17 November 2013). A 'semi-retired' batman 'controlling drones from the batcave' in 'batman vs. superman.' Under the Gun Review. http://www.underthegunreview.net/2013/11/17/a-semi-retired-batman-controlling-drones-from-the-batcave-in-batman-vs-superman/.

Cisco Systems. (2014). The internet of everything. White Paper, Available online at: http://share.cisco.com/IoESocialWhitepaper/#/.

Department of Homeland Security. (2013). Cell-all: Super smartphones sniff out suspicious substances. http://www.dhs.gov/cell-all-super-smartphones-sniff-outsuspicious-substances. Accessed 15 May 2014.

Discover. (October 2011). How to fix our most vexing problems. http://discovermagazine.com/2011/oct/21-how-to-fix-problems-mosquitoes-potholes-corpses.

Eggers, W. (15 July 2014). Social by the numbers: An interview with Sandy Pentland. Deloitte Review. http://dupress.com/articles/sandy-pentland-mit-interview/.

Epps, G. (15 March 2013). Did Rand Paul ask the wrong questions in his drone filibuster? *The Atlantic,* http://www.theatlantic.com/politics/archive/2013/03/did-rand-paul-ask-the-wrong-questions-in-his-drone-filibuster/274033/.

Foucault, M. (1977). *Discipline and punish: The birth of the prison*. LLC: Random House.

Foucault, M., & Ewald, F. (2003). *"Society must be defended": Lectures at the Collège de France, 1975–1976* (Vol. 1). New York: Macmillan.

Friedersdorf, C. (21 April 2014). Eyes over Compton: How police spied on a whole city. *The Atlantic,* http://www.theatlantic.com/national/archive/2014/04/sheriffs-deputy-compares-drone-surveillance-of-compton-to-big-brother/360954/.

Galloway, A. R. (2013). The poverty of philosophy: Realism and post-fordism. *Critical Inquiry, 39*(2), 347–366.

Gerson, A. D., Parra, L. C., & Sajda, P. (2006). Cortically coupled computer vision for rapid image search. *Neural Systems and Rehabilitation Engineering, IEEE Transactions on, 14*(2), 174–179.

Greenfield, D. (19 August 2013). The case against drone strikes on people who only 'act' like terrorist. *The Atlantic,* http://www.theatlantic.com/international/archive/2013/08/the-case-against-drone-strikes-on-people-who-only-act-like-terrorists/278744/.

Greenwald, G. (2014). *No place to hide: Edward Snowden, the NSA, and the US Surveillance State*. New York: Metropolitan Books.

Haggerty, K. D., & Ericson, R. V. (2000). The surveillant assemblage. *The British Journal of Sociology, 51*(4), 605–622.

Hill, K. (22 May 2014). Facebook wants to listen in on what you're doing. *Forbes,* May 22, http://www.forbes.com/sites/kashmirhill/2014/05/22/facebook-wants-to-listen-in-on-what-youre-doing/.

Hunt, G. (2012). Big data: Operational excellence ahead in the cloud. Presentation to the Amazon Web Services Government Summit 2011, Washington, DC, October 26. http://www.youtube.com/watch?v=SkIhHnoPpjA. Accessed 15 May 2014.

Kittler, F. (2010). Optical Media. London: Polity.

LiKimWa, R. (2012). MoodScope: Building a mood sensor from smartphone usage patterns. Doctoral dissertation, Rice University, Houston, TX.

Mack, K., & Henry, S. (17 September 2012). The future of direct mail is digital. *Campaigns and Elections Magazine,* http://www.campaignsandelections.com/magazine/1796/the-future-of-direct-mail-is-digital.

McCue, C. (2006). *Data mining and predictive analysis: Intelligence gathering and crime analysis*. Oxford: Butterworth-Heinemann.

Moglen, E. (27 May 2014). Privacy under attack: The NSA files revealed new threats to democracy. *The Guardian,* http://www.theguardian.com/technology/2014/may/27/-sp-privacy-under-attack-nsa-files-revealed-new-threats-democracy.

Nakashima, E., & Whitlock, C. (2 January 2011). With Air Force's Gorgon drone "we can see everything." *Washington Post,* http://www.washingtonpost.com/wp-dyn/content/article/2011/01/01/AR2011010102690.html.

Packer, J. (2013). Epistemology not ideology or why we need new Germans. *Communication and Critical/Cultural Studies, 10*(2–3), 295–300.

Packer, J., & Reeves, J. (2013). Romancing the drone: Military desire and anthropophobia from SAGE to swarm. *Canadian Journal of Communication, 38*(3), 309–331.

Parks, L. (2001). Satellite views of Srebrenica: Tele-visuality and the politics of witnessing. *Social Identities, 7*(4), 585–611.

Prigg, M., & Thornhill, T. (7 February 2014). Could google class find your dream date? *The Daily Mail,* http://www.dailymail.co.uk/sciencetech/article-2554133/First-Google-Glass-facial-recognition-app-launches-match-potential-couples-not-checking-sex-offenders-registry.html#ixzz3Eg3ArLmc.

Rotella, C. (12 September 2013) No child left untableted. *The New York Times,* http://www.nytimes.com/2013/09/15/magazine/no-child-left-untableted.html?pagewanted=all&_r=0.

RT.com (29 March 2014). DARPA project seeks hive mind for drones. http://rt.com/usa/darpa-drones-swarm-autonomous-985/.

Rudder, C. (28 July 2014). "We experiment on human beings!" OK Trends: Dating Research from OKCupid (Blog). http://blog.okcupid.com/index.php/we-experiment-on-human-beings/.

Sharkey, N. (2011). Automating warfare: Lessons learned from the drones. *Journal of Law, Information, and Science, 21,* 141–154.

Singer, N. (4 January 2014). "Listen to Pandora, and it listens back." *The New York Times.* http://www.nytimes.com/2014/01/05/technology/pandora-mines-users-data-to-better-target-ads.html?_r=1.
Sosadmin. (2012). Identifying you from your heart rate, from a drone? Privacysos.org. http://privacysos.org/node/694.
Virilio, P. (1989). *War and cinema: The logistics of perception*. London: Verso.
Volkova, S., Coppersmith, G., & Van Dume, B. (2014). *Inferring user political preferences from streaming communications.* Proceedings of the Association for Computational Linguistics (ACL).
Weinberger, D. (2011). *Too big to know: Rethinking knowledge now that the facts aren't the facts, experts are everywhere, and the smartest person in the room is the room*. New York: Basic Books.
Zimmer, B. (26 July 2013). The flight of 'drone' from bees to planes. *The Wall Street Journal,* http://online.wsj.com/news/articles/SB10001424127887324110404578625803736954968.
Žižek, S. (2000). *The ticklish subject: The absent centre of political ontology*. Memphis: Verso.
Zizek, S. (2011). *How to read Lacan*. London: Granta Books.

Mark Andrejevic is an associate professor of Media Studies at Pomona College, Claremont, CA, USA. He is a media scholar who writes about surveillance, new media, and popular culture. In broad terms, he is interested in the ways in which forms of surveillance and monitoring enabled by the development of new media technologies impact the realms of economics, politics, and culture. His first book, Reality TV: The Work of Being Watched (2003), explores the way in which this popular programming genre equates participation with willing submission to comprehensive monitoring. His second book, iSpy: Surveillance and Power in the Interactive Era (2007), considers the role of surveillance in the era of networked digital technology and explores the consequences for politics, policing, popular culture, and commerce. His third book, Infoglut: How Too Much Information Is Changing the Way We Think and Know, explores the social, cultural, and theoretical implications of data mining and predictive analytics. His work has appeared in a edited collections and in academic journals including Television and New Media; New Media and Society; Critical Studies in Media Communication; Theory, Culture & Society; Surveillance & Society; The International Journal of Communication; Cultural Studies; The Communication Review; and The Canadian Journal of Communication. His current work explores the logic of automated surveillance, sensing, and response associated with drones.

Chapter 3
The Political and Moral Economies of Dual Technology Transfers: Arming Police Drones

Kristin Bergtora Sandvik

Introduction

To offset declining demand from the defense sector, the drone industry is engaging in significant marketing efforts throughout the world, including emerging markets, to promote the "good" use of drones for purposes other than national security.[1] To better understand the process by which military drone technologies are being transferred to the domestic sphere, this chapter takes a close look at the emergence of what I will call the "public order drone," whose purposes encompass both firefighting and law enforcement. I am particularly interested in a subset of the public order category—namely, the armed police drone—and in the process through which police drones may be weaponized.

Given the significant public resistance to the deployment of armed police drones—and the technology's failure, so far, to attain certain technical and safety standards—some observers view the use of armed police drones as an unlikely prospect for the near future (RT 2013). I would argue, however, that the arming of police drones should be conceived of as a process, rather than as a one-time technological breakthrough or political decision. My perspective is based on the view that technology is neither inherently neutral nor passively adopted by society. Instead, society and technology are mutually constitutive; thus, the construction of technology is subject to political contestation, commercial competition, and economic considerations. More specifically, drone technology allows a specific set of political and military rationales and projects that must be investigated—not merely for their "newness" but for the power they represent and the practices they enable (Herrera 2003; Fritsch 2011; McCarthy 2013).

[1] This chapter relies on a broad conceptualization of the "drone industry": As used here, the term refers mainly to US-based military manufacturers but also to established Israeli and European military manufacturers as well as to start-up manufacturers in the USA and elsewhere.

K. B. Sandvik (✉)
Norwegian Centre for Humanitarian Studies, Peace Research Institute Oslo (PRIO), Oslo, Norway
e-mail: bergtora@prio.no

A. Završnik (ed.), *Drones and Unmanned Aerial Systems*,
DOI 10.1007/978-3-319-23760-2_3

I would argue that armed police drones have already begun to emerge from the realm of the imaginary, through the confluence of a number of factors:

- Patterns of research and development (R&D) in the drone industry and of procurement in the law enforcement sector
- Transformations of the payload
- Improvements in safety and airworthiness
- The drone lobby's promotion of military and civilian success stories
- The drone lobby's success at pushing the geographical and substantive boundaries of use
- The continued trends towards increasingly sophisticated, miniaturized, and inexpensive technology goods

Scholarship on police drones—primarily in the fields of criminology, critical war studies, and critical terrorism studies—has focused on the emergence of police drones in relation to two contexts: the militarization of policing and the securitization and pacification continuum between war and peace. In addition to building on insights from this literature, this chapter draws on other literature focusing on both military and civilian drones, media reports, explorations of the topic undertaken by civil society organizations, and the author's direct participation in the domestic and international policy debate on drone proliferation.

My reading of the process of weaponization takes political economy as its analytical starting point. As used here, the phrase *political economy* refers to the study of how political institutions, the political environment, and the economic system influence each other. In the late 2000s, as the gradual winding down of engagements in Iraq and Afghanistan intersected with a worldwide financial crisis, defense budgets tightened and drone manufacturers stepped up their efforts to open US and European civil airspace to the use of drones outside the context of violent conflict (Hayes et al. 2014). Symptomatic of this push for drone technology transfer is the fact that in the USA and elsewhere, police now use drones for general surveillance, crowd control, crime prevention, evidence gathering, and arrests. Nevertheless, drone manufacturers and their lobbyists are aware that the use of drones for law enforcement, as well as for more general civil uses, is hampered by regulations, including those that limit access to civil airspace. In their view, the result is missed economic opportunities domestically, lost market shares globally, and a lag in the cultural acceptance of the use of drones in civil airspace, including for the purpose of maintaining public order. Finally, drone manufacturers and lobbyists perceive lack of access to civil airspace as hobbling public discussion of the potential future use of armed drones to maintain public order.

Any discussion of armed police drones must acknowledge that in the sphere of military action, a prototype for such drones already exists at the interface of miniaturization and weaponization. As Tyler Wall has observed, the move to arm police drones has been concurrent with the move to arm micro-unmanned aerial vehicles (UAVs; Wall 2013). Using the fragmented information publicly available, I have sketched out a biography of the Switchblade, a microdrone with a lethal payload that has been used in Afghanistan since 2012. Any account of the Switchblade is an account of unknowns: We do not really know that the Switchblade has been deployed with lethal effects or *how* lethal it has been. Similarly, little is known about the industry's long-term goals, particularly with respect to transferring armed microdrones to the civilian market. Such technology transfers are now possible, however, and are a source of potential interest to drone manufacturers. I would argue, however, that the process through which the transfer may occur is more complicated than what is commonly envisioned in the scholarship on the "militarization of policing."

My contribution is to illuminate this process by exploring the role of moral economies in the transfer of military technology to the civilian law enforcement realm. By *moral economy,* I refer to an environment in which social expectations, cultural transactions, and emotional investments collectively create a shared understanding between the participants in an economic exchange.[2] With regard to the development of armed police drones, the drone industry faces legitimacy problems (stemming from the "drone wars") and the associated difficulty of "educating the public" about drones. To overcome these problems, the industry has had to engage in the strategic mobilization of moral economy in the public domain, principally by attempting to gain leverage in the struggle for resources—whether for access to civil airspace or funding for procurement and R&D.[3] The industry's overall objective is to create a shift within the existing moral landscape, which is characterized by negative emotions, including both fear and skepticism. By mapping the types of arguments that have been deployed to bring the armed police drone out of the realm of the imaginary, I hope to offer an inventory of the moral economy that is at play.

The media and the blogosphere are rife with reports on beneficial new applications for drones—from peacekeeping to humanitarian relief, search and rescue, border control, environmental and wildlife protection, and precision agriculture (Sandvik and Gabrielsen Jumbert forthcoming). Despite the inclusion of "good" uses (finding lost children), such reports tend to focus on "endless" uses of drones in the realm of law enforcement, particularly in light of the considerations of size, cost, and operationability.

In my view, this emphasis on open-ended possibility as a *value in itself* is an important part of the constitution of armed police drones. Moreover, at present, public discussions of such drones play out with reference to a particular type of payload—namely, nonlethal or less lethal weapons (this chapter will use the latter term)—and in two specific contexts. The first context is what I will call "innovation talk": Innovation talk is engendered by way of the public demonstration of microdrones, armed with less lethal weapons, that are being manufactured by smaller manufacturers, including start-ups. Such talk is rendered both possible and respectable under

[2] While this definition of moral economy is useful, it is difficult to gauge its origin. There are several examples of unattributed uses in scholarly works. Building on the seminal contributions of E. P. Thompson and James Scott on moral economies in pre-market and traditional societies, contemporary scholarship emphasizes the notion that there is a "moral economy" that shapes habits and norms in the economic sphere and also lends legitimacy to the constitution of the economy and markets, and to the allocation of social goods. The term is often invoked in discussions of actors who have been deprived of social justice (Thompson 1971; Scott 1976; Arnold 2001; Mau 2004; Karstedt and Farrall 2006).

[3] For a discussion of the moral economy of humanitarian drones, see Sandvik and Lohne (2014).

the guise of "critical public reflection." The second context is the production and deployment of drones that carry nonlethal weapons and that are used for specific "crisis management" purposes in countries with problematic human rights records and high levels of societal violence. In an interconnected world, news coverage of the usage and capabilities of these drones quickly travels globally.

The effectiveness of this moral economy arsenal will be evaluated in light of the criticisms raised against it by activists, lawmakers, scholars, and the media. Hence, I also include a discussion of these criticisms. In their efforts to promote the transfer of armed drone technology to the civilian realm—and to law enforcement in particular—the drone industry and its lobbyists frequently tout the advantages of drone use: specifically, precision, prevention (of further harm), and cost. The "precision, prevention, cost" argument for armed police drones echoes the rationales that have been used for the deployment of combat drones in targeted killing—rationales that critics have met by pointing to the risks associated with such uses, including high numbers of civilian casualties, increased civil resistance, and massive loss of hardware. Similarly, in the realm of armed police drones, industry representatives cite the efficiency, cost-effectiveness, and public safety value of the technologies, while critics point to potential budgeting problems, including unforeseen expenses; to the risks associated with inadequate maintenance, equipment malfunctions, and insufficient training; and to the possible dangers for socially marginalized groups.

The chapter consists of eight sections: parts 1 and 2 provide context; part 3 offers a biography of the Switchblade, an extant and operational killer microdrone; parts 4, 5, and 6 map out the moral economy of the police drone; and part 7 evaluates criticisms of the arguments on which the moral economy of the drone industry is based. Part 8 provides a brief conclusion.

The Militarization of Policing

The criminological literature has paid significant attention to the tradition of military technology shifting to domestic law enforcement and creating a paramilitary police culture (Balko 2013; Holmqvist 2014; Li 2014) According to Abigail Hall and Christopher Coyne (2014), the political economy underlying the militarization of domestic policing is premised on "crises" that prompt the government to take immediate action but that ultimately become perpetual wars—the War on Drugs and the War on Terror being the primary examples. Hall and Coyne argue that as the police engage in military-style training, acquire military weapons, and employ military tactics in everyday operations, the protective state devolves into a predatory state that undermines the rights of the populace.

Because of the entirely distinct forms of violence in which the police and the military are supposed to engage—domestic police are trained to use violence only as a last resort, whereas military forces are trained to achieve victory through combat—the militarization of domestic policing is highly problematic (Hall and Coyne 2013a). One corollary result is a shift in the conceptualization of the events and behaviors with which law enforcement is expected to deal: For example, criminality is redefined as insurgency and crime control as low-intensity conflict; in a militarized law enforcement environment, both require counterinsurgency tactics and equipment (Kraska 2007).

The "golden years" of procurement for the War on Terror saw a blossoming market for new equipment, along with concomitant "tech innovation" discourses. The continuing transfer of large-scale (and often free) surplus military equipment—both offensive and defensive—to police departments has become an important field of empirical and theoretical inquiry (Harwood 2014; Salter 2013).[4] Recently, specific concerns have emerged regarding the ways in which both the technological and techno-cultural transfer of drone technology may affect law enforcement. Public discourse about armed police drones—including some academic discourse—often places them in the context of the long history of colonial airpower (Satia 2014), with a particular focus on drones as a continuation of the police logic that has been inherent in airpower since its inception (Neocleous 2013).

Academics are not alone in their fascination with drones: their deep interest is shared by politicians, the media, and the public at large. While Wall (2013, p. 34) predicts a "repurposing" of military drones as a means of providing domestic policing, he warns against fetishizing the technology, observing that the "boomeranging" of military UAVs is but one contemporary example of the long alliance between war power and police power.[5] He suggests that the military and law enforcement have historically operated on a continuum of control. In this perspective, policing would not be "transformed" by the use of armed drones but would simply move along the spectrum of control in the direction of the military. Despite valuable contributions from Wall and other observers, however, the notion of the "good drone" has so far received little critical attention, nor has the concept of the "public order drone" been subjected to analysis as a precondition for the emergence of armed police drones.

Police Drones in Civil Airspace: Access and Proliferation Issues

While many US states and some European nations have drone legislation pending or in place (Hayes et al. 2014), industry actors consistently present the lack of general access to civil airspace as the single most important obstacle to the proliferation of commercial and government drones (Presutti 2014). There is much talk, on the part of drone manufacturers and lobbyists, of "missed opportunities": One industry study argues, for example, that failing to integrate drones into US civil airspace is costing tens of thousands of jobs and that the economic impact is more than

[4] I will note, but not delve further into, emergent criticism focusing on an alleged "emptiness" and token interdisciplinarity in the burgeoning scholarship emerging from these insights (Neocleous 2014).

[5] The boomerang effect is where control technologies deployed abroad in colonial and military campaigns "boomerang" back to the metropole to be deployed against "homefront" populations (Wall 2013, p. 34).

US$10 billion annually or US$27.6 million a day (El-Hasan 2013). As John Sifton (2012) notes, however, new civilian-operating environments for drones, especially in the law enforcement sector, tend to emerge in advance of policies and procedures to govern their use. In keeping with this view, Wall (2013) notes that the drone is becoming "normalized" in police circles through special permits and exceptions. Hall and Coyne (2013a), meanwhile, observe that military and police interests in expanding their activities intersect with the lobbying efforts of special interest groups.

In recent years, the media have reported on extensive use of drones by US law enforcement for general surveillance, crowd control, crime prevention, evidence gathering, and arrests. Police agencies in Alabama, Arkansas, California, Colorado, Florida, Idaho, Maryland, North Dakota, South Carolina, Texas, Utah, and Washington State have requested or received authorization from the Federal Aviation Administration (FAA) to use surveillance drones, including the ShadowHawk, the T-Hawk, the WASP III, and the Draganflyer X6. Local law enforcement agencies are also receiving direct help from the military and the federal government: For example, the US Air Force has assisted North Dakota police with surveillance, and Predator drones belonging to the US Border Patrol have been loaned out to unnamed sheriffs' departments (Evans 2014; Fox News 2014). The US Department of Homeland Security (DHS) provides police "with all the gadgets, hardware, and software necessary to keep everybody under surveillance," giving local police enormous amounts of data about citizens (Wolverton 2014).

The increasing acquisition and use of surveillance drones by law enforcement agencies is not limited to the USA—suchdrones have been deployed by police in Australia, Canada, New Zealand, and the UK (Fisher 2013; Bowman 2014; Greene 2014; Huffington Post 2014; Mcleod 2014) as well as in nations with fewer legal protections and higher levels of social unrest. In Brazil, for example, surveillance drones have been used in favelas and at sporting events. In Colombia, drones have been used for counterterrorism; in Mexico, to control drug trafficking; and in Russia, to control drugs, crime, and street protests (Franceschi-Bicchierai 2012; Mercopress 2012; Robbins 2013). Moreover, drones are increasingly likely to have been designed, manufactured, and exported from such countries, as domestic drone industries evolve and mature. As I will suggest later in the chapter, a future deployment of weaponized public order drones *anywhere* can lead to the normalization and legitimization of such uses in Western liberal democracies.

The Switchblade: A Biography

Wall (2013) observes that the move to weaponize police drones has been concurrent with the move to weaponize micro-UAVs. Thus, an important starting point for a discussion of armed police drones is the observation that prototypes for such drones already exist and are being deployed for military uses. Although there may be no societal or political readiness to use armed drones to impose public order, there is technological readiness.

Well-known drone models such as the Predator and the Reaper require hundreds of people to remain airborne, to produce legible information, and to effectively strike a target. However, the technologies that are part of the US military's Lethal Miniature Aerial Munition System (LMAMS, a drone category sometimes referred to broadly as "loitering munitions." This chapter also sees LMAMS as synonymous with weaponized microdrones) offer "the infantry something it never had with traditional mortar rounds: the ability to reconnoitre an enemy target before delivering a quick, precision airstrike" (McLeary 2013a). The most famous LMAMS is the Switchblade, described by its manufacturer, AeroVironment, as "the warfighter's magic bullet" (Aerovironment 2012; Defense Update n.d.). In 2012, *Time* magazine heralded the Switchblade as "one of the best inventions of the year" (Time 2012).

Over the past decade, AeroVironment has become the Pentagon's top supplier of small drones, including the Raven and the Puma. As of 2011, the firm was spending nearly US$1 million annually on lobbying, much of which focused on "expanding the use of drones beyond military uses into domestic law enforcement settings"—which AeroVironment sees as intrinsic to its future growth (Greenwald 2011). But in 2012, the wars in Iraq and Afghanistan were winding down, military spending declined, and AeroVironment began to report losses: In 2013, the value of the company's stock plummeted more than 18 % in April alone (Hennigan 2013). Nevertheless, in September 2013, the US government awarded AeroVironment US$51.4 million in contracts for the Switchblade (Smith 2013). The company says that it sees "significant opportunities" for long-term, compound growth in its Switchblade loitering munitions systems and international small-drone sales, in both defense and nondefense industries (Ventura County Star 2013).

At about 24 in. in length, weighing between 5 and 6 lb (2.7 kg), and with a price tag of somewhere between US$40,000 and US$150,000, the Switchblade—variously referred to by its admirers as the "kamikaze drone" (Barrie 2012) and the "ultimate assassin bug" (Gouré 2011)—can be transported in the trunk of a car and assembled and deployed within a matter of minutes by a single soldier or by a police officer chasing a suspect—whom it can not only find but kill (Greenwald 2011). The drone, which can be launched from a small tube and can fly either manually or autonomously, carries an explosive charge equivalent to that of a 40-mm grenade, allowing it to target lightly armored vehicles and embedded (or otherwise inaccessible) infantry positions, such as on rooftops or ridge lines (Strategy Page 2013; Barrie 2012; NavalDrones n.d.; Tarantola 2013).

The Switchblade is marketed as a quick killer: Once it has been released, its four wings spring open, its propeller starts spinning, and it can move at a speed of up to 8 knots (nearly 100 mph), providing "a powerful, but expendable miniature flying intelligence, surveillance and reconnaissance…package on a beyond line-of-sight…target within minutes." One of the Switchblade's crucial attributes is its potential for silence: According to AeroVironment, the drone can "either glide or propel itself via quiet electric propulsion," making it difficult to detect, recognize, and track (AeroVironment n.d.b).

From the Pentagon's perspective, the advantage of acquiring the Switchblade was the opportunity to counter one of the principal criticisms of the drone wars—

namely, by reducing "civilian casualties and collateral damage" (Hennigan 2012). The manufacturer, for its part, describes the Switchblade as both a "precision strike solution" (AeroVironment n.d.b) and a new type of munition to "protect US forces" (AeroVironment 2014). Importantly, the Switchblade is also marketed as a tool of urban warfare: "instead of calling in an airstrike in a densely populated urban area to clear out dug-in enemy positions," the Switchblade enables forces to strike with "pinpoint accuracy" (Tarantola 2013).

The Switchblade has been deployed in Afghanistan since late 2012—and, according to industry sources, is succeeding in its goals and has been well received in the field. An industry press article from May 2013, for example, notes that "Army officials confirmed in February that the Switchblade officially became lethal earlier this year, scoring several hits on enemy targets" (McLeary 2013a). Another source notes that by 2013, "theater came in and said, 'We need dramatically more' than the 75 initially supplied in late 2012." Colonel Pete Newell, the director of the US Army's Rapid Equipping Force, has stated that the Switchblade has "gained some notoriety of its own on both sides"—meaning even among the insurgents targeted by it (McLeary 2013b). Finally, according to another industry source, "the US-NATO military command…[in Afghanistan] says they're loving what they're seeing from it [Switchblade]," and the drone has been "very effective."

In an environment in which virtually all available information comes from the industry itself, it is difficult to get a clear view of the Switchblade's performance. Wired.com—which, along with the Danger Room, has filed a Freedom of Information Act request with the US Central Command (Ackerman 2014)—notes that there is no reliable, publicly available information about the "deployment, effectiveness, distribution or tactical employment" of the system, in particular, about "the breakdown between its surveillance missions and its lethal ones; and certainly nothing about its accuracy" (Ackerman 2014). Nevertheless, industry advocates continue to make claims not only about the drone's perceived effectiveness but about the US military's increasing demand for it. Such claims contribute to the constitution of the Switchblade—and loitering munitions in general—as feasible, protective (saves soldiers on the ground today, can save law enforcement tomorrow), and effective tools for maintaining public order.

The emergence of drones like the Switchblade means that small, tactical ground units can now orchestrate air-to-ground ambushes or flanking maneuvers from concealed positions, a capability that has several implications. First, it raises significant concerns about thresholds for the use of force (Lively 2012). Second, it changes the "boots on the ground" argument usually deployed in defense of armed drones: Instead of being used to *avoid* the use of combat troops, armed drones would be deployed *to keep boots on the ground.* [6] In practice, this means that the decision to kill could be brought down to the level of the platoon commander or even that of individual soldiers (Hennigan 2012). Such a scenario would be in sharp contrast to the deployment of the Predator and Reaper drones, which is monitored by a small army of military lawyers. Concerns about decision-making will become even more acute

[6] This paragraph builds on insights from Sandvik (forthcoming).

when set against the backdrop of past training practices for micro-UAVs: Historically, there has been no consistency in the "frequency or timing" of pilot training, which is only now being institutionalized and standardized.[7]

Industry Efforts to "Educate the Public"

The drone lobby includes both drone manufacturers and providers of supportive services (training, maintenance, and consulting; Hall and Coyne 2013a). Despite being open to "members…with an active interest in UAVs and the development of opportunities to use these systems on a routine basis for the overall benefit of mankind" (UAVS 2014), the industry's lobbying organizations perceive themselves as having a "public relations problem" (Wolverton 2012), caused by its failure to properly "educate the public" about the benefits of drone use in civil airspace. It is useful to examine some of the reasons for this state of affairs:

To begin with, the very use of the term *drone* continues to be contested. Both the military and the drone industry object to the word and have for some time insisted on a variety of alternatives, including UAV, remotely piloted aircraft (RPA), remotely piloted vehicle (RPV), and unmanned aircraft system (UAS). Critics have responded by accusing the industry of failing to recognize the existence of "legitimate issues to be debated about how unmanned systems might be used," and have observed that the industry appears to believe that a semantic shift will make the criticism go away (McNeal 2013).

In the wake of broad coverage, including human rights reporting, on civilian deaths in Afghanistan, Pakistan, and Yemen, the drone industry found itself being regarded as partially responsible for what are known as "drone wars" (Cavallaro and Sonnenberg 2012). In February 2013, Paul Applewhite, a member of the board of directors of the Association for Unmanned Vehicle Systems International (AUVSI), which represents 80 companies based in the Pacific Northwest region of the USA, participated in a congressional hearing on drones. In response to a question regarding the backlash against drones, Applewhite said, "My opinion is that the way that we're currently using drones in warfare, we're moving away from indiscriminate killing to discriminate killing" (Thalen 2013).

Yet, the image of drone manufacturers as a particular breed of war profiteers is not the only problem; another concern is the lack of attention to safety and security as the industry attempts to open civil airspace. The FAA Modernization and Reform Act of 2012, for example, which was drafted with extensive involvement from the drone lobby, makes no explicit reference to privacy. To counter an outburst of criticism, AUVSI later issued a code of conduct, which has been strongly criticized for ignoring key concerns with respect to pilot skills, privacy, and sanctions for vio-

[7] While operators of larger UAVs have their own military occupational specialties and belong to units dedicated to that purpose, Marines using the smallest UAVs have received training on a catch-as-catch-can basis (Sanborn 2012).

lating the code (Thalen 2013). Moreover, despite general support for drone wars abroad, Americans remain skeptical about the domestic use of police drones: In a 2013 poll, 58% of respondents agreed that law enforcement's use of militarized weapons, armored vehicles, and drones has gone too far (Detrick 2013).

I would suggest that because of its failure to "educate the public" (and thereby repudiate the negative perceptions of drones), the drone industry is attempting to embrace drone uses that can be regarded as serving the "public good"—an important strategy in the struggle to open civil airspace (ABANEWS 2013). By creating a shared moral economy that traverses the boundaries between military and civil airspace, the drone industry is making a concerted effort to persuade the public that drones are acceptable in civil airspace. By 2012, the drone industry in the UK had found it necessary to engage in a "long-term public relations effort to counter the negative image of the controversial aircraft"; the goal was to create "a narrative that presents the introduction of drones in the UK as part of a 'national mission'" and to demonstrate that drones "benefit mankind in general" (Gallagher 2012). That same year, the US drone lobby launched the website increasinghumanpotential.org, which emphasizes the shared moral values ("saving time, saving money and, most importantly, saving lives") and conceptions of justice inherent in civilian drone uses.

Within this "good drone" discourse, the public order (sometimes "public safety") drone occupies a particular space. The concept of the public order drone merges firefighting and policing, which is a valuable strategy for several reasons: first, firefighting has fewer negative connotations (such as association with the drone wars) and raises fewer difficult issues (such as privacy and fear of government intrusion); second, firefighting appears to offer more unambiguous advantages (such as protecting the lives of civilians and firefighters, as well as saving property and wildlife; Bennett 2012a).

The Mobilization of Moral Economies: Precision, Preemption, and Cost

Recurrent references to three advantages—preemption, precision, and price—are a central feature of the moral economy of the police drone. As noted earlier, these same factors are used as rationales for the use of combat drones (including the Switchblade) in targeted killing. The Switchblade's manufacturer, AeroVironment, is now "zeroing in on a new market—police and fire departments too small to afford their own helicopters, but big enough to have a need for overhead surveillance" (Rieland 2012).

The targeting logic of drones is shaped by what Tyler Wall and Torin Monahan (2011, p. 240) call "an actuarial form of surveillance," which is predicated on preempting harm. Independent of where or for what purpose drones are deployed, they accumulate data that are then used as the basis for risk calculations (and, where risk is deemed excessive, elimination of targeted individuals; Haggerty et al 2011). The dominant politico-military rationale for the use of drones in war is that the "drone stare"—a video feed in near real-time—allows the operator to see and strike with "surgical precision," not only minimizing civilian casualties but also eliminating the risk to one's own soldiers. This rationale is in keeping with what James Der Derian has referred to as "virtuous war," which is founded on "the technical ability and ethical imperative to threaten and, if necessary, actualize violence from a distance—with no or minimal casualties" (Der Derian 2009, p. 772). In its promotion of their candidate for the law enforcement drone, the microdrone "Qube," AeroVironment notes that the Raven, Puma, and Wasp microdrones "prove their value on the battlefield on a regular basis by providing their operators with better information to help them make better decisions." AeroVironment describes the Qube as an affordable, portable, and rapidly deployable eye in the sky, capable of providing "real-time situational awareness without putting anyone in harm's way" (AeroVironment n.d.a).

As noted by the American Civil Liberties Union (ACLU), both drones themselves and the surveillance technologies attached to them are steadily decreasing in cost and increasing in power: With mass production, these tools will become sufficiently inexpensive that a local police department will be able to fill the skies over a town with them (ACLU 2011). The Qube, for example, "will likely be marketed mainly to smaller agencies that can't afford manned aircraft" (Jones 2012), such as helicopters. According to an AeroVironment representative, "a Qube system's less than $50,000, which is about what police agencies pay for a fully equipped police cruiser…and much less than what they might pay for a helicopter, which costs $1–2 million" (CBS 2013).

The manufacturer presents the Qube as "a small unmanned aircraft that's designed to give first responders an immediate eye in the sky so they can find lost kids, they can investigate accidents, they can support disaster recovery for earthquakes in California, tornadoes in the Midwest, hurricanes in the Gulf Coast" (CBS 2013). While this sort of rhetoric explicitly emphasizes beneficial uses, the real focus is on the endless possibilities for future use. For example, in touting the Qube as just what a forward-thinking police department needs (Rieland 2012), AeroVironment uses the phrase "joining the 21st century."

I would argue that this emphasis on open-ended possibility *as a value in itself* is an important part of the ongoing and future constitution of armed drones. To defend the current use of drones by the Border Patrol, a senior DHS official argued that "it is not about the things we are doing today. It is about the things we might be able to do" (Bennett 2012b). I would also suggest that the notion of endless possibility is clearly linked to past uses of the same or similar technology for military purposes. For example, an AeroVironment representative observed, during a trade show interview, that "a lot of law enforcement are familiar with the full line of Aerovironment's unmanned products. Some of them have used them in the military or watched them" (KESQ.com 2013).

Less Lethal Killer Drones, Innovation Talk, and Emergencies

Generally, legislative provisions against weaponizing drones explicitly prohibit only "putting guns" on the drones (Brustein 2013).[8] As early as the late 1990s, however, drones were already being proposed as a means of helping to "deliver and deploy non-lethal agents" such as smoke canisters (crowd control) and steel spikes (to destroy tires) for law enforcement purposes (Murphy and Cycon 1999). Globally, at least where there are high rates of urban violence, less lethal weapons are fairly common in law enforcement—as are deaths caused by the use of such weapons. Proponents argue that weaponizing drones with less lethal weapons will reduce both collateral damage and threats to the security of police officers. Scholars need to be alert, however, to the legitimizing effect of less lethal weapons (Rappert 2003). In particular, I would argue that we must ask whether there is an inevitable logic to the sequencing of payloads, whereby less lethal payloads (such as tasers and stun guns) will eventually be complemented by lethal payloads (such as missiles and machine guns).[9]

According to one observer, there is an extensive list of possible uses for drones armed with less lethal weapons. Such drones could limit the need to resort to lethal weapons and thereby save the lives of police. They could also be deployed for crowd control, prison riots, organized attacks on police stations, and hostage situations. In hostage situations, nanodrones could listen in on the conversations of hostage takers, and "virtually silent and invisible drones armed with various less lethal weapons could enable capture and arrest" (McDougal 2013).

The drone industry has engaged in considerable flirtation with such possibilities: The ShadowHawk—which has already been purchased by local US police departments—is capable of firing rubber bullets and releasing tear gas canisters and taser projectiles; reportedly, it also has the capacity to launch grenades and to fire 12-gauge shotguns (Kindynis 2012). In a 2010 report submitted to Congress by US Customs and Border Protection (CBP), the agency raises the possibility of eventually equipping its drones with "nonlethal weapons" to "immobilize" people and vehicles trying to cross the border illegally. CBP later stated that it has no such plans (Sengupta 2013).

Although less lethal drones continue to be topics of discussion and speculation, actual demonstrations of the technologies remain rare. The exceptions occur in the realm of what I will call "innovation talk," of which I will provide two examples. The first concerns the do-it-yourself (DIY) and start-up drone scene—which, in recent years, has emerged as a field that, at least in theory, is disconnected from the military–industrial complex. In order to "provoke controversy" or "spark debates," DIY drone enthusiasts and nonmilitary drone start-ups display weaponized microdrones. Although such activities are conducted under the guise of the "public con-

[8] This section builds on Sandvik (forthcoming 2015).

[9] The issue of less lethal weapons is hugely controversial (see McCray 2012).

versation" rationale, they generate interest not only from the public but also from commercial, military, and law enforcement entities.

Despite the field's claim to offer opportunities to discuss difficult issues raised in various physical or virtual forums (such as large tech meet-ups or within the tech blogosphere), the precise nature of the discursive space occupied by DIY drone enthusiasts and start-ups remains elusive. I would propose, however, that innovation hubs (including both DIY-ers and start-ups) have both the freedom and the credibility to do what the conventional drone industry might be reluctant to undertake: namely, to "devirtualize" the YouTube feeds that depict *imaginary* armed drones by engaging in demonstrations of real hardware for wide audiences, all under the guise of contributing to "public debate."[10]

The idea of taser drones has been around for years now (not least in science fiction literature and Hollywood movies), and both governments and private companies have considered developing such technology (Watson 2014). At the 2014 South by Southwest gathering, Chaotic Moon Studios engaged in a tech demo of Chaotic Unmanned Personal Intercept Drone (CUPID), a "stun-copter" capable of delivering an 80,000-V shock—which was deployed on a human during a live demonstration. Chaotic Moon has stated that CUPID could be used by law enforcement personnel to apprehend fleeing suspects (Aamot 2014). As William Hurley, the firm's cofounder, noted, "If you imagine a S.W.A.T. raid and people running—why send officers, with gun blazing, down an alley way where they can shoot and harm an innocent person or whatever, when you could just have the drone follow them" (Watson 2014).

Chaotic Moon's expressed ambition is to stay "on the forefront of inevitably convergent technologies." Although the company has stated that CUPID could be quickly brought to production whenever a personal security or law enforcement client sees fit (Aamot 2014), they have declared that they are not seeking to commercialize the technology but to create discussion: "Chaotic Moon built CUPID to raise awareness of technology that's outpacing everything from regulatory agencies to social norms. We have no plans to develop drone type or commercialize this in any way" (Suba 2014). Nevertheless, one of Chaotic Moon's (unnamed) cofounders confirmed to Fox News that CUPID has sparked a great deal of interest from military and law enforcement agencies. Critics have observed that it seems unclear what exactly the company wants to achieve with this display (Mass 2014).

A second example of an armed, nonlethal drone is the Skunk Riot Control Copter designed by Desert Wolf, a South African company, which is designed to be used to "control unruly crowds…without endangering the lives of the protestors or the security staff." So far, the Skunk Riot has been sold to an international mining company and to police units outside South Africa. The drone has four high-capacity gun barrels capable of shooting up to 4000 paintballs, pepper spray balls, or solid plastic

[10] As a side note, the internet is home to a fascinating proliferation of phoney footage depicting functioning weaponized drones; by both depicting an ideal type and offering the promise of technical feasibility, these images make important contributions to the constitution of the armed police drone (Johnson 2012).

balls at rates of up to 80 balls per second, which will only be used in an extreme "life threatening situation." The paintballs can be used to mark people in a crowd with red (those carrying dangerous weapons) or blue (vandalizing protestors; Desert Wolf n.d.a).

The Skunk Copter is equipped with both "blinding lasers" and on-board speakers to send verbal warnings to a crowd. The drone also uses a thermal camera with night-vision capabilities and two full-high-definition video cameras to record events as they unfold. The eight powerful electric motors, with 16-in. propellers, allow the drone to lift up to 45 kg (99 lb). According to the manufacturer, the goal was to "assist in preventing another Marikana"—a reference to a 2012 strike in South Africa, in which 44 miners were killed by police (Smith 2014).

Unpacking Assumptions

According to Torin Monahan and Jennifer Mokos, an important element of the securitization process is the creation of compelling narratives to justify the surveillance systems under consideration; Mike Crang and Stephen Graham (2007) refer to such narratives as "technological fantasies" that position emergent technological systems as necessary—and effective—responses to dire threats. Such narratives are not, however, simply instrumental devices designed to achieve desired ends; they also actively shape the larger security cultures and afford them influence (Monahan and Mokos 2013).

I would propose that the potential for the future deployment of armed police drones—the potential of the technological fantasy—may lie precisely in the lack of knowledge about the battlefield effectiveness of loitering munitions, including the Switchblade. At the same time, it has been noted that the drone industry's narratives evoke a level of havoc that is extremely rare in Western liberal democracies (terrorist atrocities, nuclear meltdown, high-speed car chases), for which a technological solution—the R&D and procurement of drones—is proposed, and thereby "used to justify both substantial public expenditure and the acquisition and use of drones by domestic police forces" (Hayes et al. 2014).[11]

The very capabilities that make microdrones effective—intrusiveness and silence—make possible new forms of privacy invasion. As the ACLU has noted, in light of the crowd innovation that is characteristic of drone development, it is likely that "the technology will develop new and more advanced capabilities that have never existed for police helicopters—such as swarms, or more continuous surveillance" (Stanley 2013). The use of police helicopters has already raised privacy is-

[11] Ben Hayes and Eric Toepfer suggest that in many scenarios developed for the European Union (EU), drones look more like a solution looking for a problem than vice versa. At a drone conference, a drone manufacturer acknowledged as much to a representative from Statewatch, an independent organization that monitors civil liberties in the EU: "You're quite right. We don't actually know what the problem is; we just know that the solution is UAVs" (Hayes et al. 2014).

sues, as well as concerns about the use of lethal police airpower; however, because manned helicopters are expensive to acquire, staff, and maintain, cost has so far provided a natural limitation on their use. The ACLU have argued, however, that drones "erase natural limits" on aerial surveillance (Stanley 2013). As the developer of CUPID has noted, "This is something that's affordable for almost everybody and in the next two or three years the technology will probably cut in half, by price" (Mass 2014).

Yet, the possibility that the use of surveillance and armed drones will be more expensive than traditional methods of "keeping order" is rarely communicated: For example, an audit by the Homeland Security Department's inspector general criticized DHS for buying more Predator drones than the Border Patrol can use. Additionally, the former president of the National Border Patrol Council, the border agents' union, has complained that "The big problem is that they [drones] are more expensive than traditional methods" of patrolling (Bennett 2012b).

The same can be said of weaponized microdrones: Even as the industry competes for more contracts, an analyst with the Teal Group Corporation has suggested that LMAMS will be "purchased in limited quantities, especially given the withdrawal from Afghanistan." The analyst notes that "It [LMAMS] is very expensive per mission compared to more conventional fire support methods" (Schechter 2013). While it has been reported that the USA is attempting to lower the cost of the War on Drugs by using drones in the Caribbean (Mick 2013), it has also been reported that CBP has raided budgets of its manned aircraft to pay for drones, resulting in a cut in flight hours of surveillance planes hunting smuggling ships (Bennett 2012a). Hence, although drones are cheaper than traditional police airpower, they may not be cheaper than traditional policing—an equation that is kept out of the industry's promotional material.

Critics of the cost-effectiveness argument not only question the utility of police drones but also point to the high rate of vehicle loss, both civilian and military. Observing that "it is becoming clear that police interest in purchasing drones far outweighs their practical utility," Salter (2013) proposes that the various and conflicting rationales offered by police forces in their pursuit of public support for drones can be understood as mystifications of the desire to embody and enact the militarized "subject position" made possible by such technology—including states seeking to legitimize particular governmental strategies through crime control.

Other problems associated with drones stem from poor maintenance, inadequate coordination of deployment, inadequate training, and insufficient resources and staffing (Sengupta 2013). At CBP, for example, drone deployment has been beset by problems: As observed by a commentator, "Many of the logs indicate missions were terminated or cancelled due to undisclosed issues affecting both the aircraft and surveillance equipment" (Lynch 2014). CBP has also been criticized for the performance of its drone fleet, including the failure to help increase the seizure of drugs. This has "stained the reputation of the technology as limited and insignificant in the border patrol mission" (McDougal 2013). The CBP has also been criticized for lending out its drones to a high number of agencies and for using a more sophisticated surveillance system than foreseen (and for loaning it to other agencies;

Lynch 2014). Reports of lost drones and drone malfunction are rampant: In January 2014, the CBP grounded its drone fleet after having been forced to down a malfunctioning drone at sea (Dobuzinskis 2014). Human error is also common: UK police have crashed several drones. They have also been unable to use them due to legal restrictions and fears for public safety if the drones should fall down (Kindynis 2012; Personal Drones 2013).

Finally, the accuracy of drone surveillance and drone strikes has been heavily contested (Braun 2012). Much can and does go wrong at various points in the "kill chain" (Gregory 2011, p. 193), "for reasons ranging from pilot error and bad weather to mechanical failure" (Turse 2012). Moreover, regardless of the accuracy of the technology, the data collected by surveillance systems is subject to interpretation: As Lane DeNicola has observed, assumptions of manifest transparency and mechanical objectivity must be abandoned in favor of recognizing that the practice of airborne and satellite imagery analysis is a "significantly interpretive one" (DeNicola 2008). Finally, drones are not safe from being hacked: In 2012, college students from Texas hacked a drone in front of DHS officials (RT 2012).

There are significant differences between an armed Predator or Reaper and a loitering munitions system. The Predators and Reapers are deployed exclusively in remote battle spaces; loitering munitions *can* be deployed in such spaces, but they can also be deployed by the municipal police. Relying on grainy, soda-straw views of the world, remote drone pilots identify "targets." Criminology has a long tradition of questioning the behavior and intent of "loitering individuals"; will such individuals be seen as loitering beneath the "loitering munitions" of police departments? Salter (2013) suggests that like other examples of police militarization, drones are likely to have differential effects on marginalized or disadvantaged groups.[12]

Conclusion

By exploring the political and moral economies of armed police drones, this chapter has attempted to highlight the political, legal, and discursive process of "creating" the armed drone as an option in policing. I would argue that the creation of the armed police drone revolves around several principal factors: the militarization of policing; the interrelationship between proliferation, lack of regulation, and globalization; overt struggles over legitimacy between industry, political actors, and civil society; the careful recycling of expediency arguments from the drone wars; and, finally, the use of innovation talk.

[12] According to the ACLU, the proliferation of Special Weapons and Tactics (SWAT) teams, the arming of police with weapons of war, and the use of hyperaggressive tactics escalates the risk of needless violence, destroys property, and undermines individual liberties (ACLU 2014), particularly in poor neighborhoods and communities of color (ACLU 2013). In March 2014, provincial police in Ontario, Canada, used drones to monitor First Nations protests about the issue of missing and murdered aboriginal women (Bowman 2014).

In conclusion, I would like to suggest three directions for further research. First, the emerging political economy scholarship on drones focuses on the relationship between regulatory efforts and the development and marketing of new prototypes. I would propose that more attention be given to the moral economy side of the equation—namely, to the ways in which UAV procurements are processed, labeled, and legitimated by governments and international organizations. Second, I would recommend giving further attention to the globalization of the police drone and its deployment in authoritarian states or states in conflict, as well as to the backlash against drone policing in democratic states. Third, under one of the rationales for drone deployment, known in military circles as "force protection," drones can replace boots on the ground: What needs further examination is whether police drones are being deployed to *secure* or to *keep* boots on the ground. A related issue is whether and how the claims to surgical precision and decreased civilian casualties will play out in the policing sphere.

References

Aamot, D. (10 March 2014). Recently in controversy: Drone can tase people with 80,000 volts. *Time*. http://time.com/19076/recently-in-controversy-drone-can-tase-people-with-80000-volts/. Accessed 30 Sept 2014.

ABANEWS. (2013). The rise of the drones: Experts discuss legal implications of UAVs in civilian airspace. http://www.americanbar.org/news/abanews/aba-news-archives/2013/10/the_rise_of_the_dron.html. Accessed 30 Sept 2014.

Ackerman, S. (2014). Tiny, suicidal drone/missile mashup is part of US' Afghanistan arsenal. http://www.wired.com/2013/03/switchblade-afghanistan/. Accessed 30 Sept 2014.

ACLU (American Civil Liberties Union). (2011). Protecting privacy. In aerial surveillance: Recommendations for government use of drone aircraft. https://www.aclu.org/files/assets/protectingprivacyfromaerialsurveillance.pdf. Accessed 30 Sept 2014.

ACLU. (6 March 2013). ACLU Launches nationwide investigation into police use of military technology & tactics. https://www.aclu.org/criminal-law-reform/aclu-launches-nationwide-investigation-police-use-military-technology-tactics. Accessed 30 Sept 2014.

ACLU. (23 June 2014). War comes home: The excessive militarization of American police—Report. https://www.aclu.org/criminal-law-reform/war-comes-home-excessive-militarization-american-police-report. Accessed 30 Sept 2014.

AeroVironment. (2012). Switchblade overview. https://www.avinc.com/downloads/Switchblade_Datasheet_032712.pdf. Accessed 30 Sept 2014.

AeroVironment. (2014). Unmanned aircraft systems overview. https://www.avinc.com/downloads/AVAV_UAS_FactSheet_1-14-14.pdf. Accessed 30 Sept 2014.

AeroVironment. (n.d.a). Qube: Small UAS missions. http://www.avinc.com/uas/small_uas/qube. Accessed 30 Sept 2014.

AeroVironment. (n.d.b). Switchblade. http://www.avinc.com/uas/small_uas/switchblade/. Accessed 30 Sept 2014.

Arnold, T. C. (2001). Rethinking moral economy. *American Political Science Review, 95*(1), 85–95.

Balko, R. (2013). *Rise of the warrior cop: The militarization of America's police forces*. New York: Public Affairs.

Barrie, A. (31 May 2012). Meet the micro-drone tag team assault force. *Fox News*. http://www.foxnews.com/tech/2012/05/31/meet-micro-drone-tag-team-assault-force/. Accessed 30 Sept 2014.

Bennett, B. (28 April 2012a). Predators drones have yet to prove their worth on border. *Los Angeles Times*. http://articles.latimes.com/2012/apr/28/nation/la-na-drone-bust-20120429.

Bennett, B. (5 August 2012b). Drones tested as tools for police and firefighters. *Los Angeles Times*. http://articles.latimes.com/2012/aug/05/nation/la-na-drones-testing-20120805. Accessed 30 Sept 2014.

Bowman, J. (5 March 2014). Ontario police defend use of drone cameras over protests. *CBC News*. http://www.cbc.ca/newsblogs/yourcommunity/2014/03/ontario-police-defend-use-of-drone-cameras-over-protests.html. Accessed 30 Sept 2014.

Braun, M. (18 December 2012). Columbia drone report misses the target. *Foreign Policy*.

Brustein, J. (2 July 2013). Dear FBI: Keep guns off your drones, please. Yours, Congress. *Business Week*. http://www.businessweek.com/articles/2013-07-02/dear-fbi-keep-guns-off-your-drones-yours-congress#r=rss. Accessed 30 Sept 2014.

Cavallaro, J., & Sonnenberg, S. (2012). *Living under drones: Death, injury, and trauma to civilians from US drone practices in Pakistan*. Stanford: University and NYU.

CBS News. (16 May 2013). Drones: Eyes in the sky. http://www.cbsnews.com/news/drones-eyes-in-the-sky/. Accessed 30 Sept 2014.

Crang, M., & Graham, S. (2007). Sentient cities: Ambient intelligence and the politics of urban space. Information. *Communication and Society, 10*(6), 789–817.

Defense Update. (n.d.). Switchblade—miniature killer drone. http://defense-update.com/products/l/switchblade_31122010.html. Accessed 30 Sept 2014.

DeNicola, L. (2008). Credibility and the use of geospatial media in activism and advocacy. *Proceedings of the American Society for Information Science and Technology, 45*(1), 1–12.

Der Derian, J. (2009). *Virtuous war: Mapping the military—industrial-media-entertainment network*. New York: Routledge.

Desert Wolf. (n.d.a). Skunk riot control copter. http://www.desert-wolf.com/dw/products/unmanned-aerial-systems/skunk-riot-control-copter.html. Accessed 30 Sept 2014.

Detrick, P. (3 December 2013). Reason-Rupe poll: Americans still uncomfortable with cops and drones. *Reason.com*. http://reason.com/blog/2013/12/13/reason-rupe-poll-americans-still-uncomfo. Accessed 30 Sept 2014.

Dobuzinskis, A. (28 January 2014). US Customs grounds fleet of drones after crash at sea. *Reuters*. http://www.reuters.com/article/2014/01/28/us-usa-drone-crash-idUSBREA0R1JT20140128. Accessed 30 Sept 2014.

El-Hasan, M. (23 June 2013). Businesses see opportunity in civilian drones, but regulations stand in the way. *Los Angeles News Group*. http://www.dailyfreeman.com/general-news/20130623/businesses-see-opportunity-in-civilian-drones-but-regulations-stand-in-the-way. Accessed 30 Sept 2014.

Evans, J. (18 January 2014). The techno-militarization of America. *TechCrunch*. http://techcrunch.com/2014/01/18/the-techno-militarization-of-america/. Accessed 30 Sept 2014.

Fisher, D. (March 2 2013). Police use drones to catch criminals. *New Zealand Herald*. http://www.nzherald.co.nz/nz/news/article.cfm?c_id=1&objectid=10868674. Accessed 30 Sept 2014.

Fox News. (28 January 2014). Predator drone helps convict North Dakota farmer in first case of its kind. http://www.foxnews.com/us/2014/01/28/first-american-gets-prison-with-assistance-predator-drone/. Accessed 30 Sept 2014.

Franceschi-Bicchierai, L. (25 July 2012). Russia is stockpiling drones to spy on street protests. Wired.com.http://www.wired.com/2012/07/russia-2/. Accessed 30 Sept 2014.

Fritsch, S. (2011). Technology and global affairs. *International Studies Perspectives, 12*(1), 27–45.

Gallagher, R. (2 February 2012). Surveillance drone industry plans PR effort to counter negative image. *The Guardian*. http://www.theguardian.com/uk/2012/feb/02/surveillance-drone-industy-pr-effort. Accessed 30 Sept 2014.

Gouré, D. (19 October 2011). Unmanned aircraft: Bringing a switchblade to a knife fight. Lexington Institute. http://www.lexingtoninstitute.org/unmanned-aircraft-bringing-a-switchblade-to-a-knife-fight/#sthash.yJCbbLrm.dpuf. Accessed 30 Sept 2014.

Greene, A. (1 March 2014). AFP using drones to investigate major crime as questions raised over privacy. *ABC News Australia*. http://www.abc.net.au/news/2014-03-01/afp-using-drones-to-solve-crime/5292096. Accessed 30 Sept 2014.

Greenwald, G. (12 December 2011). The growing menace of domestic drones. *Salon*. http://www.salon.com/2011/12/12/the_growing_menace_of_domestic_drones/. Accessed 30 Sept 2014.

Gregory, D. (2011). From a view to a kill: Drones and late modern war. *Theory, Culture & Society, 28*(7–8), 188–215.

Haggerty, K. D., Wilson, D., & Smith, G. J. D. (2011). Theorizing surveillance in crime control. *Theoretical Criminology, 15*(3), 231–237.

Hall, A. R., & Coyne, C. J. (2013a). The militarization of US domestic policing. *Independent Review, 17*(4), 485–504.

Hall, A. R., & Coyne, C. J. (2014). The political economy of drones. *Defence and Peace Economics, 25*(5), 445–460.

Harwood, M. (14 August 2014). One nation under SWAT. TomDispatch.com. http://www.tomdispatch.com/post/175881/tomgram%3A_matthew_harwood%2C_one_nation_under_swat/. Accessed 30 Sept 2014.

Hayes, B., Jones, C., & Toepfer, E. (2014). *Eurodrones Inc*. London: Transnational Institute and Statewatch.

Hennigan, W. J. (11 June 2012). Pentagon soon to deploy pint-sized but lethal switchblade drones. *Los Angeles Times*. http://articles.latimes.com/2012/jun/11/business/la-fi-kamikaze-drone-20120611. Acessed 30 Sept 2014.

Hennigan, W. J. (4 April 2013). Drone maker AeroVironment sputters as defense spending dries up. *Los Angeles Times*. http://articles.latimes.com/2013/apr/04/business/la-fi-0404-aerovironment-troubles-20130404. Accessed 30 Sept 2014.

Herrera, G. L. (2003). Technology and international systems. *Millennium: Journal of International Studies, 32*(3), 559–593.

Holmqvist, C. (2014). *Policing wars: On military intervention in the twenty-first century. Houndsmill*. Basingstoke: Palgrave Macmillan.

Huffington Post. (27 March 2014). UK police "Skyranger" drones to patrol skies above Gatwick Airport after major disasters. http://www.huffingtonpost.co.uk/2014/03/12/uk-police-drones-gatwick_n_4949862.html. Accessed 30 Sept 2014.

Johnson, R. (24 April 2012). Sorry, this awesome flying submachine gun is a hoax. *Business Insider*. http://www.businessinsider.com/watch-this-flying-submachine-gun-wreak-havoc-on-everything-from-mannequins-to-cars-2012-4#!CmKqA. Accessed 30 Sept 2014.

Jones, J. H. (30 September 2012). Tiny drones promoted for police use. *U-T San Diego*. http://www.utsandiego.com/news/2012/sep/30/tiny-drones-promoted-for-police-use/. Accessed 30 Sept 2014.

Karstedt, S., & Farrall, S. (2006). The moral economy of everyday crime markets, consumers and citizens. *British Journal of Criminology, 46*(6), 1011–1036.

KESQ.com. (13 February 2013). Drone technology could be coming to a police department near you. http://www.kesq.com/kesq/Drone-technology-could-be-coming-to-a-Police-Department-near-you/19098288. Accessed 30 Sept 2014.

Kindynis, T. (14 October 2012). Contours of control, eyes in the sky: The rise of the police drones. Ceasefire. http://ceasefiremagazine.co.uk/eye-sky-primer-police-drones/. Accessed 30 Sept 2014.

Kraska, P. B. (2007). Militarization and policing—Its relevance to 21st century police. *Policing, 1*(4), 501–513.

Li, S. (15 August 2014). The evolution of police militarization in Ferguson and beyond. *The Wire*. http://www.thewire.com/national/2014/08/the-evolution-of-police-militarization-in-ferguson-and-beyond/376107/. Accessed 30 Sept 2014.

Lively, T. (20 March 2012). The Switchblade UAS: Is everyone now a sniper? *SOFREP*. http://sofrep.com/4754/the-switchblade-uas-is-everyone-now-a-sniper/. Accessed 30 Sept 2014.

Lynch, J. (14 January 2014). Customs & border protection loaned predator drones to other agencies 700 times in three years according to "newly discovered" records. Electronic Frontier

Foundation. https://www.eff.org/deeplinks/2014/01/newly-discovered-drone-records-show-customs-border-protection-flew-its-predator. Accessed 30 Sept 2014.

Mass, W. (25 March 2014). CUPID drone has 80,000-volt taser dart. *New American*. http://www.thenewamerican.com/usnews/crime/item/17917-cupid-drone-has-80-000-volt-taser-dart. Accessed 30 Sept 2014.

Mau, S. (2004). Welfare regimes and the norms of social exchange. *Current Sociology, 52*(1), 53–74.

McCarthy, D. R. (2013). Technology and "the International" or: How I learned to stop worrying and love determinism. *Millennium: Journal of International Studies, 41*(3), 470–490.

McCray, R. (5 March 2012). Tasers no longer a non-lethal alternative for law enforcement. American Civil Liberties Union. https://www.aclu.org/blog/criminal-law-reform/tasers-no-longer-non-lethal-alternative-law-enforcement. Accessed 30 Sept 2014.

McDougal, C. (2013). From the battlefield to domestic airspace: An analysis of the evolving roles and expectations of drone technology. *PublicINReview, 1*(2), 92–102.

McLeary, P. (7 May 2013a). Special report: Precision strike; US Army expanding soldier attack options. *Defense News*. http://www.defensenews.com/print/article/20130507/DEFREG02/305060011/Special-Report-Precision-Strike. Accessed 30 Sept 2014.

McLeary, P. (12 February 2013b). US Army wants more switchblades: Remote-control munitions are small, lethal. *Defense News*. http://www.defensenews.com/article/20130212/DEFREG02/302120017/U-S-Army-Wants-More-Switchblades. Accessed 30 Sept 2014.

McLeod, P. (2 January 2014). Nova Scotia RCMP to get drones. *Chronicle Herald*. http://thechronicleherald.ca/novascotia/1177077-nova-scotia-rcmp-to-get-drones. Accessed 30 Sept 2014.

McNeal, G. S. (2 March 2013). Just call it a drone. *Forbes*. http://www.forbes.com/sites/gregorymcneal/2013/03/02/just-call-it-a-drone/. Accessed 30 Sept 2014.

MercoPress. (10 September 2012). Rio police begin to use drones to control drugs and crime in the city's shanty towns. http://en.mercopress.com/2012/09/10/rio-police-begins-to-use-drones-to-control-drugs-and-crime-in-the-city-s-shanty-towns. Accessed 30 Sept 2014.

Mick, J. (29 April 2013). US turns to drones to cut cost of "War on Drugs" in the Caribbean. *Daily Tech*. http://www.dailytech.com/US+Turns+to+Drones+to+Cut+Cost+of+War+on+Drugs+in+the+Caribbean/article30444.htm. Accessed 30 Sept 2014.

Monahan, T., & Mokos, J. T. (2013). Crowdsourcing urban surveillance: The development of homeland security markets for environmental sensor networks. *Geoforum, 49*, 279–288.

Murphy, D. W., & Cycon, J. (1999). Applications for mini VTOL UAV for law enforcement. In enabling technologies for law enforcement and security. International society for optics and photonics. http://www.dtic.mil/dtic/tr/fulltext/u2/a422459.pdf. Accessed 30 Sept 2014.

NavalDrones. (n.d.). Switchblade UAS. http://www.navaldrones.com/switchblade.html. Accessed 30 Sept 2014.

Neocleous, M. (2013). Air power as police power. *Environment and Planning D: Society and Space, 31*(4), 578–593.

Neocleous, M. (2014). *War power, police power*. Edinburgh: Edinburgh University Press.

Page, S. (6 September 2013). Weapons: Troops demand more Switchblade. https://www.strategypage.com/htmw/htweap/articles/20130906.aspx. Accessed 30 Sept 2014.

Personal Drones. (3 November 2013). The rise and fall of UK Mereyside police drone. http://www.personal-drones.net/the-rise-and-fall-of-uk-mereyside-police-drone/. Accessed 30 Sept 2014.

Presutti, C. (25 February 2014). Drones at work worldwide, but US still lacks laws. *Voice of America*. http://www.voanews.com/content/drones-at-work-worldwide-but-us-still-lacks-laws/1859180.html. Accessed 30 Sept 2014.

Rappert, B. (2003). *Non-lethal weapons as legitimizing forces? Technology, politics, and the management of conflict*. London: Frank Cass.

Rieland, R. (15 October 2012). Drawing the line on drones. Smithsonian.com. http://www.smithsonianmag.com/innovation/drawing-the-line-on-drones-72749356/?no-ist=. Accessed 30 Sept 2014.

Robbins, S. (8 January 2013). Colombia develops UAV technology to fight drugs, rebels, illegal mining. *Diáologo*. http://dialogo-americas.com/en_GB/articles/rmisa/features/regional_news/2013/08/01/colombia-drones. Accessed 30 Sept 2014.

RT. (27 June 2012). Texas college hacks drone in front of DHS. http://rt.com/usa/texas-1000-us-government-906/. Accessed 30 Sept 2014.

RT. (14 February 2013). Feds: No armed drones in US, but forget about privacy. http://rt.com/usa/us-privacy-aircraft-drone-243/. Accessed 27 Sept 2014.

Salter, M. (2013). Toys for the boys? Drones, pleasure and popular culture in the militarisation of policing. *Critical Criminology, 22*(2), 167–177.

Sanborn, J. K. (11 September 2012). Marines stand up schools for small UAV training. *Defense News*. http://www.defensenews.com/article/20120911/C4ISR01/309110007/Marines-Stand-Up-Schools-Small-UAV-Training. Accessed 30 Sept 2014.

Sandvik, K. B. (forthcoming). Arming police drones: A global perspective (unpublished manuscript on file with author).

Sandvik, K. B., & Jumbert, M. G. (Eds.), (Forthcoming.) The good drone. Surrey: Ashgate Publishing.

Sandvik, K. B., & Lohne, K. (2014). The rise of the humanitarian drone. *Millennium: Journal of International Studies, 43*(1), 145–164.

Satia, P. (2014). Drones: A history from the British Middle East. *Humanity: An International Journal of Human Rights, Humanitarianism, and Development, 5*(1), 1–31.

Schechter, E. (14 April 2013). Kamikaze drones: Miniature munitions for dismounted troops. *Army Times*. http://www.armytimes.com/article/20130814/NEWS04/308140013/Kamikaze-drones-miniature-munitions-dismounted-troops. Accessed 30 Sept 2014.

Scott, J. C. (1976). *The moral economy of the peasant: Rebellion and subsistence in Southeast Asia*. New Haven: Yale University Press.

Sengupta, S. (3 July 2013). US border agency allows others to use its drones. *New York Times*. http://www.nytimes.com/2013/07/04/business/us-border-agency-is-a-frequent-lender-of-its-drones.html?_r=1&. Accessed 30 Sept 2014.

Sifton, J. (27 February 2012). A Brief History of Drones. *The Nation*. http://www.thenation.com/article/166124/brief-history-drones. Accessed 30 Sept 2014.

Smith, R. (2013). AeroVironment reaps $51.4 million in Switchblade UAV contracts in 9 days. *sUAS News*. 9 September. http://www.suasnews.com/2013/09/24892/aerovironment-reaps-51-4-million-in-switchblade-uav-contracts-in-9-days/. Accessed 30 Sept 2014.

Smith, D. (20 June 2014). Pepper-spray drone offered to South African mines for strike control. *The Guardian*. http://www.theguardian.com/world/2014/jun/20/pepper-spray-drone-offered-south-african-mines-strike-control. Accessed 30 Sept 2014.

Stanley, J. (8 March 2013). We already have police helicopters, so what's the big deal over drones? American Civil Liberties Union. https://www.aclu.org/blog/technology-and-liberty-criminal-law-reform/we-already-have-police-helicopters-so-whats-big-deal. Accessed 30 Sept 2014.

Suba, R. (13 March 2014). This CUPID drone doesn't shoot love arrows—It tases you with 80,000 volts of pure electricity. *Tech Times*. http://www.techtimes.com/articles/4280/20140313/this-cupid-drone-doesnt-shoot-love-arrows-it-tases-you-with-80-000-volts-of-pure-electricity.htm. Accessed 30 Sept 2014.

Tarantola, A. (5 September 2013). America's kamikaze drone makes the skies way less friendly. *Gizmodo*. http://gizmodo.com/americas-kamikaze-drone-makes-the-skies-way-less-frien-1227821895/all. Accessed 30 Sept 2014.

Thalen, M. (22 February 2013). Drone lobbyist makes shocking comment about "indiscriminate killing" policy. *Examiner.com*. http://www.examiner.com/article/drone-lobbyist-makes-shocking-comment-about-indiscriminate-killing-policy. Accessed 30 Sept 2014.

Thompson, E. P. (1971). The moral economy of the English Crowd in the eighteenth century. *Past & Present: A Journal of Historical Studies, 50,* 76–136.

Time. (31 October 2012). Best inventions of the year. http://techland.time.com/2012/11/01/best-inventions-of-the-year-2012/slide/the-switchblade-drone/. Accessed 30 Sept 2014.

Turse, N. (16 January 2012). The crash and burn future of robot warfare. *Huffington Post*. http://www.huffingtonpost.com/nick-turse/the-crash-and-burn-future_b_1208644.html. Accessed 30 Sept 2014.

UAVS (Unmanned Aerial Vehicle Systems) Association. (2014). About UAVS. http://www.uavs.org/aboutuavs. Accessed 30 Sept 2014.

Ventura County Star. (5 March 2013). Drone maker AeroVironment's third-quarter earnings fall short. http://www.vcstar.com/business/drone-maker-aerovironments-third-quarter-earning. Accessed 30 Sept 2014.

Wall, T. (2013). Unmanning the police manhunt: Vertical security as pacification. *Socialist Studies/Études socialistes, 9*(2), 32–56.

Wall, T., & Monahan, T. (2011). Surveillance and violence from afar: The politics of drones and liminal security-scapes. *Theoretical Criminology, 15*(3), 240.

Watson, S. (24 March 2014). Military, police declare interest in taser drone. Infowars.com. http://www.infowars.com/military-police-declare-interest-in-taser-drone/. Accessed 30 Sept 2014.

Wolverton, J. (3 July 2012). Drone group drafts code of conduct citing public relations problem. *New American*. http://www.thenewamerican.com/usnews/item/11948-drone-group-drafts-code-of-conduct-citing-public-relations-problem. Accessed 30 Sept 2014.

Wolverton, J. (30 January 2014). Local police adopt military technology and tactics. *New American*. http://www.thenewamerican.com/usnews/item/17515-local-police-adopt-military-technology-and-tactics. Accessed 30 Sept 2014.

Kristin Bergtora Sandvik S.J.D., Harvard Law School, is a senior researcher at the Peace Research Institute Oslo (PRIO) and the director of the Norwegian Centre for Humanitarian Studies. Her research takes a socio-legal approach to technology, international law, and humanitarian action. Her work has appeared in, inter alia, *Disasters, the International Journal of Refugee Law, Refugee Survey Quarterly, Millennium: Journal of International Studies, the International Committee of the Red Cross (ICRC) Review, and the International Federation of Red Cross and Red Crescent Societies (IFRC) World Disasters Report 2013*. She is the coeditor of "the Good Drone," forthcoming at Ashgate 2015.

Part II
Drones Between Privacy and Security

Chapter 4
The (F)utility of Privacy Laws: The Case of Drones?

Primož Gorkič

The Problem

Consider the following hypothetical example: one morning, the citizens of Townsville wake up only to receive a letter from their local law enforcement. In the letter, they learn that the law enforcement is about to start using several drones. The drones will operate at an altitude of 300 m and will have state-of-the-art surveillance equipment and data-processing capabilities. These will include thermal imaging cameras, high-resolution cameras, GPS tracking capabilities, long-range microphone, automated license plate recognition system, face recognition capabilities, international mobile subscriber identity (IMSI) catcher and a wide-band real-time connection to available personal data records. They will be used for surveillance, identification and traffic control purposes (for detailed account on the range of available instruments, see e.g. Takahashi 2012, p. 86 et seq.; Hiltner 2013, p. 400 et seq.).

How should an average citizen feel when confronted with such a development? While some may feel that the application of drones will contribute to the safety of their communities, others may feel less inclined to accept such practices. They may consider that the application of drones will provide authorities with information they were previously capable of retrieving only by "classic" judicially supervised measures, such as search of premises or orders to produce communication traffic data; or, at least, by measures that each individual was aware of, for example, stopping and identifying an individual, a vehicle, etc. In short, some would feel that the application of drones severely impacts those areas of their lives that they feel should be left alone and undisclosed to others, including law enforcement agencies.

In legal terms, application of drones raises yet unresolved questions of privacy protection. Those that consider drones a threat to privacy will no doubt seek an adequate legal framework which would enable them to face such threats to privacy. While they may instinctively feel that drones raise privacy issues, an attempt at a legal analysis

P. Gorkič (✉)
Faculty of Law, University of Ljubljana, Poljanski nasip 2, SI-1000 Ljubljana, Slovenia
e-mail: primoz.gorkic@pf.uni-lj.si

A. Završnik (ed.), *Drones and Unmanned Aerial Systems*,
DOI 10.1007/978-3-319-23760-2_4

seems more demanding. At this point, it is important to note that we have yet to reach a comprehensive definition of privacy. The concept seems utterly elusive (see, for example, the analysis in Solove 2007). However, a privacy-centred analysis of using drones for law enforcement purposes may not require such an in-depth approach. In the field of law enforcement, the reasonable expectation of privacy appears to be the leading concept helping us to recognize and regulate measures that intrude into the personal lives of others.

This enables us to pose a deceivingly simple question: can we rely on the concept of reasonable expectation of privacy to regulate the application of drones?

The Katz Test

The general contours of the reasonable expectation of privacy test are well known. In *Katz v. United States* (389 U.S. 347 (1967)), the US Supreme Court departed from its previous trespass-based approach to IV Amendment, established in *Olmstead v. United States* (277 U.S. 438 (1928)). Under the Katz rule, trespass was no longer required to find a IV Amendment violation. Instead, the court held:

"/…/ the Fourth Amendment protects people, not places. What a person knowingly exposes to the public, even in his own home or office, is not a subject of Fourth Amendment protection/…/ But what he seeks to preserve as private, even in an area accessible to the public, may be constitutionally protected/…/

/…/ once it is recognized that the Fourth Amendment protects people—and not simply "areas"—against unreasonable searches and seizures, it becomes clear that the reach of that Amendment cannot turn upon the presence or absence of a physical intrusion into any given enclosure."

Later, the court adopted the position held by Harlan, J., in his concurring opinion on Katz, and the Katz test developed its twofold structure. In order to decide whether a violation of IV Amendment took place, it is necessary to establish (i) whether a person has exhibited an actual (subjective) expectation of privacy and (ii) that such an expectation is recognized by the society as reasonable.

The application of these principles to regulating the use of drones by law enforcement agencies raises several trying questions. To somewhat simplify the nature of drones and their usage, it seem reasonable to say that they are airborne devices with state-of-the-art technical gadgets and data-processing capabilities (see also Takahashi 2012, p. 81 et seq.). In this sense, they operate "out in the open", collecting and processing data that appears publicly accessible—or, at least, accessible from a public vantage point. By adopting the reasonable expectation of privacy test, we therefore need to make at least two inquiries. First, should we recognise as reasonable any expectations of privacy that a person may hold as s/he steps onto the boardwalk or drives away in his/her car? And second, are we capable of responding to technological advances that support the all-pervasive surveillance practices and draw the thin line that helps us in recognising the (un) lawfulness of privacy intrusions? Despite several attempts in (mostly) US case law, the reasonable expectation of privacy test seems to be of little help in tackling these modern-day privacy issues (see also Kerr 2004; Takahashi 2012, p. 108 et seq.)).

Aerial Surveillance

There are several relevant cases that can help us clarify the extent of privacy protection provided by applying reasonable expectation of privacy test in cases of airborne observation and surveillance. Even so, they offer little added value to the state of privacy safeguards.

In *Ciraolo v. California* (476 U.S. 207 (1986)), the US Supreme Court produced guidelines on airborne naked-eye observation and surveillance practices by law enforcement agencies. In Ciraolo, the police engaged in warrantless aerial observation of suspect's fenced home backyard at an altitude of 1000 ft. In deciding whether the safeguards under the IV Amendment apply, the court had to consider, whether such activities amount to a "search". If so, the police activities were illegal due to a lack of court warrant.

The majority opinion of the court held that such activities do not constitute a search. Applying the Katz test, the court found that the applicant did, indeed, manifest "his own subjective intent" to maintain privacy—by raising a fence around his backyard. The main thrust of the opinion, however, focused on whether such an expectation can reasonably prohibit any government observation. Clearly, raising a fence around one's backyard effectively prevents the police (or anybody else) from observing the backyard. However, in court's opinion, the IV Amendment does not "require law enforcement officers to shield their eyes when passing by a home on public thoroughfares", provided that they do so from a public vantage point, that they have the right to be there and that their activities are clearly visible. Finding that the police conducted their observation in publicly navigable space, in physically nonintrusive manner, the court found no violation of the IV Amendment.

To what extent do the court's findings in Ciraolo still apply to modern-day drones? In some aspects, the Ciraolo is certainly still important. The court found it relevant to stress that the police was within its rights when it conducted its observation from a publicly navigable space. Clearly, this calls for a comprehensive regulation of drone's usage. In some aspects, however, the Ciraolo hardly applies. For instance, drones and their hi-tech gadgets far surpass the capabilities of naked-eye observation. Cameras that can be employed at long distances no longer make drone surveillance "clearly visible" to those under observation. As a whole, the court's reasoning is lagging behind the rapid progress of nonintrusive means of observation made possible by technological progress.

Interestingly enough, the majority in Ciraolo emphasized the physically nonintrusive nature of the observation. After Katz, it appeared that the physical aspect of police conduct might play a lesser role in determining violations of the IV Amendment (see also Kerr 2004, p. 831 et seq.). The shift in Katz could have been significant: abandoning the Olmstead trespass test could signal the shift towards individual-, rights-oriented understanding of search and seizure laws. Instead, the court, not only in Ciraolo, remains focused on the nature (intrusiveness) of police conduct, that is, the manner the government chose to expose individual's conduct (see Rubenfeld 1989). While it may have been crucial to maintain such a focus when physical intrusiveness was at the heart of search and seizure, technological progress requires a change in perspective (see also Talai 2014, p. 766 et seq.). Indeed, it appears the court's majority—in Ciraolo, at least—chose to ignore the warnings of Brandeis J., dissenting in *Olmstead v. United States*:

"Ways may someday be developed by which the Government, without removing papers from secret drawers, can reproduce them in court, and by which it will be enabled to expose to a jury the most intimate occurrences of the home. Advances in the psychic and related sciences may bring means of exploring unexpressed beliefs, thoughts and emotions."

These same warnings were echoed by Powell J., dissenting, in Ciraolo, with little effect. The court's majority failed to recognise that police conduct in Ciraolo amounted to an equivalent of a search of the applicant's home. Were they to abandon the conduct-oriented approach and adopt a rights-based understanding of the IV Amendment, they should be able to recognise that the information police collected though aerial observation was the equivalent of the information they would collect upon entering and physically searching the suspect's house and its backyard. From this perspective, the majority's opinion that the police are not required "to shield their eyes" when passing by a home on public thoroughfares is delightfully distracting. The police, as well as anybody else, are indeed required to shield their eyes—when their curiosity is checked by a clear demonstration of individual's expectations to be left alone.

The court's failure to take into account the marvels of technological progress in Ciraolo, however, hardly appears accidental. In *Dow Chemical Co. v. United States* (476 U.S. 227 (1986)), decided at the same time as Ciraolo, the court faced a similar set of facts. In Dow Chemical, too, a law enforcement agency engaged in aerial observation of an industrial compound, but with a difference: They employed sophisticated aerial high-definition photography and took photographs of the applicant's business premises. The court's majority found no violation of the IV Amendment. Its argument was twofold. First, the court chose to draw a distinction between "intimate activities associated with family privacy and the home" and "areas within an industrial manufacturing complex". The latter, following the court's reasoning, does not provide an individual with the same set of safeguards under the IV Amendment. Instead, the court compared the industrial complex area with "open fields", applying a doctrine that restricted the safeguards under the IV Amendment, due to a lack of reasonableness of any expectations an individual may have formed beyond the curtilage of his/her home (see, for example, *Oliver v. United States*, 466 U.S. 170 (1984)).

The first part of the reasoning is clearly individual oriented and rights based. It applies to both natural and legal persons, arguing that one cannot expect privacy protection under the IV Amendment while on such business premises. The argument makes no reference to the nature of the conduct of the law enforcement agency. There is, however, an important second part of the court's majority's reasoning. In this second part, the court, to some extent, follows the same rationale as in Ciraolo: its argument turns towards the manner of the government conduct. The observation in Dow Chemical, too, took place from a publicly navigable airspace.

The court, however, had to deal with an important distinction: the observation was conducted by a sophisticated aerial photography camera. The court noted that the law enforcement agency was

"not employing some unique sensory device that, for example, could penetrate the walls of buildings and record conversations in Dow's plants, offices, or laboratories, but rather a conventional, albeit precise, commercial camera commonly used in map-making/…/

It may well be/…/ that surveillance of private property by using highly sophisticated surveillance equipment not generally available to the public, such as satellite technology, might be constitutionally proscribed absent a warrant. But the photographs here are not so revealing of intimate details as to raise constitutional concerns. Although they undoubtedly give EPA more detailed information than naked-eye views, they remain limited to an outline of the facility's buildings and equipment. The mere fact that human vision is enhanced somewhat, at least to the degree here, does not give rise to constitutional problems."

How far-reaching is this reasoning? Undoubtedly, the court took great care to stay within the particularities of the case. At best, the majority's reasoning leads to more questions than answers. What is clear is that no privacy protection will be afforded under the open-fields doctrine. Other issues, however, remain poorly answered. Most importantly, how to respond to technological progress and the opportunities it provides for law-enforcement surveillance practices? On the one hand, the court appears prepared to recognise that the use of technology can lead to privacy intrusions equivalent to physical searches and physically intrusive surveillance. On the other hand, the court stressed that the technology used by the government was common, conventional and readily accessible.

Advanced Technologies

In Ciraolo and Dow Chemical, the nature of police conduct appears to be of paramount relevance. Police observation always took place in a publicly accessible airspace. This appears as the prevailing circumstance, regardless of whether the object of surveillance was "the curtilage" of individual's home or not. Only in Dow Chemical does the court address the issue of technology-supported observation that is crucial to drones and their modern-day application. The question we need to ask is fairly simple: Would the decision in Dow Chemical be different, if law enforcement used technology "not generally available to the public"? This seemingly fleeting remark in the court's reasoning hardly allows for easy predictions on how the court would respond to drone surveillance in a technology-infested society (see also Molko 2013, p. 1325).

As Kerr (2004) observed, the court—even when applying the Katz test—upheld property-oriented approach in determining IV Amendment violations. The decisions in *United States v. Knotts* (460 U.S. 276 (1983)) and *United States v. Karo* (468 U.S. 705 (1984)) are witness to such an approach. In Knotts, the police used electronic beepers without a warrant to gain data on suspect's movements in public, while in Karo the police installed such beepers in a way that enabled them to verify that the tracked object was inside suspect's house. In this part, the actions of the police constituted a warrantless search of the suspect's apartment.

Both Knotts and Karo set the stage for decisions in *Kyllo v. United States* (533 U.S. 27 (2001)), and *United States v. Jones* (132 S. Ct. 945 (2012)). Both Kyllo and Jones tackle the question of new technologies, but differ in one important aspect: In Kyllo, the court had to decide whether warrantless thermal imaging of suspect's home amounted to a search. In Jones, the court faced the problem of warrantless long-term use of GPS tracking device to monitor the movement of suspect's car. Surprisingly, in both cases, the court ruled there was a violation of the IV Amendment.

Why surprisingly? After a history of property-oriented interpretations, the court in Kyllo took the interpretation of the Katz test an inch further:

"We think that obtaining by sense-enhancing technology any information regarding the interior of the home that could not otherwise have been obtained without physical "intrusion into a constitutionally protected area,"/…/, constitutes a search—at least where (as here) the technology in question is not in general public use."

As Kerr (2004) observed, the impact of Kyllo was smaller than expected. The courts applied Kyllo mainly in order to strengthen the protections of privacy in one's home. The US Supreme Court, however, was mainly concerned with the impact of "property-defeating" (Kerr) technologies:

"/…/ just as a thermal imager captures only heat emanating from a house, so also a powerful directional microphone picks up only sound emanating from a house—and a satellite capable of scanning from many miles away would pick up only visible light emanating from a house. We rejected such a mechanical interpretation of the Fourth Amendment in Katz, where the eavesdropping device picked up only sound waves that reached the exterior of the phone booth. Reversing that approach would leave the homeowner at the mercy of advancing technology, including imaging technology that could discern all human activity in the home. While the technology used in the present case was relatively crude, the rule we adopt must take account of more sophisticated systems that are already in use or in development."

The actual scope of Kyllo, however, is uncertain. First, the court's position does not apply to use of advanced technologies in open fields (comp. Talai 2014, p. 762). The position of the court in Dow Chemical remains undisturbed, given that the court found that an industrial complex cannot be regarded as "home". Second, Kyllo also appears to differ substantively from Ciraolo, given that Ciraolo dealt with naked-eye observation. One can argue, however, that Ciraolo, too, constituted a technology-supported (aerial) observation that allowed the police to gain knowledge they could not otherwise obtain without physical intrusion. If Kyllo is understood as applicable to all property-defeating technologies, it applies to facts such as in Ciraolo as well. And third, Kyllo exposes the key deficiency in the Katz test: use of property-defeating technologies that amount to an equivalent of a house search will constitute a search as long as "the technology in question is not in general public use". It appears as though the court (as before in Dow Chemical) is setting the stage for a technology-infested society where privacy is no longer an issue. Later, Alito J., concurring with the decision in Jones, appears very much aware of this, when stating:

"/…/ the Katz test rests on the assumption that this hypothetical reasonable person has a well-developed and stable set of privacy expectations. But technology can change those expectations. Dramatic technological change may lead to periods in which popular expectations are in flux and may ultimately produce significant changes in popular attitudes. New technology may provide increased convenience or security at the expense of privacy, and many people may find the trade-off worthwhile. And even if the public does not welcome the diminution of privacy that new technology entails, they may eventually reconcile themselves to this development as inevitable/…/"

To paraphrase, in a world where every smartphone is equipped with a thermal imaging sensor, the court's holding in Kyllo may no longer be relevant.

In the context of drone surveillance, Kyllo (for the time being) draws the line when it comes to the use of technologies that allow obtaining knowledge about the intimacy of one's home. It applies to thermal imaging, use of long-range microphones and modern laser appliances. It may apply even to use of so-called IMSI catchers when used for locating cellphones inside one's home (Gorkič 2014). In short, under Kyllo, use of such technologies (as long as they are not in general use) to obtain insight into one's home should amount to a search under IV Amendment—even when they are used from a "public thoroughfare" (see Talai 2014, p. 764).

Before we analyse the conceptual drawbacks of the Katz test, we need to take a look at Jones. In Jones, the court addressed the issue of long-term use of GPS tracking device, attached to suspect's vehicle in order to monitor its movements. In Knotts and Karo, the court found that using a tracking device to monitor the movement of a container (as long as in public accessible areas) does not amount to a search. Open-fields doctrine, too, supports such a conclusion. In Knotts and Karo, however, the court had to decide whether using a tracking device constitutes a search under the Katz test. In Jones, the court returned to seemingly abandoned trespass test. In this way, the fact that the government placed a tracking device onto suspect's vehicle constitutes a search under the trespass test. By focusing on government trespassing conduct, open-fields doctrine (relying on Katz test) no longer applies. The fact that the court in Jones relied on trespass doctrine, however, makes Jones inapplicable to the problem of drones. Flying an aircraft over one' home simply does not amount to trespass (Talai 2014, p. 761).

* * *

At this point, we can try and answer the two questions we began with. First, should we protect individual's privacy expectations beyond his/her home in the context of drone surveillance? It appears that the answer is no, at least following the line of reasoning adopted by the US Supreme Court. And second, can the Katz test adequately set the limits to the ever-expanding usage of advanced technologies? Here, the answer, too, in the long term, is negative, even when applied to situations of remote sensing wall-penetrating equipment. The moment such equipment becomes ubiquitous, the Katz test will be of little value.

A Note on the Katz Test and Its Drawbacks

Still, the US Supreme Court seems determined to pursue Katz in situations "involving merely the transmission of electronic signals without trespass", as put in Jones. Talai, too, is optimistic when stating that the problem of drones can be tackled by Katz test (Talai 2014, p. 762).

Those debating the Katz test in the context of advanced technologies are highly critical of its ability to effectively pose limits to technology-supported surveillance. Takahashi (2012, p. 107 et seq.) sees the use of drones as a possible "catalyst" leading to a new paradigm of privacy. His analysis of US Supreme Court case law on IV Amendment shows that it hardly provides a satisfactory answer to questions of long-term surveillance and the ability to minimize the scope of surveillance, due to multimodality of the drones surveillance. Schlag (2012, p. 15 et seq.) points out that the use of drones—even in the "open fields"—can lead to collection of intimate information about an individual, without his knowledge—due to the technological advances that enable drones to operate autonomously and inconspicuously. Others, for example Ghoshray (2009), warn that the use of drones for domestic surveillance is the ultimate threat to the "inherent sanctity" of individual's home. More important, however, is the argument that privacy needs to extend beyond the "physical dwelling" to include "individual's digital confines of the home-like community of connected individuals" within which an individual may exercise his right to be left alone (Ghoshray 2009, pp. 596–597).

These reflections on the state-of-play of privacy laws in the context of drone surveillance signify that the applicable doctrines under the IV Amendment may not be satisfactory. Their common denominator is no doubt a shift from "conduct-oriented" nature of the trespass and the Katz tests towards an "individual-centred" understanding of privacy-related issues.

Calls for a shift towards a different interpretation of the IV Amendment are nothing new. Limits of the Katz test when used to regulate collecting and processing publicly accessible data about an individual have been well noted. Solove (2007, pp. 765–766), for example, points out two factors that are an inherent part of surveillance practices: the aggregation of data and the chilling effect. Aggregating bits of data leads to insight about our "personalities and activities", by putting together small, seemingly unimportant pieces of information. The "chilling effect" results from the facts that surveillance, even of legal activities, may result in inhibiting individual's conduct, in particular by chilling free speech and free association (Solove 2007, p. 765). While Solove wrote primarily in the context of surveillance and data mining practices, same considerations apply to the context of drone surveillance. Only by recognising that the aggregation of accessible pieces of information is, indeed, a privacy problem, can drones-related privacy issues be adequately addressed.

Rubenfeld, too, presents a different interpretation of the IV Amendment, focusing on the security of personal life (Rubenfeld 2008, p. 127 et seq.) rather than the conduct of law enforcement agencies. His understanding of the protections afforded under the IV Amendment can be labelled as "anti-totalitarian" (Rubenfeld 2008,

p. 129). His reliance on Mill leads Rubenfeld to stress the importance of freedom, in personal life, to "defy public norms", that is, to speak and act as one chooses (Rubenfeld 2008, p. 128). Similar to Solove, Rubenfeld reshapes the issues of privacy and presents them as political issues. Focusing on the individual's personal security of private life allows Rubenfeld to understand personal life a "collective good", as including "personal communications and other interactions among individuals, even when those communications and other interactions occur outside the home" (Rubenfeld 2008, p. 130). Each individual's private life "depends in part on other's having that capacity as well" (Rubenfeld 2008 p. 131).

These approaches can be more effective when applying the IV Amendment to new methods of surveillance. Their strongpoint is, no doubt, their focus on the individual's personal life. Law-enforcement surveillance practices will invoke concerns under IV Amendment, when they provide an insight into person's personal life, regardless of their conduct or the accessories used, free from the restraints under the trespass rule or the Katz test and its exceptions, such as open-fields doctrine. There is, however, a downside: it remains uncertain to what extent they can be introduced into current IV Amendment case law.

The mosaic theory can be understood as an attempt to overcome the limits of the Katz test. The term "mosaic theory" is used by Kerr (2012, p. 313) to identify a new approach in interpreting the term "search" under the IV Amendment. In contrast to a traditional, sequential approach, a step-by-step analysis of government's conduct (Kerr 2012, p. 315 et seq.), using the mosaic theory a "search" can be construed as a "collective sequence of steps" (Kerr 2012, p. 313) that, taken as a whole, "can lead the Government to learn intimate details" about suspect's life (Kerr 2012, p. 328). As put in *United States v. Maynard* (615 F.3d 544 (D.C. Cir.)) "when it comes to privacy,/…/ the whole may be more revealing than the parts."

The concurring opinions in Jones, too, offer some support to those predicting that the problem of drone surveillance could be more effectively addressed by the mosaic approach (see, for example, Talai 2012, p. 765, 778 et seq.). Alito J., for example, held that

"relatively short-term monitoring of a person's movements on public streets accords with expectations of privacy that our society has recognized as reasonable/…/ But the use of longer term GPS monitoring in investigations of most offenses impinges on expectations of privacy. For such offenses, society's expectation has been that law enforcement agents and others would not—and indeed, in the main, simply could not—secretly monitor and catalogue every single movement of an individual's car for a very long period/…/"

The mosaic theory does, indeed, offer a helping hand in tackling advanced methods of surveillance. It can be applied to drone surveillance as well. It does not rely on the general availability of surveillance technologies, nor does it confine itself to the boundaries of "home and its curtilage". There are important questions that remain however, which the judicial process of developing the mosaic approach, at least in Kerr's opinion (2012, p. 350 et seq.; also Kerr 2004), cannot answer adequately or in time (for some legislative proposals, see Thompson 2013, p. 18 et seq.).

Does Europe Fare Any Better?

In the European context, courts are less reliant on the Katz test. The European Court of Human Rights (ECtHR) has integrated the Katz test into its case law on Article 8 of the European Convention of Human Rights (ECHR), for example in *Halford v. The United Kingdom* (no. 20605/92, 25 June 1997). Later on, however, the ECtHR distanced itself from the test. This is particularly evident in cases where the Katz test does not yield satisfactory results, for example in cases concerning surveillance in public. In *P.G. and J.H. v. The United Kingdom* (no. 44787/98, 25 Sept 2001), the ECtHR, for example, held:

"Since there are occasions when people knowingly or intentionally involve themselves in activities which are or may be recorded or reported in a public manner, a person's reasonable expectations as to privacy may be a significant, although not necessarily conclusive, factor. A person who walks down the street will, inevitably, be visible to any member of the public who is also present. Monitoring by technological means of the same public scene (for example, a security guard viewing through closed circuit television) is of a similar character. Private life considerations may arise, however, once any systematic or permanent record comes into existence of such material from the public domain."

Later on, the ECtHR adhered to this position when faced with cases concerning the use of GPS tracker in *Uzun v. Germany* (no. 35623/05, 2 September 2010). In comparison to Jones, there is no indication that the ECtHR attached any importance to the trespassory nature of the conduct of German law enforcement bodies. In considering whether the use of GPS trackers intruded upon applicant's personal life, the ECtHR did stress that

"GPS surveillance is by its very nature to be distinguished from other methods of visual or acoustical surveillance which are, as a rule, more susceptible of interfering with a person's right to respect for private life, because they disclose more information on a person's conduct, opinions or feelings."

At the same time, the fact that GPS surveillance allows for a systematic collection and storage of data on applicant's whereabouts and movements in the public sphere, resulting in a pattern analysis of applicant's movements, the ECtHR nevertheless found that the use of GPS trackers interferes with the right to private life under Article 8 of the ECHR. Adding to this, the ECtHR, while analysing its previous case law, found no reason to differentiate between covert and overt methods of surveillance.

Under the ECtHR case law, the use of drones would constitute an interference with individual's right to private life. While Uzun addressed surveillance practices in public places, there is little doubt that the ECtHR would address problems of wall-penetrating technologies in a similar manner. Property-related considerations and the issue of all-presence of the applicable technologies, which arise under the Katz test, are simply not important; considerations under Article 8 of the ECHR are, in contrast to IV Amendment case law, very much individual and rights oriented.

Article 8 of the ECHR case law does not, however, (yet) address the problem of general, unfocused surveillance in public sphere. The judgement in Uzun deals with a surveillance practice that is individual focused. The use of GPS tracker allows for systemic collection, storage and processing of data about a particular individual. It does not, however, address the problem of a general surveillance in the public sphere.

Recent judgements of the Court of Justice of the European Union (ECJ), resting on Articles 7 and 8 of the Charter of Fundamental Rights of the European Union, may, however, provide some insight into arguments related to such generalized surveillance practices.

The use of drones as a general surveillance tool raises similar issues as those addressed by the ECJ in its Digital Rights judgement (C-293/12, C-594/12, 8 April 2014), namely, the aggregation of data and the concern of a chilling effect (Solove). Even though the court's judgement dealt with the retention of communication traffic data, the similarities between drone surveillance and traffic data retention and processing are striking. Traffic data

"allow very precise conclusions to be drawn concerning the private lives of the persons whose data has been retained, such as the habits of everyday life, permanent or temporary places of residence, daily or other movements, the activities carried out, the social relationships of those persons and the social environments frequented by them,"

as the court put it the aforementioned judgement.

Similarities can also be drawn with respect to the wide-ranging nature of privacy interference due to drone surveillance. The court, following Advocate General and consequently the reasoning of the German Constitutional Court, considered the interference due to traffic data retention and processing "particularly serious"—partly due to the fact that

"data are retained and subsequently used without the subscriber or registered user being informed is likely to generate in the minds of the persons concerned the feeling that their private lives are the subject of constant surveillance."

The possibilities of long-range, in effect, covert drone surveillance must raise similar concerns. What the ECJ is essentially referring to are the dangers of a chilling effect due to "feelings of constant surveillance" that can result from dragnet type of data collection and processing. There are no reasons why such considerations would not be applicable to some of the drone surveillance practices.

Conclusion

Comparing the US case law and the jurisprudence of the two European courts reveals a fundamental difference in the approaches. These differences have been well noted by Jacoby in her comparative analysis of privacy protections in the context of technical surveillance under US and German law (Jacoby 2007, p. 479 et seq.). She, too, notes the overreliance of the US courts on the nature of law enforcement conduct and the ubiquity of the technology used. In contrast, the German approach is more concerned with the protection of privacy itself, focusing on the dignity of the individual and the level of intimate details revealed to law enforcement agencies (Jacoby 2007, p. 484). She concludes that the US law, applying the Katz test, is ill-equipped to deal with privacy issues that arise in the context of surveillance using advanced technologies (see Jacoby 2007, p. 493).

These characteristics in the German approach very much coincide with the positions of the ECtHR and the ECJ. Their emphasis on the level of intimate details revealed to law enforcement agencies offers a significantly greater level of safeguards against privacy intrusions, when compared to the mainstream US approach. Furthermore, any attempts to develop alternative interpretations of the IV Amendment may very well be destined to fail. The views of those US scholars that argue for alternative solutions may be quite difficult to implement due to the nature of the right to privacy in the US law. Jacoby (2007, p. 491) argues that the right to privacy in the US context is understood as a negative right, requiring the government to refrain from unreasonable warrantless searches and seizures, with no duty of the government to actively preserve the individual's private sphere. In contrast, the alternative solutions, proposed by US scholars, resemble very much the views adopted by the German law and by the two European courts. The "European" approach sees the right to privacy as a positive right: The government is required to "foster and uphold the privacy sphere" (Jacoby 2007, p. 491). Hence, we should add, any judicial consideration of privacy intrusions must take into account the quality and quantity of information collected and processed in the course of law-enforcement surveillance. And how should the concerned people of Townsville proceed in challenging the use of drones? Their attempts at preserving their privacy interests by relying on the reasonable expectation of privacy may very well prove futile. On the other hand, their arguments should be more effective in a European context.

Without a doubt, drone surveillance makes the debate on privacy protections in a technology-permeated society more interesting than ever before. While "dataveillance" (Clarke) appears distant and vague ("out of sight, out of mind"), drone surveillance should represent a more tangible threat to the private life of an average citizen. To this moment, legal regulation of surveillance practices has proven quite ineffective. Perhaps a reaction to the use of drones can spark a more vigorous response to the challenges our technological society brings for the protection of the right to privacy.

References

Ghoshray, S. (2009). Domestic surveillance via drones: Looking through the lens of the fourth amendment. *Northern Illinois University Law Review, 33*, 580–600.

Gorkič, P. (2014). Sodobni prikriti preiskovalni ukrepi, prvič: Lovilec IMSI. *Odvetnik, 65*(2), 47–53.

Hiltner, P. J. (2013). The drones are coming: Use of unmanned aerial vehicles for police surveillance and its fourth amendment implications. *Wake Forest Journal of Law and Policy, 3*(2), 397–415.

Jacoby, N. (2007). Redefining the right to be let alone: Privacy rights and the constitutionality of technical surveillance measures in Germany and the United States. *Georgia Journal of International & Comparative Law, 35*(3), 433–493.

Kerr, O. S. (2004). The fourth amendment and new technologies: Constitutional myths and the case for caution. *Michigan Law Review, 102,* 801–888.

Kerr, O. S. (2012). The mosaic theory of the fourth amendment. *Michigan Law Review, 111,* 311–354.

Molko, R. (2013). The drones are coming! Will the fourth amendment stop their threat to our privacy? *Brooklyn Law Review, 78,* 1279–1699.

Rubenfeld, J. (1989). The right of privacy. *Harvard Law Review, 102,* 737–807.

Rubenfeld, J. (2008). The end of privacy. *Stanford Law Review, 61,* 101–162.

Schlag, C. (2012). New privacy battle: How the expanding use of drones continues to erode our concept of privacy and privacy rights. *Journal of Technology Law & Policy, 13,* 1–22.

Solove, D. J. (2007). "I've got nothing to hide" and other misunderstandings of privacy. *San Diego Law Review, 44,* 745.

Takahashi, T. (2012). Drones and privacy. *The Columbia Science and Technology Law Review, 14,* 72–114.

Talai, A. B. (2014). Drones and jones: The fourth amendment and police discretion in the digital age. *California Law Review, 102,* 729–780.

Thompson, R. M. (2013). Drones in domestic surveillance operations: Fourth amendment implications and legislative responses. congressional research service.

Primož Gorkič After attaining BA in law in September 2002, Primož Gorkič enrolled in postgraduate studies in civil and commercial law at the Ljubljana Faculty of Law, later substituted for studies in criminal law. He attained his doctoral degree in November 2009. Finishing his traineeship with the Ljubljana Higher Court in November 2004, he applied for the position of assistant researcher at the Ljubljana faculty. He passed the national bar exam in April 2005. From October 2007 till November 2009, he worked as a part-time advisor to the Supreme Court of Slovenia. He is currently working as an assistant professor with the Chair for Criminal Law at Ljubljana Faculty of Law and as a part-time researcher at the Institute of Criminology at the Ljubljana faculty of law. His research focuses on criminal procedure, evidence law and human rights in the context of criminal justice. He teaches criminal procedure law (undergraduate and graduate studies) and criminal Investigation (graduate studies). He currently presides over the Slovenian Association for Criminal Law and Criminology.

Chapter 5
Re-bordering the Peripheral Global North and Global South: Game of Drones, Immobilising Mobile Bodies and Decentring Perspectives on Drones in Border Policing

Sanja Milivojevic

> *Without making a sound, invisible sets of eyes patrol our southwest border.... Virtually undetectable by naked eye, unmanned areal vehicles—better known as UAVs or drones—silently patrol the border, looking for potential terrorists, drug smugglers, and illegal immigrants. (Longmire 2014, p. 75)*

Introduction

Border security, asylum seekers and undocumented migrants are undoubtedly one of the key issues in a colourful Australian political landscape in the past decade (see McCulloch and Pickering 2012; Weber and Pickering 2011). Elections are won or lost based on political parties' perceived (in)ability to secure Australian borders, especially from undocumented migrants coming to Australia by boat (Manne 2010). The process of separating kinetic elites from kinetic underclass (Adey 2006) in Australia is formalised through series of interventions in the area of border management, many of which include surveillance technologies deployed at the border and beyond. In May 2010, then Opposition leader Tony Abbot announced that the Coalition government is interested in pursuing unmanned areal vehicles (UAVs)—commonly known as drones—in order to 'regain control of our borders' (The Guardian 2013). Given the importance of border control in both political and social context in Australia, it is hardly surprising that the issue of finding a technological 'solution' for asylum seekers' 'problem' emerged as a critical element of the Coalition's 2013 elections defence policy. Surprisingly, a few days before the elections, Mr. Abbot, now the prime minister of Australia, backed away from this election promise, stating that he is 'not making specific commitment [about purchasing drones] at this time' (The Guardian 2013). However, in February 2014, the newly elected

S. Milivojevic (✉)
School of Social Sciences, University of New South Wales, Sydney, Australia
e-mail: s.milivojevic@unsw.edu.au

A. Završnik (ed.), *Drones and Unmanned Aerial Systems*,
DOI 10.1007/978-3-319-23760-2_5

Australian government led by Mr. Abbott announced that '[u]p to seven giant unmanned aircraft would be bought for $3 billion under a government plan that would transform the country's ability to patrol its borders' within 6 years (Stewart 2014a).

Elsewhere across the Western world, border policing and regulating mobile bodies are also a priority. In Europe, policymakers and political parties are 'competing for electoral support with promises to restrict unwanted migration' (Boswell 2003, p. 621) amidst growing conviction that the European Union (EU) cannot cope with (largely exaggerated) immigration inflow from non-EU states (Sassen 2006). The ongoing process of re-bordering and fortifying external borderlands includes a range of hi-tech security and surveillance technologies (Neal 2009; Milivojevic 2013) developed to strengthen EU flanks and prevent the arrival of migrants and asylum seekers to the Fortress Europe. The borders in Europe are rapidly transforming into segregation zones built, expanded and maintained through state-of-the-art initiatives such as *frontières extérieures* (FRONTEX), European border surveillance system (EUROSUR), Schengen Information System (SIS), Visa Information System (VIS) and European Dactyloscopy (Eurodac). Such development resulted in creating a 'cyber-fortress Europe' (Guild et al. 2008)—(im)migration exclusion zones based on hi-tech mechanisms of social segregation. Surveillance from the sky is quickly emerging as a preferred method for strengthening risky borderlands of the Mediterranean and a troubled Southeast frontier of the EU. With a budget over £ 320 million set for drone development (Waterfield 2014), we are witnessing 'an emerging EU drone policy' (Hayes et al. 2014) that gradually spills across the border to the neighbouring countries of Global South. Drones now patrol not only maritime borders of the EU, UK, Australia and the deadly US–Mexico border but also green borders of Turkey, Ukraine, Chile and Serbia (Stein 2014; ABC News 2014c; Press Online 2011; Cupolo 2013)—new custodians of the order at the peripheral borders of Global North.

Such development, however, has generated limited interest in the criminological literature. While surveillance studies, sociology and other disciplines in social sciences increasingly look into the use of drones, criminological inquiry rarely explores the topic (for notable exemptions see works of Tyler Wall, Torin Monahan, Kevin Haggerty, Dean Wilson and Gavin Smith). Even when criminologists engage with the issue, the use of drones in military context sits at the top of the agenda, while the border policing aspect of drone use is rarely discussed (again, for notable exemptions see works of Peter Andreas, Katja Franko Aas and Hille Koskela). In addition, the focus on use of drones in border policing both in academia and in the media discourse has been mostly US- and/or Global North-centric.

This chapter fills this gap by decentring the border–drone scholarship. It looks at the process of re-bordering of both Global North and Global South through case studies of Australia and Serbia. The chapter highlights the development of security and surveillance border control technologies in both countries, especially in the context of the usage of drones in border policing. Its purpose is not to offer a comparative analysis of two case studies; rather, the chapter seeks to illuminate the issue of drone use in border policing from opposing sides of the borderland. The chapter presents preliminary findings from a research project that looks at mobility and border control in Western Balkans in the context of the EU integration, conducted in the period August–December 2013. It also draws on the analysis of media reporting on drones in Serbia and Australia in the past 2 years, in order to explore predominant narratives around the usage of surveillance technologies in border control. This chapter argues that, although the rhetoric in the use of drones in two contexts is largely different, the military rhetoric underpins both military and non-military use of drones. Finally, the chapter begins to examine the impact of surveillance-based, drone-led border policing on mobile bodies caught in the 'drone stare' (Wall and Monahan 2011). I first turn to the state of knowledge in relation to the intersection of border management and technology in the twenty-first century.

Towards 'An Exhaustive Surveillance': Border Anxieties, Risk Management and Technology Advancements in Border Policing

Ceyhan (2008) reminds us that security technologies comprise technologies of the living (that include developments in genetics and identification via body part prints), optical and electronic technologies (laser, glass fibre networks) and information and communication technologies. The implementation of security technologies is largely based on three sets of logic: security logic of identification of risks; a logic of management flows of goods, people and transportation; and a logic of ambient intelligence that seeks to improve the quality and comfort of our daily lives (Ceyhan 2008). The first two logics underpin the development of security/surveillance border policing initiatives in the broader context of globalisation. A crisis over boundaries (Berman 2003) strengthened by ever-growing inventory of 'threats' to national security and perceived (in)ability of nation states to regulate mobile populations is just one of many anxieties that characterise the globalised world; yet, this anxiety is arguably one of the most potent ones, with significant policy, legal and social outcomes within Global North and Global South. The ongoing process of re-bordering (Popescu 2012) sees borders renewed and enabled/fortified, both terrestrially and virtually. Apparent failure in traditional migration control policies prompted the quest for 'more sophisticated, flexible, and mobile devices of tracking, filtration, and exclusion' (Vukov and Sheller 2013, p. 225). Surveillance as 'the focused, systematic and routine attention to personal details for purposes of influence, management, protection or direction' (Lyon 2007, p. 14) became the backbone of 'crimmigration' (Aas 2011; Welch 2012) border policing strategies.

In a software-based modernity (Bauman 2000), the decision on who is in and who is out, who can cross and who is to be immobilised is—while still in the hands of humans—increasingly based on data captured, processed and analysed by a range of non-human technological advancements. The various degrees of permeability of contemporary borders are defined and enforced by machines, both at border crossings and beyond. Information and communication technologies, as Broeders (2009,

p. 34) reminds us, 'play an important role in widening the possibilities to document, codify and store information on the activities of subjects and citizens'. Contemporary 'surveillance assemblage' (Haggerty and Ericson 2000)—a range of systems and techniques, from biometrics, video and wireless surveillance, scanning devices, facial recognition technology, motion detectors and fingerprinting—is designed to increase efficiency in border policing (Neal 2009; Milivojevic 2013). The ultimate goal of such development, to borrow from Foucault (cited in Amoore et al. 2008), is to achieve 'an exhaustive surveillance' of mobile populations. Indeed, overarching total surveillance in crime control is spreading across Global North, unchallenged and almost universally recognised as a best practice in combating transnational crime.

'Risk management' strategies that aim to perform social sorting have been interwoven into numerous cross-border surveillance systems (or 'surveillance fantasies'—Aas 2011, p. 332), driven by an ever-going pursuit for security (Zedner 2009). The site for the development of 'smart' surveillance technologies (Kenk et al. 2013) is the borderland—the border itself as the fundamental political institution—but also terrestrial and digital spaces deep inside nation-states that act as sites for border policing interventions. The surveillance state, as Broeders (2009) notes, performs the watch both internally (within its borders) and externally (across borders and in countries of transit and origin). Security technologies deployed in the context of border control penetrate deep into the private, domestic sphere of mobile (and immobile) populations' lives in countries of origin and transit.

Surveillance from the sky is, arguably, of paramount importance for policymakers and law enforcement. For the purpose of this chapter, drones[1] are defined as aircrafts without a pilot on board.[2] They come in all shapes and sizes[3] and are perceived as 'ideal instruments for…ISR (intelligence, surveillance and reconnaissance) missions' (Barry 2013, p. 2). Drone's 'Three Ds' capability—to be Dull (work long hours and engage in repetitive tasks), Dirty (the resilience to toxicity) and Dangerous (not having a human price tag attached if something goes wrong) makes them a red-hot favourite for both military and non-military deployment (Barry 2013).

Like many other technological advancements, the development of drones is associated with science fiction[4]. While dating back to World War I (Salter 2013; Wilson 2012), the expansion of drones is linked to the development of the military–industrial complex in the late twentieth century (Barry 2013). The first military use of drones was for surveillance purposes in the 1990s Balkan wars and especially in

[1] A range of synonyms are used to describe drones, such as UAVs and remotely piloted vehicles (RPVs), while unmanned areal systems (UAS) or remotely piloted vehicle systems (RPVS) define systems comprising drones and land-based control stations (Hayes et al. 2014).

[2] This does not include landlocked sea-based vehicles that have also been developed (Hayes et al. 2014).

[3] A common categorisation is light (between 20 and 150 kg), medium (150–600 kg) and large (over 600 kg) (Hatzigeorgopoulos 2012).

[4] Clark (2014, p. 231) reminds us that '[t]he first major 20th century anti-utopian novel—25 years before Orwell's 1984—imagined drones (aeros) as the means by which the government observed and repressed the population'.

the 1999 bombing of Serbia (Hayes et al. 2014; Becker 2013). A non-military use of drones followed: The US border agency Customs and Border Protection (CBP) signalled its interest in drones in the late 1990s, with first drones deployed along the borderlands in 2003 (Barry 2013). The EU is trying to integrate high-tech security technologies in border policing for quite some time now (see Milivojevic 2013; Hayes and Vermeulen 2012; Aas 2011). While drones entered the EU policy discourse in 2002, it was in 2012 that the first formal EU strategy on the use of drones emerged (Hayes et al. 2014). Currently, 16 of the 27 EU member states use drones for military and non-military purposes, while more than 400 UAV systems are deployed in at least 21 EU nation states (Hayes et al. 2014). Frontex—the EU's border police agency—uses drones in joint operations with EU member states and has recently expressed the interest to acquire its own drones in the near future (Suilleabhain 2013; Hayes et al. 2014). As envisioned by EUROSUR (a pan-European border surveillance system) and Frontex, drones are expected to become a backbone of surveillance and monitoring of external borders of the EU in the near future, with the budget for overall research on drones growing to a mind-blowing €3.8 billion (Hayes et al. 2014). The debate on the use of drones is heated, as they can potentially assist in border policing, but also fire missiles and kill from a distance.

The Drone Dilemma: Deadly Weapons, Pushback Tools or Rescuers at High Seas?

As Hayes et al. (2014, p. 8) point out, '[f]ew technologies have captured the media's attention like drones'. In recent years, drones have been increasingly associated with remote killings or drone strikes that have caused thousands of (military and civilian) casualties in Pakistan, Afghanistan, Yemen and Somalia (Hayes et al. 2014; see also Wall and Monahan 2011). Perceived as the ultimate 'smart weapons', use of drones in the military context provides an advantage in the information warfare; however, drones are increasingly seen as 'automated surveillance-military killing machines' (Wilson 2012, p. 274). The predominant narrative around the military use of drones is the one of a hunt, in which superior, technology-equipped enforcers are out there to 'hunt and kill' the enemy. Given that many victims of drone strikes are civilian casualties, the use of drones in war is extensively scrutinised by the media and scholars (Kilcullen and McDonald Exum 2009; Wall and Monahan 2011; Orr 2011). Importantly, the use of drones results in a drone stare, surveillance 'that abstract people from contexts, thereby reducing variation, difference, and noise that may impede action or introduce moral ambiguity' (Wall and Monahan 2011, p. 239). People are, thus, reduced to targets, dead bodies to numbers and civilian victims to collateral damage and 'bug splats' (Not A Bug Splat 2014).

The pursuit for security in border policing resulted in adapting military technology for non-military purposes (Hayes et al. 2014). The intersection of national security (and border security as one of its key priorities), risk management of mobile bodies and security technologies resulted in a series of developments in border

Fig. 5.1 General atomics. (Website: www.ga.com)

policing, with funding for drones in border management now surpassing funding for any other drones (Hayes et al. 2014). Perceived as cheap and efficient solution for the border security 'problem', drones have increasingly been used to detect and immobilise undocumented migrants. Hundreds of types of drones currently patrol the skies of both Global North and Global South, the smallest ones barely the size of a small bird, while the biggest ones rival the Boeing 737 (Farber 2014). And while military use of security and surveillance technologies has been smeared by ethical considerations, their use in domestic (surveillance and border policing) context has an almost unanimous support by political actors and participants in the debate.

The language used to describe security technologies deployed at the border resembles reading a sci-fi novel, intertwined with Greek or Roman mythology. Since 2012, the US border agency CBP uses a border surveillance drone system called VADER (Vehicle and Dismount Exploitation Radar)—the name of one of the greatest villains in modern cinematography—Lord Darth Vader from Star Wars (Becker 2013). One cannot help but notice a heavyweight, military-style language with a plethora of acronyms used to describe projects and initiatives in relation to border control (such as GLOBE (Global Border Environment), An Interoperable Approach to European Union Maritime Security Management (OPERAMAR), TALOS (European Union Border Protection System), Protection of European Seas and Borders Through the Intelligent Use of Surveillance (PERSEUS), Open Architecture for UAV-Based Surveillance System (OPARUS), Sea Border Surveillance (SEABILLA), Smart Unattended Airborne Sensor Network for Detection of Vessels Used for Cross Border Crime and Irregular Entry (SUNNY), Low Time Critical Border Surveillance (LOBOS), Collaborative Evaluation of Border Surveillance Technologies in Maritime Environment by Pre-operational Validation of Innovative Solutions (CLOSEYE) and others),[5] complemented with logos that resemble military-style insignia. The aircrafts themselves carry mythological and/or military-style hunting names, such as 'Predator', 'Avenger' or 'Triton'. Pictures provided by drone manufacturers often portray drones as an essential element to the border security paraphernalia (Fig. 5.1).

[5] See Hayes et al. 2014 for more details; also interview with no. 11, NGO, Hungary.

Recently, we see a potential shift in semantics around the use of drones in the border policing context, with 'Predator' drones replaced by 'Guardians' (Longmire 2014, p. 76). In that sense, as drone industry public relations (PR) campaigns indicate, drones are described as inventions that can 'benefit mankind in general' (Gallagher 2012). Border policing is identified as one of the key positive roles drones might play in the future.[6] The concept of 'drones for human rights' (Sandvik and Lohne 2014) is based on the notion that drones can save lives. Indeed, the Australian Defence Ministry argues that 'unmanned aircraft[s]...have the ability to swoop low and identify asylum-seeker boats and if necessary drop a life raft to boats in distress' (Stewart 2014a). Similarly, Zdravko Kolev, Frontex research officer argues that

> [Drones are] of particular interest of Frontex and EU Member states for improving the capacity to detect and track small and unseaworthy vessels, which are being used on a regular basis for irregular migration and cross-border crime.... [Drones] technology might offer great potential by improving the areal surveillance capacity resulting in more lives saved. (Cited in Hayes et al. 2014, p. 71)

Such rhetoric was especially strengthened after the tragedy at Lampedusa in October 2013. The president of the European Commission, Jose Manuel Baroso, commented that

> [w]e need also to strengthen our capacity for search and rescue, and our surveillance system to track boats, so that we can launch a rescue operation and bring people to safe grounds before they perish. I think the kind of tragedy we have witnessed here so close to the coast should never happen again. Our initiative 'Eurosur' is meant to do that. (Cited in Heller and Jones 2014)

Such narrative is seldom contested. Critics argue that rather than rescuing migrants in distress drones serve to hunt migrants and refugees (Hayes et al. 2014), while surveillance is 'a key component of the conditions that have led to over 14,000 documented deaths at the EU's maritime borders over the last 20 years' (Heller and Jones 2014, p. 9). They remind us that under EUROSUR rules there is no requirement for EU member states to assist vessels in distress (Hayes et al. 2014; Heller and Jones 2014). The efficiency of drones in intercepting mobile bodies at the border is also questioned,[7] as well as its use in rescue missions. As Ska Keller, a German Member of Parliament (MP) in the European Parliament argues,

> [d]rones are very expensive and they don't help. Even if a drone detects a vessel, it can't do anything for them. You need to have actual people there, and having a drone doesn't guarantee that. (Cited in Hayes et al. 2014, p. 71)

Nevertheless, the expansion of drones seems unstoppable, within and beyond the borderlands of Global North.

---

[6] Along with law enforcement, firefighting and road traffic monitoring (Gallagher 2012).

[7] According to the CBP, in 2012 the drones assisted in apprehending 143 undocumented border crossers at the US–Mexico border (out of 365,000 apprehensions; Replogle 2013).

Managing 'Risky' Populations: Drones and (Im)Mobile Bodies in Global South and Global North

Case Study 1

Serbia: The European integrations in the 1990s initiated a pursuit for unified border management within the EU. While the external borders have largely been fortified by the Schengen Accords (1995), the process of the EU enlargement delegated the responsibility for border management to the ascending countries of Central and Southeast Europe. In 2003, the EU launched the European Neighbourhood Policy, with cross-border cooperation between EU and non-EU countries at the top of the agenda (Popescu 2012). The key aim of the policy is to establish a 'circle of friends' (Aas 2005) that serves to 'keep as many asylum seekers as possible away from external European borders and to reduce the numbers to whom refugee status is granted' (Freedman 2007, p. 136). In this context, one of the key objectives is to 'build up surveillance resources of third world countries, particularly in Eastern Europe and Northern Africa, and enable them to take over some of the EU's surveillance labor' (Aas 2011, p. 333). The 'external dimension' in justice and home affairs (as it is known in the context of EU; Boswell 2003) sees countries of origin and countries of transit as new key sites of surveillance, detection and immobilisation of undocumented border crossers.

Non-EU, bordering and countries in the process of the EU accession are assigned with the onus to protect the integrity and external borders of the EU. This process of offshoring or, as Pickering (2011) calls it, 'contracting out' of border policing is developed to counterbalance perceived vulnerability of EU after the abolishment of internal borders (Boswell 2003). Yet, such 'cooperation' is often not voluntary (see Aas 2005). Across the borderlines of Global South, drones are deployed to identify and pre-empt mobility of undocumented migrants prior to border crossing (Hayes and Vermeulen 2012). As I mentioned earlier, Frontex is currently considering purchasing its own drones to further enhance its 'common pre-frontier intelligence picture', in which surveillance is performed on the outside of the walls of the Fortress Europe (Suilleabhain 2013).

Countries of former Yugoslavia that joined the EU (first Slovenia in 2004 and then Croatia in 2013) are accountable for the integrity of the Southeast frontier of the "Fortress Europe". This burden is increasingly shared with its non-EU, neighbouring countries (Bosnia and Herzegovina and Serbia). The Republic of Serbia is in the process of EU accession since the fall of Slobodan Milosevic in 2000. The country gained the official member status in 2012 and opened official negotiations for EU membership in 2013 (SEIO 2014). During the 1990s, Serbia was predominantly country of origin for undocumented migrants and asylum seekers. Since 2009, however, the country faces a large influx of people from the Middle East and Northern Africa (MENA) region (Frontex 2013). In 2012, more than 13,900 migrants were registered in Serbia, a growth of 34 % compared to 2011. Given that not all migrants are registered, this number potentially represents 'no more than 30 % of all migrants that transit through Serbia' (respondent 9, international non-governmental organisa-

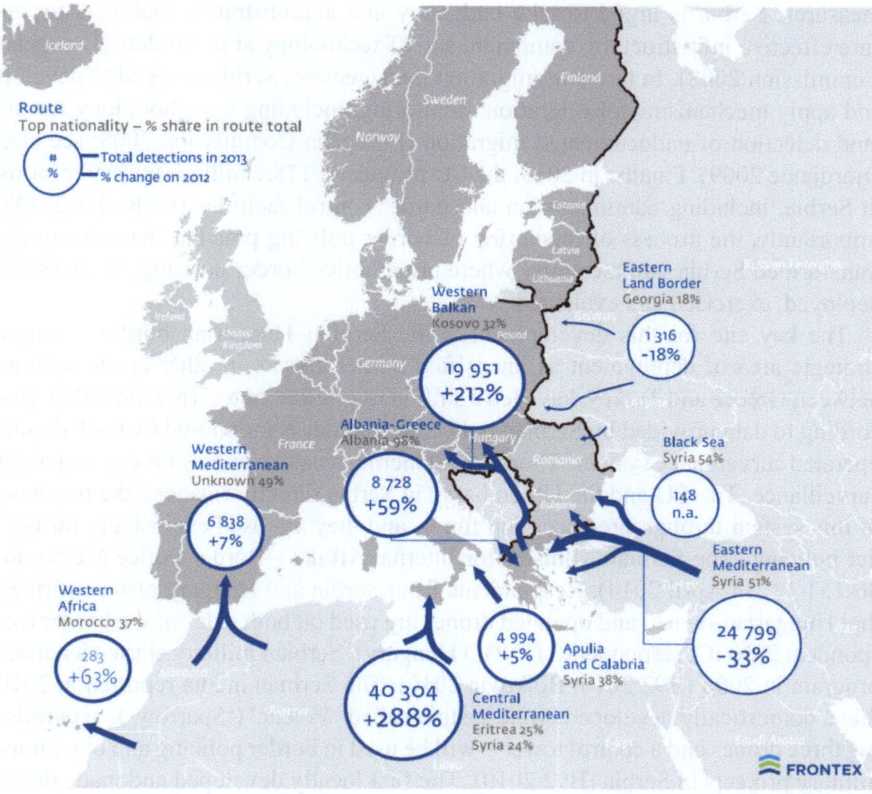

Fig. 5.2 Detections of illegal border crossings in 2013 with percentage change from 2012 by route and top nationality detected. (Frontex 2014)

tion (INGO), Serbia; for more estimates see APC/CZA 2013). A Belgrade-based non-governmental organisation (NGO) working on issues of migration and asylum estimates that at least 20,000 people transited through Serbia in 2012 (respondent 2, NGO, Serbia), while over 13,000 people that passed through Serbia in the first 6 months of 2013 sought asylum in Hungary (mostly from Kosovo, Pakistan, Afghanistan and Syria and other MENA countries—respondent 10, NGO, Hungary). In its 2014 Annual Risk Analysis, Frontex identified a sharp rise in detection of irregular migration reported on the Hungarian–Serbian border (from 6400 in 2012 to 19,500 in 2013) as one of the three key changes in 2013. This Western Balkans route is now, according to Frontex, the third largest route for irregular migration in Europe (after the central Mediterranean and eastern Mediterranean route; Frontex 2014, p. 8) (Fig. 5.2).

The Serbian Government is under constant pressure to undertake a number of initiatives relating to the border and migration management and to harmonise its border policies and legislation governing the movement of persons with EU framework (Renner and Trauner 2009; see also Milivojevic 2013). Amongst other measures, Serbia is urged to take budgetary and administrative measures to ensure effective infrastructure, equipment and IT technology at its borders (European Commission 2008). In terms of migration management, Serbia is urged to develop and apply mechanisms for migration monitoring, including a methodology for inland detection of undocumented migration (European Commission 2008; see also Djordjevic 2009). Finally, in 2013, the EU provided €178.7 million for key reforms in Serbia, including administration and border control facilities (EUROPA 2013). Importantly, the process of offshoring of border policing practices has effectively transformed Serbia into a country where pre-emptive border policing practices are deployed, exercised and evaluated.

The key site for this development is the Serbian–Hungarian border—'a new strategic area of deployment as intensified border checks in other areas, such as between Greece and Turkey, have led a shift in migration routes' (Martin 2013). According to data provided by Serbian police, this border is monitored by the balloon-operated surveillance system 'Thor-1' with thermo-visual cameras for day and night surveillance. The EU and the US embassy in Serbia directly financed the purchase of this system through pre-accession funds, and they are used exclusively for border policing (The Serbian Ministry for Internal Affairs—Border Police Memo no. 06-151/13, 14 April 2014). Respondents from Serbia and Hungary also confirmed that Hungarian-owned and operated drones are used on both sides of the border (respondent 2, NGO; respondent 11, NGO Hungary). Serbian military started its drone program in 2008 (B92 2010; Bojkovic 2010). The Serbian media reported in 2010 that a domestically developed drone system called 'Vrabac' ('Sparrow')—comprising three drones and a control tower—will be used in border policing and other non-military projects in Serbia (B92 2010). The first locally developed and made drone called 'Pegasus 011' was (somewhat enthusiastically and using the familiar sci-fi rhetoric) announced as a new custodian of the Serbian–EU border:

> It looks like it came from a science-fiction movie. It has an almost 20/20 vision…It is fast, precise, elegant and mostly invisible. … [Pegasus 011] is planned to be a flagship of Serbian military, but will be of great use in…border policing. (Blic Online 2011)

However, I was unable to confirm whether Serbian drones are used in border policing. Yet, as one respondent argues, these developments—whether actual or futuristic—fail to deter mobile bodies from their migratory journeys:

> Migrants who cross (the Serbian–Hungarian) border tell us they don't see police patrols very often. This is interesting: they say, they ask us 'Where can we buy mosquito repellent?' When we ask 'Why you need this?' they say, 'You know, when we cross border, it is woods, it's flat, we cross and there are a lot of mosquitos. So they can see us. They have little flying objects and can see everything and uncover us'… So, [border police] sees them, crossing border, all sweaty, because of mosquitos. But they now put insect repellent and manage to cross the border. (Respondent 2, NGO)

Case Study 2: Australia

At Global North located at the geographical South, mobility has long been redefined as a security issue (Pickering and Weber 2006; Wilson and Weber 2008; McDonald 2011). Australian border policies have been developed in the context of 'punitive pre-emption' (Wilson and Weber 2008), in which mobile bodies are detected, immobilised and removed before they reach the physical border. The focus of the intervention are 'boat people'—a pejorative term used since 1970s to describe asylum seekers coming to Australia by boat (Phillips and Spinks 2013). The 'hard-line' policies have been tightened over the years by both Labour and Coalition governments, with the introduction of the mandatory immigration detention for unauthorised boat arrivals, the excision of external territories from the migration zone ('the Pacific Solution'), offshore processing for asylum seekers and temporary protection visas (Phillips and Spinks 2013). While such approach to asylum subjected Australian governments to extensive international scrutiny as it violates Australia's international obligations (especially the Refugee Convention to which Australia is a signatory), the punitive policies continued based on mainstream public opinion that asylum seekers who arrive by boat are not genuine refugees and that they need to be treated more harshly (Dorling 2014; see also Pickering 2008).

The interception of mobile bodies at sea and 'push back, tow back' policy is a new 'height' in Australian government's pursue for non-permeable borders for mobile bodies coming from Global South. So-called Operation Sovereign Borders, implemented just 11 days after the Coalition's victory in the 2013 federal election, sends a clear message to undocumented migrants and asylum seekers coming to Australia by boat: 'No way: You will not make Australia home' (ACBP 2014). While technically a civilian-led law enforcement operation, in 'Operation Sovereign Borders' the Australian government 'has increasingly adopted military structure, language and methodology' (Leslie and Corcoran 2014). The role of the Australian Defence Force (ADF) in border policing is an important one: The 'Operation RESOLUTE' is the 'ADF's contribution to the whole-of-government effort to protect Australia's borders and offshore maritime interests' (ADF 2014). The creation of 'a single frontline operational border agency'—the Australian Border Force (to be implemented from 1 July 2015; DIBP 2014)—suggests the complete fusion of non-military and military segments of border policing, the one needed when the nation is at war. At the same time, asylum seekers are renamed as 'transferees' and 'customers of law-breaking people smugglers' (Keane 2014).

Australia is developing technologies for surveillance of space and individuals, with a particular emphasis on surveillance of its coastal waters. While human-operated aircrafts have been deployed in maritime surveillance for over 40 years (ABC News 2014a), a new pair of 'eyes in the sky' will be joining them soon. As Prime Minister Tony Abbot claims, drones are necessary as 'Australia has responsibility for something like 11% of the world's oceans [so] it's very important that we've got a very effective maritime surveillance capability'. The Australian government purchased up to seven giant Triton drones, a 40-m-wingspan UAV that can be airbound for 30 h and fly 16,000 km (Mannix 2013). The system will be operational in 2017–2018 (Veiszadeh and Le May 2013). In the debate that can be described as 'asylum panic', drones appear to be a moderate solution, when compared with calls for missile-equipped boats in border policing (Veiszadeh and Le May 2013). Interestingly, given a price tag between AU $ 1–3 billion (ABC News 2014b) and while drones for military use are subjected to rigorous public scrutiny, drones in the context of border policing are not making the headlines Down Under.

In the past 2 years, leading Australian newspapers and news broadcasting agencies *(The Age* and *Herald Sun*—Melbourne, *The Sydney Morning Herald (SMH)* and *The Daily Telegraph*—Sydney, *The Australian* and the *Australian Broadcasting Corporation (ABC)*—national) reported extensively on the use of drones in the military context ($n=326$). The reports predominantly covered the use of drones by the Obama administration in Pakistan, Iraq and Afghanistan, with headlines such as 'Uncle Sam's latest lethal toy' (*The Daily Telegraph*, 11 January 2014), 'War by remote control' (*The Australian*, 6 July 2013), '"Cool war" a click away' (*The Australian*, 26 April 2013) but also 'Drones are here to stay but with what consequences' (*The SMH*, 29 June 2014), 'US counts the cost of a deadly idea' (*The SMH*, 12 February 2013), 'Most of the dead are women and children' (*The Australian*, 20 November 2012), 'A dirty business' (*The Age*, 1 August 2013) and 'Drones: the killing is cheap but the price high' (*The Age*, 31 May 2013). As these headlines indicate, the debate on the use of drones in the military context is profoundly polarised along moral, ideological and political lines.

The use of drones in non-military context is significantly under-reported with several of articles reporting on the issue in the past 2 years ($n=42$). The headlines that describe the use of drones in border policing context are more subtle, such as 'Under the wings of a big bird' (*The Australian*, 15 February 2014), 'Australian Defence Force looks at drones to replace border patrol aircraft' (*ABC*, 26 July 2013), 'Drones on the agenda as Abbott flies in to help state leader' (*The Age*, 14 March 2014), 'Drones are forces' eyes on the sky' (*The SMH*, 3 May 2013), and 'Australia moves to buy $ 3 b spy drone fleet' (*ABC*, 4 September 2012). They also seem to indicate a bipartisan support for such deployment in border policing ('Surveillance drones program dodges cutbacks', *The Australian*, 27 October 2012), as drones are perceived as potential 'answer to our border security issues' (Stewart 2014b).

I argue, however, that, while distinctively different, both military and border policing narratives around use of drones are underpinned by military rhetoric, especially the notion of a hunt. The key argument is that 'eyes in the sky' can indeed assist in locating, immobilising, apprehending, removing or eliminating the threat, the 'risky' populations—whether they are enemy combatants or asylum seekers and whether they are at sea (Australia) or at the fringe of Global North (Serbia):

> [The drones can] intercept [asylum seekers], deal with them as SOLAS requires, safety of life at sea, not necessarily turn them around but turn them around in circumstances where it's adjudicated safe to do so. (Then Opposition defence spokesman David Johnston, cited in Henderson 2013)

The human cost, clearly visible in the drone strikes debate is carefully hidden in border policing narratives or replaced with the notion or combating transnational crime and/or 'rescuing' migrants in distress:

> 'It's highly likely that we're going to see more asylum seekers coming to Australia, there's going to be the possibility of increased transnational crime, there's going to be the possibility of increased illegal activities…. [With drones] you can really get down to rowing boat sizes. The quality of the imagery is quite phenomenal.' (Kym Bergman, the editor of *Asia Pacific Defence Reporter* and a former defence industry executive, cited in Corcoran 2012)

> 'Aerial surveillance is critical, not just from the point of view of conducting interceptions, but also potentially being able to act proactively in terms of search and rescue situations as well.' (Then Shadow Immigration Minister Scott Morrison, *ABC* 2013)

The outcome of both military and border policing use of drones, however, is much the same: Mobile bodies are gradually caught in the drone stare, removed from contexts in which acknowledging the broader milieu of vulnerability and victimisation largely generated by actions of nation states is essential to understand the plight of mobile populations for mobility and agency. The drone stare reduces mobile bodies to targets, numbers, collateral damage or transferees, lawbreakers or illegals that need to be detected, immobilised and removed, with help and under the watchful gaze of new, non-human 'eyes in the sky'.

Concluding Remarks

As Kenk et al. (2013) note, '[t]he explosion of global travels has led to the emergence of a new border architecture, which seeks to respond effectively to new demands—facilitating mobility while managing the risk associated with the cross-border travel'. The process of separating mobility rich from mobility poor (Wilson and Weber 2008)—wanted from unwanted mobile bodies—is formalised by series of interventions in the area of border management, most of which now involve security and surveillance technologies at and beyond the border. This contemporary surveillant assemblage, seen as effective and benign instrument for further stratification of global mobility, is a new tool for violation of basic human rights such as the right to seek asylum and the right to mobility. Surveillance, particularly surveillance from the sky, is increasingly becoming the preferred method for fortifying risky borderlands and the backbone of 'Crimmigration'-based border policing initiatives in both Global North and Global South.

The pursuit for security in border policing resulted in adapting military technology for non-military purposes (Hayes et al. 2014). The fusion of military and non-military features in border policing gradually changes the nature of border management. Perceived as a cheap and efficient solution for the border (and national) security 'threats', military-based 'solutions' such as drones are increasingly used to detect and immobilise mobile bodies prior to and at the border. Embedded in the intersection of sci-fi and ancient mythology, drones move us a step closer to the Foucault's notion of exhaustive surveillance of mobile populations. Drones have recently been identified as 'a large step closer to a surveillance society in which our every move is monitored, tracked, recorded, and scrutinised by the authorities' (Constantini 2012). Yet, while the use of drones in the military context has been plagued with controversy in the media and public discourse and tarnished by ethical considerations, in the border policing setting a presence of invisible 'eyes in the sky' remains largely unchallenged and desired. Also importantly, while military use of drones creates waves of discontent in media and public debate, drones used in border policing are largely out of the spotlight.

Underpinning both narratives, however, is the notion of a hunt, enabled by the drone stare. Indeed, drones represent perhaps the most visible element of a surveillance society but one that can be categorised as a 'sense-and-detect' and not a 'rescue' tool (Kenk et al. 2013). This supposedly non-lethal technology is increasingly becoming a superpowerful weapon in border control; one that aims not only to identify but also to physically immobilise mobile bodies—or 'targets of interest' (Bump 2013). A recent shift in semantics, the argument that drones indeed can save lives is just a metaphor, an illusion that conceals the real purpose for drone development in border management. Voices that challenge such deception need to be both louder and greater in numbers. Importantly, the human costs of such interventions, carefully hidden in border policing debate, need to be clearly outlined.

Surveillance state performs its watch both within and across its borders. While preventing cross-border crime (such as terrorism, human smuggling and trafficking in people) is purportedly the main drive for development of such policies (see Carrera 2007), the goal is to keep the unwanted 'Other' on the other side of the border. The process of offshoring border control mechanisms to countries of origin and countries of transit is of utmost importance in creating 'buffer zones' or the 'cordon sanitaire' of the EU (Morrison and Crosland 2000, p. 47). In this new world order, the accountability for the preservation of borders is delegated to those lower on the global kinetic scale, nations that produce the mobile bodies or enable their transit towards Global North. At the same time, developed nations engage in punitive pre-emption strategies that are deployed, exercised and evaluated across Global South, in which drones play an important role. Seen as a moderate solution for a (largely invented) border security problem, drones are here to stay. They will continue to play an important role in detecting and immobilising the 'Other' across the borderlands, often with a bipartisan support of key political actors in the debate. To problematise such state of affairs, the drone stare needs to be deconstructed, with broader social contexts of vulnerability, exploitation, victimisation and human rights violation positioned in the very centre of the drone debate. This is even more important given that mobile bodies refuse—and will keep to refuse—to be 'petrified and immobilised by the drone stare' (Wall and Monahan 2011, p. 250).

Without making a sound, invisible sets of eyes patrol our southwest border.... Virtually undetectable by naked eye, unmanned areal vehicles—better known as UAVs or drones—silently patrol the border, looking for potential terrorists, drug smugglers, and illegal immigrants. (Longmire 2014, p. 75)

References

Aas, K. F. (2005). "Getting ahead of the game": Border technologies and the changing space of governance. In E. Zureik & M. Salter (Eds.), *Global surveillance and policing: Borders, security, identity*. London: Routledge.

Aas, K. F. (2011). "Crimmigrant" bodies and bona fide travellers: Surveillance, citizenship and global governance. *Theoretical Criminology, 15*(3), 331–346.

ABC News. (3 July 2013). Senior customs officials question opposition's drone strategy. http://www.abc.net.au/pm/content/2013/s3795312.htm. Accessed 22 Aug 2014.
ABC News. (21 February 2014a). New P-8 Poseidon surveillance aircraft to boost border protection. http://www.abc.net.au/news/2014-02-21/p-8-surveillance-aircraft-to-boost-border-protection/5274976. Accessed 22 Aug 2014.
ABC News. (13 March 2014b). Triton drone surveillance fleet to be based at Edinburgh air force base in Adelaide. http://www.abc.net.au/news/2014-03-13/drone-fleet-to-be-based-in-adelaide-triton-edinburgh/5317262. Accessed 10 June 2014.
ABC News. (12 September 2014c). Ukraine crisis: European security watchdog says it will use drones to monitor Russia-Ukraine ceasefire. http://www.abc.net.au/news/2014-09-12/osce-will-use-drones-to-monitor-russia-ukraine-border/5738786. Accessed 16 Oct 2014.
ACBP. (2014). Operation sovereign borders. http://www.customs.gov.au/site/operation-sovereign-borders.asp. Accessed 24 Aug 2014.
Adey, P. (2006). "Divided we move": The dromologics of airport security and surveillance. In T. Monahan (Ed.), *Surveillance and society: Technological politics and everyday life* (pp. 195–208). New York: Routledge.
ADF. (2014). Border protection. http://www.defence.gov.au/op/borderprotection/general.htm. Accessed 24 Aug 2014.
Amoore, L., Marmura, S., & Salter, M. (2008). Editorial: Smart borders and mobilities: Spaces, zones, enclosures. *Surveillance and Society, 5*(2), 96–101.
APC/CZA. (2013). *Analiza—tražioci azila u Srbiji i sprski azilanti u Evropi*. Beograd: APC/CZA.
B92. (1 May 2010). Prva srspka bespilotna letelica. http://www.b92.net/info/vesti/index.php?yyyy=2010&mm=05&dd=01&nav_id=428205. Accessed 10 June 2014.
Barry, T. (2013). Drones over the homeland: How politics, money and lack of oversight have sparked drone proliferation, and what we can do. International Policy Report, April 2013. http://www.ciponline.org/images/uploads/publications/IPR_Drones_over_Homeland_Final.pdf. Accessed 10 June 2014.
Bauman, Z. (2000). *Liquid modernity*. Cambridge: Polity Press.
Becker, A. (2013). New drone report: Our border is not as secure as we thought. *The Daily Beast*. http://www.thedailybeast.com/articles/2013/04/04/new-drone-report-our-border-is-not-as-secure-as-we-thought.html. Accessed 2 June 2014.
Berman, J. (2003). (Un)popular strangers and crises (un)bounded: Discourses of sex-trafficking, the European political community and the panicked state of the modern state. *European Journal of International Relations, 9*(1), 37–86.
Blic Online. (2011). Srpska bespilotna letelica. http://www.blic.rs/Vesti/Drustvo/262372/Srpska-bespilotna-letelica. Accessed 17 Oct 2014.
Bojkovic, T. (25 January 2010). "Vrabac"—novi adut Vojske Srbije. *Politika*. http://www.politika.rs/rubrike/Drustvo/Vrabac-novi-adut-Vojske-Srbije.lt.html. Accessed 10 June 2010.
Boswell, C. (2003). The "external dimension" of EU immigration and asylum policy. *International Affairs, 79*(3), 619–638.
Broeders, D. (2007). *Breaking down anonymity: Digital surveillance of irregular migrants in Germany and the Netherlands*. Amsterdam University Press, Amsterdam.
Broeders, D. (2009). *Breaking down anonymity: Digital surveillance of irregular migrants in Germany and the Netherlands*. Amsterdam: Amsterdam University Press.
Bump, P. (2 July 2013). The border patrol wants to arm drones. *The Wire*. http://www.thewire.com/national/2013/07/border-patrol-arm-drones/66793/. Accessed 10 June 2014.
Carrera, S. (2007). The EU border management strategy: Frontex and the challenges of irregular immigration in the Canary islands. CEPS Working Document no. 261, March.
Ceyhan, A. (2008). Technologization of security: Management of uncertainty and risk in the age of biometrics. *Surveillance and Society, 5*(2), 102–123.
Clark, R. (2014). Understanding the drone epidemic. *Computer Law & Society Review, 30*(3), 230–246.

Constantini, C. (1 May 2012). U.S. border patrol increases use of unmanned drones for surveillance. *The Huffington Post*. http://www.huffingtonpost.com/2012/05/01/us-border-patrol-increase_n_1467196.html. Accessed 10 June 2014.

Corcoran, M. (4 September 2012). Australia moves to buy $3b spy drone fleet. *ABC News*.

Cupolo, D. (2013). Drone use soars in Latin America, remains largely unregulated. http://upsidedownworld.org/main/international-archives-60/4615-drone-use-soars-in-latin-america-remains-widely-unregulated. Accessed 22 Oct 2014.

DIBP. (2014). Australian border force. http://www.immi.gov.au/About/Pages/australian-border-force.aspx. Accessed 22 Aug 2014.

Djordjevic, B. (2009). The European policy on the management of migration flows and Serbia role's within it. *Western Balkans Security Observer, 12,* 78–91.

Dorling, P. (8 January 2014). Australians want boat arrivals treated more harshly: Poll. *The Sydney Morning Herald*. http://www.smh.com.au/federal-politics/political-news/australians-want-boat-arrivals-treated-more-harshly-poll-20140108-30 g97.html. Accessed 25 Aug 2014.

Engbersen, G. (2001). The unanticipated consequences of panopticon Europe. Residence strategies of illegal immigrants. In V. Guiraudon & C. Joppke (Eds.), *Controlling a new migration world* (pp. 222–246). London: Routledge.

EUROPA. (2014). EU to grant 178.7 million EUR to support key reforms in Serbia. 23 December. http://europa.eu/rapid/press-release_IP-13-1311_en.htm. Accessed 11 June 2014.

European Commission. (2008). Visa liberalisation with Serbia: Road map. http://www.esiweb.org/pdf/White%20List%20Project%20Paper%20-%20Roadmap%20Serbia.pdf. Accessed 10 July 2014.

Farber, H. (2014). Eyes in the sky: Constitutional and regulatory approaches to domestic drone deployment. *Syracuse Law Review, 64*(pp. 1–48).

Freedman, J. (2007). *Gendering the international asylum and refugee debate*. New York: McMillan.

Frontex. (2013). Western Balkans: Annual risk analysis 2013. Warsaw.

Frontex. (2014). Annual risk analysis 2014. Warsaw.

Gallagher, R. (2 February 2012). Surveillance drone industry plans PR effort to counter negative image. *The Guardian*. http://www.theguardian.com/uk/2012/feb/02/surveillance-drone-industy-pr-effort. Accessed 3 June 2014.

Guild, E., Carrera, S., & Geyer, F. (2008). The Commission's new border package: Does it take us one step closer to a 'cyber-fortress' Europe? Centre for European Policy Studies Policy Brief. http://www.ceps.be/book/commissions-new-border-package-does-it-take-us-one-step-closer-cyber-fortress-europe. Accessed 24 July 2013.

Haggerty, K., & Ericson, R. (2000). The surveillant assemblage. *British Journal of Sociology, 51*(4), 605–622.

Hayes, B., & Vermeulen, M. (2012). Borderline: The EU's New Border Surveillance Initiative. Henrich Böll Foundation. http://www.boell.de/sites/default/files/DRV_120523_BORDERLINE_-_Border_Surveillance.pdf. Accessed 10 June 2014.

Hayes, B., Jones, C., & Toepfer, E. (2014). EURODRONES Inc. Statewatch. http://www.statewatch.org/news/2014/feb/sw-tni-eurodrones-inc-feb-2014.pdf. Accessed 3 June 2014.

Heller, C., & Jones, C. (2014). Eurosur: Saving lives or reinforcing deadly borders? *Statewatch, 23*(3–4), 9–12.

Henderson, A. (11 April 2013). Drone patrols for asylum boats could lead to legal minefield. *ABC*.

Keane, J. (5 July 2014). Operation sovereign borders. *The Conversation*. http://theconversation.com/operation-sovereign-borders-28831. Accessed 24 Aug 2014.

Kenk, V., Krizaj, J., Struc, V., & Dobrisek, S. (2013). Smart surveillance technologies in border control. *European Journal of Law and Technology*. http://ejlt.org/article/view/230/378. Accessed 20 July 2014.

Kilcullen, D., & McDonald Exum, A. (17 May 2009). Death from above, outrage down below. *The New York Times*. http://www.agriculturedefensecoalition.org/sites/default/files/file/drones_517/517V_4_2009_Pakistan_Drones_Death_From_Above_Pakistan_Outrage_Down_Below_Kilcullen_May_19_2009_NYTimes.pdf. Accessed 25 Aug 2014.

Koslowski, R. (2005). Real challenges for virtual borders: The implementation of US-VISIT. Migration Policy Institute, Washington D.C.
Leslie, T., & Corcoran, M. (2014). Operation sovereign borders: The first six months. http://www.abc.net.au/news/interactives/operation-sovereign-borders-the-first-6-months/. Accessed 18 Aug 2014.
Longmire, S. (2014). *Border insecurity*. New York: Macmillan.
Lyon, D. (2007). *Surveillance studies: An overview*. Cambridge: Polity Press.
Manne, R. (September 2010). Comment: Asylum seekers. *The Monthly*.
Mannix, L. (20 March 2013). The drones are coming: Border protection fleet for Adelaide. *Crikey*. http://www.crikey.com.au/2013/03/20/the-drones-are-coming-border-protection-fleet-for-adelaide/. Accessed 10 June 2014.
Martin, M. (2013). "Trust in Frontex": The 2013 work programme. Statewatch. http://www.statewatch.org/analyses/no-219-frontex-work-programmes.pdf. Accessed 11 June 2014.
McCulloch, J, & Pickering, S. (2012). *Borders and crime: Pre-crime, mobility and serious harm in an age of globalization*. Hampshire: Macmillan.
McDonald, M. (2011). Deliberation and resecuritization: Australia, Asylum-Seekres and the normative limits of the Copenhagen School. *Australian Journal of Political Science, 46*(2), 281–295.
Milivojevic, S. (2013). Borders, technology and mobility: Cyber-fortress Europe and its emerging Southeast frontier. *Australian Journal of Human Rights, 19*(3), 99–120.
Morrison, J., & Crosland, B. (2000). The trafficking and smuggling of refugees: The end game in European asylum policy. http://www.unhcr.org/3af66c9b4.html. Accessed 17 July 2014.
Neal, A. (2009). Securitization and risk at the EU border: The origins of FRONTEX. *Journal of Common Market Studies, 47*(2), 333–356.
Not A Bug Splat. (2014). A giant installation targets predator drone operators. http://notabugsplat.com/. Accesssed 16 October 2014.
Orr, A. (2011). Unmanned, unprecedented, and unresolved: The status of american drone strikes in Pakistan under international law. *Cornell International Law Journal, 44*, 729–752.
Phillips, J., & Spinks, H. (2013). Boat arrivals in Australia since 1976. http://www.aph.gov.au/about_parliament/parliamentary_departments/parliamentary_library/pubs/bn/2012–2013/boatarrivals. Accessed 23 Aug 2014.
Pickering, S. (2008). The new criminals: Refugees and asylum seekers. In C. Cunneen & T. Anthony (Eds.), *The critical criminology companion* (pp. 169–179). Sydney: Hawkins Press.
Pickering, S. (2011). *Women, borders and violence*. New York: Springer.
Pickering, S., & Weber, L. (2006). *Borders, mobility and technologies of control*. The Netherlands: Springer.
Popescu, G. (2012). *Bordering and ordering in the twenty-first century: Understanding borders*. Lanham: Rowman & Littlefield Publishers.
Press Online. (19 October 2011). Prvi let bespilotne letelice proizvedene u Srbiji. http://pressrs.ba/sr/vesti/vesti_dana/story/2387/Prvi+let+bespilotne+letelice+proizvedene+u+Srbiji.html Accessed 17 Oct 2014.
Renner, S., & Trauner, F. (2009). Creeping EU-membership in Southeast Europe: The dynamics of EU rule transfer to the Western Balkans. *Journal of European Integration, 31*(4), 449–465.
Replogle, J. (23 April 2013). Senate immigration bill calls for a drone-patrolled border. *Fusion Net*. http://fusion.net/abc_univision/story/senate-immigration-bill-calls-drone-patrolled-border-12867. Accessed 10 June 2014.
Salter, M. (2013). Toys for the boys? Drones, pleasure and popular culture in the militarisation of policing. *Critical Criminology, 22*, 163–177.
Sandvik, K., & Lohne, K. (2014). The promise and perils of "disaster drones". Humanitarian Practice Network. http://www.odihpn.org/humanitarian-exchange-magazine/issue-58/the-promise-and-perils-of-disaster-drones. Accessed 22 Aug 2014.
Sassen, S. (2006). Europe's migrations: The numbers and the passions are not new. *Third Text, 20*(6), 635–645.

SEIO. (2014). Istorijat. http://www.seio.gov.rs/србија-и-еу/историјат.6.html. Accessed 11 June 2014.

Stein, A. (2014). Turkey's missile programs: A work in progress. http://www.edam.org.tr/Media/Files/208/Turkey%20Missile%20Programs.pdf. Accessed 16 Oct 2014.

Stewart, C. (15 February 2014a). $3 billion plan for drone force to patrol our borders. *The Australian.* http://www.theaustralian.com.au/national-affairs/policy/bn-plan-for-drone-force-to-patrol-our-borders/story-e6frg8yo-1226827786156#. Accessed 2 June 2014.

Stewart, C. (15 February 2014b). Under the wings of a big bird. *The Australian.*

Suilleabhain, A. (2013). Already pushing legal boundaries, EU's Border Agency Now Seeking Drones. Global Observatory, 27 March. http://theglobalobservatory.org/analysis/465-already-pushing-legal-boundaries-eus-border-agency-now-seeks-drones.html. Accessed 10 June 2014.

The Guardian. (2 September 2013). Drone plan shot down in Coalition defence policy launch. http://www.theguardian.com/world/2013/sep/02/coalition-drone-plan-defence-policy. Accessed 25 Aug 2014.

Veiszadeh, E., & Le May, R. (2013). Threats of missiles, drones to stop boats. *The Sydney Morning Herald.* http://news.smh.com.au/breaking-news-national/threats-of-missiles-drones-to-stop-boats-20130411-2hmjj.html. Accessed 22 July 2014.

Vukov, T., & Sheller, M. (2013). Border work: Surveillant assemblages, virtual fences, and tactical counter-media. *Social Semiotics, 23*(2), 225–241.

Wall, T., & Monahan, T. (2011). Surveillance and violence from afar: The politics of drones and liminal security-scapes. *Theoretical Criminology, 15*(3), 239–254.

Waterfield, B. (12 February 2014). EU "spent £ 320 million on surveillance drone development". *The Telegraph.* http://www.telegraph.co.uk/news/worldnews/europe/eu/10632262/EU-spent-320-million-on-surveillance-drone-development.html. Accessed 3 June 2014.

Weber, L., & Pickering, S. (2011). *Globalization and borders: Death at the global frontier.* UK: Macmillan.

Welch, M. (2012). The sonics of crimmigration in Australia: Wall of noise and quiet manoeuvring. *British Journal of Criminology, 53*(4), 324–344.

Wilson, D. (2012). Military surveillance. In K. Haggerty & D. Lyon (Eds.), *Routledge handbook of surveillance studies* (pp. 269–276). New York: Routledge.

Wilson, D., & Weber, L. (2008). Surveillance, risk and preemption on the Australian border. *Surveillance & Society, 5*(2), 124–141.

Zedner, L. (2009). *Security.* London: Routledge.

Dr. Sanja Milivojevic is a Senior Lecturer in criminology at the University of New South Wales, School of Social Sciences. Sanja holds an LL.B. and an LL.M. from Belgrade University's law school and a Ph.D. from Monash University. Her research interests are trafficking in people and transnational crime, borders and mobility, security technologies, surveillance and crime, sexting, gender and victimisation and international criminal justice and human rights. Sanja has worked as a researcher on various projects with the Institute for Criminological and Sociological Research in Belgrade and Victimology Society of Serbia. She is a recipient of national and international grants including the Criminology Research Council's grant for project on young people and sexting (with Associate Professor Murray Lee, Associate Professor Thomas Crofts, Dr. Alyce McGovern and Dr. Michael Salter). Sanja is a New South Wales (NSW) representative at the Australian and New Zealand Society of Criminology's Committee of Management and editorial board member for the journals *Temida* (Serbia) and *The Human Rights Defender* (Australia). She has been a visiting scholar at Belgrade University's Faculty for Special Education and University of Zagreb law school (2013) and a Public Interest Law Fellow at Columbia University's law school (2001–2002). Sanja has participated in over 30 international and domestic conferences and has published in English and Serbian. Her latest book *Sex Trafficking: International Context and Response* with Dr. Marie Segrave and Professor Sharon Pickering has been published by Willan Publishing.

Chapter 6
Deploying Drones in Policing Southern European Borders: Constraints and Challenges for Data Protection and Human Rights

Luisa Marin and Kamila Krajčíková

Introduction

One of the main achievements of the European integration process is the free movement of persons, which has been completed by the Schengen process. Schengen has meant the removal of internal border checks, on the one hand, and the strengthening of external border controls and the gradual introduction of an integrated management system for external borders (Article 77, TFEU, 2012), on the other hand. In the past few years, border policing[1] has developed more and more as an autonomous policy field, according to the treaty framework, and steps have been taken toward shaping integrated border management (IBM; Mungianu 2013).

While the EU and the member states have increased their efforts in policing their external borders, the persistent poverty and the recent political instabilities

Though the chapter is the product of a common conception, sections 'Introduction', 'Current Instruments and Practices: The Challenges Ahead', 'The Legal Framework for Border Surveillance and Its Constraints', 'The Limitations for Drone Surveillance from Privacy and Data Protection' and 'The Challenges for Human Rights Created By EU's Cooperation with Third Countries in the Area of Border Management' can be attributed to Luisa Marin and sections 'Actors in Border Surveillance', 'The Deployment of Drone Technology in Border Surveillance' and 'General Aviation Law Issues' are attributed to Kamila Krajčíková. The conclusions are common.

[1] In this chapter, the terms "border policing" and "border surveillance" are used interchangeably.

L. Marin (✉)
Department of Law, School of Law, Economics and Governance, Utrecht University, The Netherlands
e-mail: l.marin@uu.nl

Institute for Governance Studies, University of Twente, The Netherlands

K. Krajčíková
Heinrich-Heine-Universität Düsseldorf, Düsseldorf, Germany
e-mail: kamila.krajcikova@gmail.com

A. Završnik (ed.), *Drones and Unmanned Aerial Systems*,
DOI 10.1007/978-3-319-23760-2_6

of the North African and Middle East regions have increased the phenomenon of boat migration, that is, migrants who decide to risk their lives in order to reach one (Northern) European state in (often unseaworthy) vessels; lacking the possibility of reaching Europe legally, migrants resort to illegal networks of smugglers. The North African coast is the main departure point for irregular boat migration. Although the maritime border crossings account for about 7 % of overall migration to the EU, irregular migration by sea is always high on the agenda of political priorities of the European Union Agency for Fundamental Rights (2013). Moreover, irregular migration by sea is problematic because of the high death toll of migrants. In 2011, the Mediterranean was declared as the most deadly water area for refugees in the world by the United Nations High Commissioner for Refugees (UNHCR 2012), and in 2013 the dramatic figures for lives lost at sea picked up again. The island of Lampedusa is the symbol of this phenomenon, and the numerous tragic accidents of migrants in the Mediterranean Sea (some known and others simply unknown) have turned the *mare nostrum* of the Romans into a *mare monstrum* (monster sea).[2]

Against the background of this contemporary social and human phenomenon, irregular migration policy is securitised and the external borders are framed in member states' and EU's policies as the gateways, the sources of (external) threats to their (internal) security. Border surveillance in contemporary politics (not only in the EU but also in the USA and Australia, to make some comparisons) is increasingly seen and practiced as a preventive policy (Collett 2011), requiring the deployment of all the most up-to-date technological means (Dijstelbloem and Meijer 2011; Besters and Brom 2010).

It is commonly acknowledged by scholars that the Europeanisation of national migration policy is caused by national failures in the domain (Boswell and Geddes 2011), and that the European migration policy is best explained by securitisation theories, according to which migration and migrants are framed, in political discourses (Weaver 1995), by security actors (Bigo 2001) and through practices (Balzacq 2008) as security threats. The framing of migrants as security threats has determined that member states and the EU react to defend the internal security from those alleged external threats. So, if globalisation has turned the world into a "global village" where goods, capitals and information circulate across the globe, it could not help decreasing persistent poverty in the southern part of the globe; instead, states have consolidated their interest in regulating the human dimension of globalisation, that is, human mobility. This policy, aiming at controlling the overall phenomenon of human migrations also by increasingly regulating legal migration and, consequently, fighting against illegal migration has attracted a number of criticisms, captured by the image describing Europe as a fortress.

Within this process, border surveillance has gained importance too. The EU and member states are investing in technological applications, ranging from biometrics to databases and drones, in order to deploy the most effective technological means in the attempt to face the security threats allegedly coming from outside. Drones are

---

[2] Estimates on migrants' deaths at sea are uncertain, but according to the Fortress Europe blog, run by journalist Gabriele del Grande, in the past few decades about 20,000 people have lost their lives in the attempt to cross the Mediterranean Sea from Africa to Europe.

part of this process, variably framed as the EU's digital fix (Besters and Brom 2010), or the transformation of Europe into a high-tech fortress, through militarisation of border surveillance (Marin 2011). Drones, more neutrally called unmanned aerial vehicles (UAVs) or remotely piloted aircraft systems (RPAS), are first aircrafts and, second, intelligence, surveillance (target-acquisition) and reconnaissance (ISR or ISTAR) machines. As such, they can perform surveillance, monitoring and intelligence operations, usually carried out by public agencies tasked with border control functions. In the USA, drones have already been used for monitoring the southern borders since 2004 and, recently, some European states, such as Italy, have deployed them too (Marin 2015). In the past few years, several European and international agencies have joined forces participating in EU-funded research programmes exploring and preparing the transfer of drone technology from the military domain to the non-military one (Hayes et al. 2014). In this context, Frontex regularly holds demonstrations and workshops concerning drone technology for border surveillance, cooperating closely with the world's industry giants, and specifically with companies like Israel Aerospace Industries, Thales Group, Aerovision or Lockheed.

Thanks to the surveillance technologies they can carry, drones can contribute to the attainment of the objectives of EU border controls, that is, to reduce the number of migrants illegally entering the EU, to prevent undocumented migration and thus contribute to the fight against cross-border crime. Drones can provide information to border guards present on the ground or at sea and therefore contribute to making border surveillance a proactive policy rather than a reactive one. These ground and sea patrols, thanks to the information acquired by drones, could then take control of migrants and, in the case of migration by sea, could support them in case of distress, taking them to the closest port, as well as re-direct them to international seas or to the authorities of cooperating third countries (TCs), if bilateral agreements so provide.

The aim of this chapter is to explore the recent developments constituted by the deployment of drone technology in border surveillance. The chapter first introduces the actors, policies and practices in the sphere of border management, specifically in the area of border surveillance (1); then, it discusses the use of drone technology for border surveillance, looking not only at the potentialities but also at its current shortcomings (2); it will then move toward the regulatory framework enabling the safe deployment of drones under aviation law and on border surveillance (3) and present the constraints represented by fundamental rights and data protection rules on drone technology and the challenges drones represent for the human rights of migrants (4), before concluding that drone technology might entail a further securitisation of border surveillance, together with a shift toward preventive border surveillance (5).

Border Policing in the EU: Actors, Practices and Critical Aspects

Actors in Border Surveillance

As stated in Article 77(1)(c) of the Treaty on the Functioning of the European Union (TFEU), the EU is mandated to develop the gradual introduction of an integrated management system for external borders. Border management in the European perspective has several actors: on the one side, the member states and their national bureaucracies and, on the other side, the European agency Frontex.

Considering the first group, that is, the particular member states, the countries most directly concerned by the recent migratory trends are the southern European states, such as Portugal, Spain, Italy, Malta, Cyprus and Greece; they have to administer an important maritime border neighbouring with third (i.e. non-EU) countries and they are (geographically) the first countries to be faced with the problem of irregular and dangerous migration by sea. Border surveillance practices vary across the member states and are obviously limited by the legal framework, and to some extent are also influenced by governments' political interests and goals. Since the management of a state's borders in the EU means the management of the external borders of the EU, the policies applied in these member states involve consequences for the rest of the EU. However, as explained above, the Schengen integration process requires member states to be the first "guardians" of the common external borders of all the member states of the EU. Therefore, the individual states are still the first and most important actors in the area of border surveillance, and, according to the legal framework, they bear the responsibility of border controls within the EU.

Traditionally, border surveillance is the task of border guards, the first actors on the frontline of borders. Usually, European states have specific bodies within their law enforcement systems, but in this respect there are many variations as to their nature and features, as institutions are the consequence of historical processes within states. So, for example, in German-speaking countries, border surveillance is usually vested with the police, whereas in other member states it is bestowed on military police corps such as the French *Gendarmerie*, the Dutch *Royal Marechaussee*, the Spanish *Guardia Civil*, and the Italian *Carabinieri*. Alongside them, border surveillance is traditionally also a competence of agencies tasked with the control of customs, such as the French *Douane* and the Italian *Guardia di Finanza*.

In the past few years, however, a new trend has been emerging in Southern European states, that is, the militarisation of border surveillance as part of a broader process of securitisation of borders and migration (Bigo 2001; Hayes 2006). Alongside border guards, other state agencies have been tasked with border surveillance, often as a consequence of the characterisation of migration flows as a security threat and as an emergency, requiring the deployment of exceptional forces. An interesting example is provided by the operation *Mare Nostrum*, carried out by the Italian Navy, in cooperation with the Ministry of Internal Affairs. In this operation, border surveillance and the management of irregular migration were characterised as humanitarian actions, and therefore this humanitarian rationale has been used to justify the deployment of warships and other military assets. Alongside this, states are progressively deploying other assets, such as drones, and also satellites, in order to map human movements in the sea. The flourishing of a civilian security industry after the fall of the Iron Curtain and the end of the Cold War is also supported by the interests of the defence industry, which is finding in these civilian missions a new and growing market, where dual-use technologies can be exploited and tested and their deployment can be made cost-efficient thanks to mutualisation.

Another increasingly influential actor in border management is Frontex, the European Agency for the Management of Operational Cooperation at the External

Borders of the Member States of the European Union, operational since 2005. The mission of Frontex is the coordination of operations mainly carried out by single states. Its core activities are joint operations, pilot projects and risk analysis. Another crucial task of Frontex is research and development (R&D); the agency regularly organises workshops and demonstrations of new technological developments informing the member states on research on border management techniques and technologies (Regulation No 1168/2011 2011). Frontex, with the help of the European Commission (EC), supports research concerning border surveillance that aims at "improving detection of irregular migration and cross-border crime as it occurs between border crossing points (BCPs)" (Frontex 2014a). This research should also help member states to identify and address any vulnerability at their borders as well as "investigate the possibilities of automated data mining and media analysis as a part of the creation of an EU-wide intelligence picture" (Frontex 2014a).[3]

However, the use of drone technology by Frontex itself raises a lot of concerns among scholars. The agency has been criticised for its lack of transparency. Many of its activities are not made public, even when the sensitivity of the contained data is not an issue (Léonard 2010). In the past, NGOs as well as scholars have shown reasons for concern on the compliance of Frontex-coordinated operations with the legal provisions on human and migrants' rights. The agency has been accused of engaging in push-back operations, carried out on the basis of bilateral agreements between a member state and a TC.[4] In the joint operations (JOs) Hera II and Hera III, the bilateral Spain–Senegal and Spain–Mauritania agreements enabled interception and diversion of migrants' boats (Heijer 2011; Marin 2014). Later on, the agency did not engage actively in push-back operations, but it claimed success and effectiveness for the JO Nautilus, carried out "in parallel" with the operations carried out by Italy in the same area on the basis of national bilateral agreements with Libya, where interceptions also took place. This raised criticism among scholars: though not directly involved with such practices, the agency seemed to suggest complementary of its and member states' controversial practices (Moreno-Lax 2014; Marin 2014).

[3] In computer sciences, data mining is the practice of examining large pre-existing databases in order to generate new information (Oxford University Press 2014). Media analysis constitutes the monitoring of open and media sources and analysing its effects on its audience, trends as well as reliability when reporting news (Dictionary Central 2012).

[4] These diversion and interception practices were carried out, for example, by Italy or Spain on the basis of bilateral agreements with third countries (Italy–Libya, Spain–Morocco/Mauritania/Senegal) (European Union Agency for Fundamental Rights 2013; Heijer 2011). These agreements also allowed the member state to join the national border patrols and participate in the surveillance of those third states' territorial waters (Heijer 2011). The criticism of these practices mainly involves the fact that these activities are actually against international refugee law, that is, they constitute a breach to the non-refoulement principle (Marin 2011) that no state should "expel or return ("refouler") a refugee in any manner whatsoever to the frontiers of territories where his/her life or freedom would be threatened on account of his/her race, religion, nationality, membership of a particular social group or political opinion" (Convention and Protocol Relating to the Status of Refugees 1951, p. 30). These practices were challenged before the European Court of Human Rights in the case of *Hirsi Jamaa and Others v. Italy* (Case of Hirsi Jamaa and Others v. Italy 2012).

However, in spite of the importance it acquired in a relatively short period, it would be wrong to see Frontex as a supranational actor. On the opposite side, member states are represented in the control body of the agency, its management board. The agency relies on the member states' (mainly technical) resources and their voluntary participation in JOs. Hence, the agency is claimed to undergo some limitations in its cooperation capacity. In addition, it claims that search-and-rescue operations fall outside the mandate of border surveillance. The case of the Greek tanker *Salamis* is illustrative. The *Salamis* rescued migrants from near the Libyan coast. This tanker was not allowed access into the territorial waters of Malta and found itself in an impasse not wanting to go back to Libya and not being allowed to enter Malta or Greece. To avoid a possible humanitarian crisis on board the tanker, the ship was allowed to disembark migrants in Italy after political negotiations, with the EC claiming no powers in the process (Balzan 2013; Balzan and Dalli 2013). Frontex was criticised for not taking action in this and similar cases; the agency claimed not to be responsible for search-and-rescue operations and that these actions lie solely under the authority of particular member states (Regulation EU 1052/2013 2013), in spite of the growth of the agency's competencies and means. With the Frontex Regulation reforms, the agency has been allowed to deploy rapid-reaction intervention teams, and now European Border Guard Teams, to provide enhanced technical and operational assistance in case of an urgent and exceptional situation, or to purchase its own equipment.

Current Instruments and Practices: The Challenges Ahead

Frontex has a strong operational dimension that shows itself clearly on the operational-to-administration expenses ratio, which is 67:33 (Marin 2011; Frontex 2013a). However, there are some doubts about the proportionality of the allocation of these resources and its effectiveness. Most of the operations of Frontex are carried out at sea borders although migration by sea only constitutes a marginal part of overall entries to the EU and is driven by structural factors like the economic and political situation in the migrants' states of origin (Marin 2011; Czaika and De Haas 2013).

Integrated border management (IBM) sees cooperation with TCs as one of its pillars. For this purpose, Frontex has been given powers to coordinate action between TCs and the member states (Regulation No 1168/2011 2011). This follows the argument that for efficient and successful operations at borders against unauthorised migration and cross-border crime, actions have to be coordinated between the countries of destination and the countries of origin. Frontex has already signed a few working arrangements with TCs that involve development of activities between the agency and TCs in the fields of information exchange and risk analysis, training and R&D related to border management, and development and coordination of joint operational measures and pilot projects on border control (Frontex 2014d).[5]

[5] Frontex has already signed working arrangements with 18 countries: Albania, Armenia, Azerbaijan, Belarus, Bosnia and Herzegovina, Canada, Cape Verde, Croatia, Georgia, Macedonia, Mol-

Nevertheless, Frontex is bound by EU law, which also means its fundamental rights provisions, when cooperating with TCs. It is, however, questionable how clear and transparent cooperation with North African countries can be (Fink 2012). These are the main departure points of unauthorised migration in the Mediterranean, but the political and human rights situation there is becoming worse rather than better.

According to the European legislation, member states are tasked to carry out border surveillance, and Frontex coordinates their actions in this regard. But what is border surveillance? In its Article 2(11), the Schengen Borders Code (SBC) defines border surveillance as "the surveillance of borders between border crossing points and the surveillance of border crossing points outside the fixed opening hours, in order to prevent persons from circumventing border checks" (Regulation (EC) No 562/2006 2006). However, at the southern sea borders, law enforcement authorities find several challenges: for example, how to detect and track small boats, high personnel costs, and efficiency and surveillance capability of border patrols (Frontex 2014b). Since the Mediterranean is one of the busiest seaways in the world, the core of these challenges is the problem of how to monitor this extensive area of 2.5 million km^2 and collect information on what is happening, also in remote places. The current technology is often perceived as insufficient: for example, first, weather conditions and clouds affect satellite images, pictures taken by satellites are not real-time and can only be taken when the satellite is directly overhead; second, the automated identification system for boats is only mandatory for commercial vessels and fishing boats over 16 m, and infrared cameras can scan shores only up to the distance of about 1.2 km (Frontex 2013b).

This suboptimal situation creates a demand for more effective and efficient technological equipment to counter these challenges, such as the deployment of drone technology according to the US example. Frontex has already set a few criteria that drone technology has to satisfy for it to be regarded as the effective solution to counter the challenges concerning border surveillance. Accordingly, drone manufacturers have to prove to Frontex that it can persistently monitor large areas, function in all weather conditions, identify suspect vessels and human presence, provide real- and near-real-time operational data and be integrated into existing surveillance networks, and mainly that drones are a cost-efficient alternative to manned aircraft and can be integrated into the normal airspace (Kolev 2012; Beugels 2011).

However, the most debated aspect in the perspective of drone deployment by Frontex or other member states is the actual impact of this technology on border management, the fundamental rights of migrants and privacy rights, that is, how much the deployment of drones changes traditional border management and its legal standards. Currently, the information on the situation in the Mediterranean collected by Frontex and the Schengen countries is shared through the European Border Surveillance System (EUROSUR). EUROSUR is a technological network whose aim is to sustain better border management of the EU's external borders developed by Frontex with the member states' border authorities (European Commission 2013b).

---

dova, Montenegro, Nigeria, Russia, Serbia, Turkey and Ukraine. Further working arrangements are being negotiated with Libya, Morocco, Senegal, Egypt, Brazil and Tunisia (Frontex 2014c).

It features several national situational pictures, a "European Situational Picture" (ESP) and a common pre-frontier intelligence picture (CPIP) on what is happening on and outside the EU borders, including information on prevention of unauthorised migration and cross-border crime. The information is gathered with the help of various technological means, for example, satellites, ship board monitoring systems, drones, ground sensors, etc. This network should enhance cooperation between national border control agencies, as well as between them and Frontex, and promote surveillance of the EU's external borders and therefore improve border protection and save migrants' lives (Regulation EU 1052/2013 2013). However, although the supporters of EUROSUR claim its humanitarian potential, Frontex has no search-and-rescue competence. The operational area of the recently established JO Triton south of Sicily is a telling example (Frontex 2014d).[6] It is therefore questionable how EUROSUR will contribute to more safety at sea preventing loss of migrants' lives in boat accidents and whether humanitarian reasons are the main objective of EUROSUR.

Having explained the current policies and practices of border surveillance, the chapter will now explain the benefits of drones in border surveillance and present current projects in the domain.

The Deployment of Drone Technology in Border Surveillance

Drones now appear to be the technology that can contribute to increasing effectiveness in border surveillance. According to the special report produced for NATO on UAVs, drones have operational advantages as they are able to lower personnel costs and are more expandable as they can stay airborne much longer than a human crew, as well as carrying the necessary technology for monitoring areas and detecting suspicious vessels, for example, infrared cameras, mobile phone jammers, thermal imagining devices and video cameras (Nolin 2012).

Advocates for increased border surveillance claim not only that drones will be beneficial for preventing and combating cross-border crime and irregular migration but also that increased surveillance will fulfil a humanitarian mission (by providing information on accidents, thereby enabling search-and-rescue (SAR) operations) (Regulation (EU) No 1052/2013 2013). However, drones are first of all aircraft, supporting sophisticated sense-and-detect technology; therefore, it is out of the question that, thanks to drones, public authorities will know more about what happens at sea. But the question is: Is this really the heart of the problem? Is there a need to know more precisely what happens at sea? Will the numerous calls for rescue by migrants' boats that have been left unanswered in the past few years be

[6] According to the (then) Commissioner for Home Affairs, Ms Cecilia Malmström, JO Triton should be seen as complementary rather than alternative to *Mare Nostrum* (Nielsen 2014).

answered by drones?[7] Or do these failures to answer calls indicate that SAR duties are now hindered by controversies on the port of disembarkation, which also means taking responsibility for providing assistance to these persons, and also giving them access to asylum systems?

Actual use of drone technology in the EU for the purposes of border surveillance is still limited. Among the southern member states, Spain and Italy dispose of drones that can be used for border surveillance. In 2010, Italy bought 12 UAVs from the US: 6 MQ-1 Predators and 6 MQ-9 Reapers or Predator Bs. Since October 2013, some of them have been used within the operation *Mare Nostrum* for monitoring the high seas and searching for migrant vessels.[8] However, because of the high costs involved, the Italian government has discontinued the operation and has asked for the participation of European states in a new Frontex coordinated operation (Scherer 2014). Consequently, JO Triton was started by Frontex to protect Italian sea borders from migration, focusing on migratory flows from Tunisia, Libya, Egypt and Turkey, but it clearly pointed out that the SAR operations are the competence of the member states (European Commission 2014a).

Other states too are on the frontline and getting ready to deploy UAVs at their borders. Spain has also considered using drones for sea border surveillance, mainly in the Strait of Gibraltar and over the Canary Islands (Purvis 2011). It also carried out several tests within the PERSEUS project in 2013 (Plunkett 2013). The Spanish authorities support or are actively engaged in several projects on drones financed under the FP7 (e.g. PERSEUS, CLOSEYE, TRITON and AEROCEPTOR). However, under the current Spanish regulation, the deployment of drones for civil purposes (including commercial or professional drones) is not allowed. Therefore, military drones can only be used under special conditions (Howell 2014).[9] Similarly, Greece is also in the process of purchasing a drone suitable for day and night border surveillance (Souliotis 2014; New Europe Online 2014; Mamakouka 2014).

Alongside member states, the European agency Frontex has also several times confirmed its interest in drones for improving the search-and-rescue and the reaction capability of the member states and the agency; in addition, drones would contribute to providing additional information to the EUROSUR network (Laitinen 2013). The agency regularly participates in various initiatives regarding R&D as well as organising practical demonstrations of and tests on drone technology (Frontex 2014a). The former executive director of Frontex, Ilkka Laitinen, in an interview for *EUobserver* admitted that the drone "seems to be a reliable and cost-effective means for surveillance" (Nielsen 2013a). The head of R&D at Frontex, Edgar Beugels, admitted that Frontex is interested in drone technology for border

[7] See two emblematic cases among many: the case of the left-to-die boat and the Salamis case of summer 2013.

[8] The Italian Navy has not disclosed how many drones are deployed for border surveillance. The operation *Mare Nostrum* is an Italian border surveillance and "humanitarian operation to save human lives" in the Mediterranean; the assets deployed confer a clearly military nature on it (Ghelli 2013; Day 2013).

[9] This is not the case of Italy, which has also enacted regulations enabling drones to fly in its skies for civil purposes, thanks to resorting to "smart segregation" (ENAC 2013).

surveillance, though “remotely piloted aircraft is just one of the technologies we are looking into. At this moment we do not know if this technology is a technology that we could potentially give to the border guard community” (Nielsen 2013b).

This uncertainty about the real deployment of drones has multiple reasons. First, current European aviation law does not allow drones to fly in nonsegregated airspace.[10] Therefore, the preparation of an appropriate legal framework, providing guarantees on safety, should precede the actual deployment of drones. Second, there are significant concerns on the cost-effectiveness, efficiency, added value and technological limitations of drones for use in border surveillance, in addition to worries about the threats drones pose to fundamental rights (Laitinen 2013; Nielsen 2013b; Frontex 2014b, 2012a). The American experience offers in this respect a lesson that cannot be ignored. In the USA, several criticisms have been raised on the deployment of drone technology in border surveillance: Though in the political debate there is a never-ending rush toward drone technology, in 2012 the fleet was unused for 63 % of the time. This was due to a lack of budget for drone operations, and associated costs of maintenance and drone-related equipment (Sternstein 2012). So the lesson for the policymakers and public agencies is that clear benchmarks and criteria are necessary in order to measure the cost-effectiveness of drones. This requires measuring not only the costs of the aircraft but also their operations and their maintenance. Right now, research on drones is, to some extent, segregated, protected from the public debate on military drones on which by contrast there is media coverage and debate on the desirability of UAVs and on the ethical implications of drones and of robotics. In a sense, the civilian deployment of drones remains unquestioned, while (defence) industries and public agencies prepare, research and manufacture new drones in a technocratic process, where the fundamentals of the debate remain unexplored (Hayes et al. 2014).

More precisely, looking at the process itself, other considerations might arise. For example, Frontex has hosted several workshops on drone technology. In 2009 and 2010, it conducted workshops on “RPA and Land Border Surveillance” in Imatra, Finland.[11] Further workshops were hosted in Sofia, Bulgaria, and Warsaw, Poland, in 2012 and 2014. These workshops are aimed at closing the communication gap between the producers and end users and bringing together the border management authorities, research institutes, universities and industry to exchange views, experiences and needs (Frontex 2014b, 2012a). Demonstrations aimed at presenting the performance of drones in various conditions are another type of event that Frontex regularly organises. At these sessions, European and international drone manufacturers were invited to present the latest developments in their industry and demonstrate their capabilities for border surveillance.[12] Criticisms have been raised

---

[10] Nonsegregated airspace is the airspace open to all civil air transport. Current aviation law prohibits the deployment of fully automated drones in nonsegregated airspace (Hayes et al. 2014).

[11] This workshop included live demonstrations of mini-RPAs (Patria MASS, Rafael Orbiter, SIM Skyeye, Selex ASIO) and Aerostats (Skystar 180) (Kolev 2012).

[12] Among the companies that attended the workshops in the past were, for example, Thales (UK), Aerovisión (Spain), Scotty Group (Austria), Israel Aerospace Industries (Israel), L–3 Communi-

on the generosity of Frontex for having subsidised companies to allow them to participate in demonstrations where member states' officials act as potential customers. For example, in 2011, Frontex paid 30,000 € to Lockheed Martin UK Integrated Systems & Solutions for participating in a demonstration of optionally piloted vehicle (OPV) UAVs in Aktio (Greece). The subsidies to the participating companies for this demonstration session varied between 10,000 and 198,000 € (Fotiadis and Ciobanu 2013; Hayes et al. 2014).

To conclude, the past few years have witnessed interest in and research on the transfer of drone technology from the military to the civil domains; in particular, Frontex and member states' ministries have engaged in the transfer of UAV technology to border surveillance, which is the focus of this chapter. Considering the high number of projects on different aspects connected with the deployment of drones, together with the actual experiences of drones' deployment in border surveillance by some member states, it can be concluded that public authorities have a long-term interest in using drone technology for the protection and policing of their external borders.

The Legal Framework for the Deployment of Drones Outside Military Spheres

The deployment of drones for border surveillance in Europe triggers general and specific questions. Under general questions, we indicate those concerning the nature of drones as aircraft, whereas the specific questions refer to drones in the sphere of border surveillance.

The following section will present the problems underlying the deployment of drones with reference to aviation standards.

General Aviation Law Issues

As recalled above, a UAV or an RPA is itself an aircraft, and as such it has to be accommodated into the airspace, according to the current regulatory requirements, and, to some extent, postulating new regulatory frameworks.

This is a preliminary set of issues that precedes and, to some extent, hinders the deployment of drone technology in the civil domains, and raises questions on whether the necessary and adequate legal framework concerning the deployment of drones in the civil airspace is in place. Within this framework, drones have to meet technological standards of airworthiness, safety and liability in case of accidents.

---

cations (USA), AeroVironment (USA), Selex (Italy), Safran Group-Sagem, Inmarsat (UK), Diamond Aircraft (Austria), Altus (USA) and Lockheed Martin (UK).

However, experts admit that the EU has an unclear legal framework regarding the flight of drones in nonsegregated airspace as well as the fundamental rights and privacy concerns raised by the application of drones. All this, accompanied by the complexity and multimodality of these surveillance systems that integrate a high range of other technologies and capabilities, hinders the current deployment of drones in Europe.

The EU has the competence to regulate only drones above 150 kg. Under this limit, drones are subject to the regulation of the national aviation authority. For an application of drones for the surveillance of large areas such as the Mediterranean, the Union will have to secure a mutual recognition of standards across the member states as is already done in other areas of the internal market. However, on its plans for the introduction of drones into the European airspace, the EC has consulted different stakeholders (e.g. industry and national authorities) since 2009 (European Commission 2014b).

The International Civil Aviation Organization (ICAO)[13] is also working on providing an international regulatory framework for the deployment of drones. ICAO states clearly that the civil market with drones will remain limited until an appropriate regulatory framework is in force. Their deployment to the civil domain is also delayed in the USA. However, across Europe and in the international airspace, drones can currently be deployed only as far as the operator maintains a visual contact with the drone, that is, there is visual line of sight (VLOS)[14]. For deployment beyond the VLOS, drones have to be operated in segregated airspace or must be equipped with a sense-and-avoid system that is safe for use. Unfortunately, sense-and-avoid systems need further development, as there is currently no system worldwide to be considered as safe for use. Moreover, the EU law prohibits the flight of fully automated unmanned drones in commercial airspace (UK Civil Aviation Authority: Directorate of Airspace Policy 2014; Hayes et al. 2014).

The EU plans to finish the revision of the regulation concerning airspace technology and its deployment by 2016. This revision should therefore integrate drones too into national airspace from that year. The EC has specified that the future regulatory framework concerning drones "should reflect the wide variety of aircraft and operations, keep rules proportionate to the potential risk and contain the administrative burden for industry and for the supervisory authorities" (European Commission 2014c). However, for now, it can be concluded that the European airspace is legally not yet prepared for a large-scale accommodation of drones into the civil airspace. From the perspective of aviation law, their use nowadays can therefore be assumed as a precipitate act that creates concerns about respect for an adequate legal framework ensuring airworthiness, safety and liability.

[13] ICAO is an UN agency created in 1944. It works with global industry and aviation organizations on development of international Standards and Recommended Practices (SARPs), which are then used by national states for the development of their legally binding national civil aviation legislation (ICAO 2014).

[14] VLOS differs across states. Usually, it means an area of about 500 m horizontally and about 120 m vertically (UK Civil Aviation Authority: Directorate of Airspace Policy 2014).

The next section will focus on problems specific to the use of drone technology for border surveillance purposes.

The Legal Framework for Border Surveillance and Its Constraints

As stated in Article 6 of the SBC, border control should be fully in line with respect to human dignity, not discriminatory, and be proportionate to the objectives pursued; border surveillance, understood as part of border control (Recital 8, Regulation (EC) No 562/2006 2006), must therefore respect this requirement. Article 12 of the SBC stipulates that border surveillance serves to "prevent unauthorized border crossing, to counter cross-border criminality and to take measures against persons who have crossed the border illegally". Furthermore, according to Article 12(4) of the SBC, surveillance may be "carried out by stationary or mobile units…the aim of such surveillance being to apprehend individuals crossing the border illegally. Surveillance may also be carried out by technical means, including electronic means". It is questionable whether drones can be considered as a legitimate tool for border surveillance if we interpret them as technical means, drones being first of all aircraft.

Right away it can be stated that drone deployment pursues legitimate objectives, for example, stopping cross-border crime, preventing unauthorised migration, saving human lives and protecting national security and order. The question is whether drones constitute the least restrictive measure that could be chosen to achieve these objectives and whether there is some imbalance between the costs of drone technology in border surveillance and its benefits.[15] Moreover, the surveillance and the information acquired will have to be processed by stakeholders in risk analysis. Frontex will therefore remain one of the important players as one of its main tasks is conducting risk analysis. Hence, it is possible to argue that drone technology might entail a shift from reactive border surveillance to proactive or preventive border surveillance.

The use of surveillance technology (e.g. drones) at legal border crossing points[16] is proportionate because states monitor their borders and people expect to be monitored there as these are highly public places (Kenk et al. 2013). Surveillance between the border crossing points can also be considered legitimate as it is perceived as responding to a need to prevent and discourage persons from circumventing border checks (Article 12(2), Regulation (EC) No 562/2006 2006). Nevertheless, even when border control is carried out outside the EU borders, the SBC still applies, as do the obligations to protect fundamental rights. The waiving of Article 7 of the SBC on border checks on persons, which according to the definition in Article 2 of the SBC should be carried out at border crossing points at the EU external borders, is provided in Annex VI of the SBC. It specifies that at sea borders, checks can be carried out even during crossings or, upon the ship's arrival or departure,

[15] This is an applied principle of proportionality (Sauter 2013).

[16] A border crossing point is any crossing point authorised by the competent authorities for the crossing of external borders (Regulation (EC) No 562/2006 2006).

in the territory of TCs (Point 3.1.1. Annex VI, Regulation EC No 562/2006 2006). The execution of these border checks should fully respect human dignity and the principles of proportionality and non-discrimination, according to Article 6 of the SBC. Alongside EU legal obligations, EU states are bound to respect the European Convention on Human Rights (ECHR) even when they act outside the EU territory (Case of Hirsi Jamaa and Others v. Italy 2012).

On the other side, drones will definitely help to locate migrants' and other boats in distress at sea because they are able to cover a much larger area than a deployed manned vessel or aircraft. Nevertheless, as drones are just sense-and-detect technology, they will not solve the core questions of search-and-rescue operations, that is, who is responsible for rescuing a boat with migrants in distress[17] and where to disembark migrants.

According to the international maritime law, there is an obligation to render assistance to a boat in distress and it applies to all vessels, both commercial and governmental (IMO 1985); however, the nearest place of safety remains a controversy among some states, such as Malta and Italy. According to the European Union Agency for Fundamental Rights (FRA) report, this point is a cause of problems in the Mediterranean: disembarkation entails taking care of the migrants, providing them with the possibility to apply for asylum because of lack of responsibility sharing — the case of the tanker *Salamis* offers an example. Shipmasters of civil and commercial ships are also concerned about their obligation to rescue migrants at sea because they have to report distress cases to the national authorities and then await their arrival (which can take a long time). Sometimes they have to start a rescue operation on their own, with the consequences this entails, and face problems on where to disembark the migrants: states are usually reluctant to allow the disembarkation of unauthorised migrants in their ports. Similar problems are faced by military vessels often cruising the Mediterranean: the case of the left-to-die boat offers a tragic example.[18] For commercial boats, taking part in rescue operation is very costly and dangerous as migrants might try to get onto the boat anyway. Therefore, many of them try to avoid migrants' vessels and do not report them to the authorities. The former Italian Prime Minister, Enrico Letta, presenting the operation *Mare Nostrum*, also stressed that identification of migrants' vessel in distress by an Italian vessel (or even drone) does not automatically mean that these migrants will be taken to an Italian port; this will be decided upon where the operation takes place (Day 2013).

Overall, it seems that drone technology will increase surveillance, but how the latter will increase the security of migrants is not yet clear. The political issues that have hindered SAR operations until now are still there. Southern states denounce

[17] Distress as defined by the SAR Convention is a "situation wherein there is a reasonable certainty that a person, a vessel or other craft is threatened by grave and imminent danger and requires immediate assistance" (IMO 1985).

[18] In 2011, a Spanish naval force rescued about 100 migrants and it took 5 days of negotiations till those people could be disembarked in Tunisia. The FRA report found that migrants often report that they encounter a military vessel during their journey but are mostly turned away without the migrants' vessel being reported or assistance being rendered (FRA 2013).

a lack of solidarity and burden-sharing concerning rescue operations and reception of the migrants. Drones will definitely help to locate migrants' and other boats in distress at sea but will not decide, regarding the situation, whether these should be rescued or not. Therefore, one can conclude that drone technology might entail a further securitisation of border surveillance, together with a shift toward preventive border surveillance.

The Deployment of Drones in Border Surveillance: An Outlook on the Limitations from and Challenges for Human Rights

The Limitations for Drone Surveillance from Privacy and Data Protection

Among the main challenges implied by the deployment of drones in the Mediterranean are the privacy and data protection rights of all people found within the sight of the drone at sea (e.g. fishermen, tourists and migrants), since drones can collect and transmit specific information to the drone operator about what happens at sea, including visual data.[19] Drones, as a sense-and-detect technology, do nothing but increase the conflict between security and privacy (Marin 2014). But what is the legal framework regulating the operation of drones?

In the EU legal order, the privacy and the protection of personal data[20] have fundamental rights status. According to Articles 7 and 8 of the Charter of Fundamental Rights of the European Union, legally binding since 2009, individuals have, respectively, the right to privacy and the right to the protection of their personal data. The latter is based on the principles of consent, the principle of specific, explicit and legitimate purpose for the collection of data, the right of access to data and right to rectification. Case law has interpreted the current legal framework as also including the right to the deletion of data, or right to be forgotten (Case Google Spain 2014). Legislative reforms are introducing other rights, such as data portability and data breach notification. Besides the now fully fledged legally binding Charter of the EU, the ECHR, in its Article 8, states that everyone has the right to have their

[19] In the case of *Peck v. UK*, the ECtHR stated that: "The monitoring of the actions of an individual in a public place by the use of photographic equipment which does not record the visual data does not, as such, give rise to an interference with the individual's private life" (Case of Peck v. The United Kingdom 2003).

[20] According to Data Protection Directive 95/46/EC, personal data are defined as "any information relating to an identified or identifiable natural person; an identifiable person is one who can be identified, directly or indirectly, in particular by reference to an identification number or to one or more factors specific to his physical, physiological, mental economic, cultural or social identity" (Directive 95/46/EC 1995).

private and family life, home and communications respected, similarly to Article 7 of the Charter.

Alongside these provisions of a constitutional nature, enshrining data protection among the fundamental rights of the EU's legal order, there are European regulations binding in all member states that further specify the constitutional provisions of the Charter: These are the Data Protection Directive 95/46/EC (hereinafter DPD) and the Framework Decision 2008/977/JHA on the protection of personal data processed in the framework of police and judicial cooperation in criminal matters (hereinafter the FD) (Directive 95/46/EC 1995; Council Framework Decision /977/JHA 2008). Last but not least, Regulation 2001/45/EC completes this legal framework, covering the processing of personal data by EU institutions and bodies (Regulation No 45/2001 2001).

The DPD has paramount importance at the EU level. Its scope covers "the processing of personal data wholly or partly by automatic means" and by contrast does not cover "processing operations concerning public security, defence, State security (including the economic well-being of the State when the processing operation relates to State security matters) and the activities of the State in areas of criminal law" (Articles 3 and 13, Directive 95/46/EC 1995). The scope of the FD instead covers member states' activities in the context of the judicial cooperation in criminal matters, the former third pillar of EU law, and hence extends data protection to that area, otherwise explicitly excluded from the scope of the DPD.

As mentioned before, the deployment of drone technology to civil applications challenges these legal frameworks and the rights they protect: drones can collect and transmit specific information to the drone operator about what happens at sea, including visual data.[21] It has been argued that the application of the DPD to border surveillance is not clear (Finn and Wright 2012). Indeed, according to the DPD, member states can restrict the scope of the obligations and rights protected by the DPD, if this is necessary in order to protect, among other things, national security, public security and for "prevention, investigation, detection and prosecution of criminal offences (…)" (Article 13, para. 1, letters a), c), d), Directive 95/46/EC 1995). However, any limitation of EU rights has to be based on the rule of law and has to be applied in a proportionate manner. Second, when member states act so, they are nevertheless bound to respect fundamental rights (Case ERT 1991). Third, considering we are also dealing with rights protected under the European Convention of Human Rights, according to that convention, as interpreted by the Court of Strasbourg, restrictions to fundamental rights have to be based on the rule of law, and have to be necessary in order to protect public order values (such as national security and public safety) and prevent disorder or crime. Alongside this, Frontex being an EU agency, its operations are covered by Regulation 45/2001/EC, as well as by application measures adopted by its management board, as stated in Article

[21] In the case of *Peck v. UK*, the ECtHR stated that: "The monitoring of the actions of an individual in a public place by the use of photographic equipment which does not record the visual data does not, as such, give rise to an interference with the individual's private life" (Case of Peck v. The United Kingdom 2003).

11 of the Frontex-reformed regulation. Therefore, even if one cannot invoke the right to privacy at the border, as the border is the legal and physical space where states exercise control to maintain their sovereignty, the individual can invoke and is entitled to data protection at the border as an autonomous fundamental right. This European human right is defined in its fundamentals by the DPD, currently in the process of being reformed into the General Data Protection Regulation. For this reason, the Meijers Committee had suggested to postpone the enactment of the EUROSUR Regulation till the adoption of the General Data Protection Regulation, which has not been done (Meijers Committee 2012).

Given this general legal background, enshrining data protection as an autonomous fundamental right, it should not be surprising that the challenges of border surveillance for data protection were discussed during the negotiation of the EUROSUR Regulation, as it has been acknowledged that extensive surveillance carried out in the context of border surveillance potentially infringes upon data protection rights. The text as agreed through the legislative process provides for specific rules for data protection in the EUROSUR system. The most sensitive issue concerns cooperation of Frontex with third country authorities (TCA) in the framework of border surveillance. For this reason, "[a]ny exchange of personal data in the European situational picture and the CPIP should constitute an exception. It should be conducted on the basis of existing national and Union law and should respect their specific data protection requirements". The European Regulation (No 45/2001) on data protection is applicable in case more specific provisions do not apply (Recital 13, Regulation No. 1052/2013 2013). It is here stressed that the first aim of EUROSUR is not to exchange personal data; if this happens in the European situational picture or in the CPIP, it is limited to personal data concerning ship identification numbers. When the national situational picture is used for processing of personal data, those data shall be processed in accordance with the European DPD and the FD, and important national provisions on data protection. All in all, Recital 13 and Article 13 of the EUROSUR Regulation anchor the processing of personal data to the European regulation on data protection. Other provisions of the EUROSUR Regulation stress the duty to comply with fundamental rights, and with data protection requirements (Article 2(4) and Article 18, Regulation (EU) No 1052/2013 2013).

However, a serious challenge to data protection comes from the cooperation of member states with neighbouring TCs. Article 20 provides that "[a]ny exchange of personal data with third countries in the framework of EUROSUR shall be strictly limited to what is absolutely necessary for the purposes of this Regulation. It shall be carried out in accordance with Directive 95/46/EC, Framework Decision 2008/977/JHA and the important national provisions on data protection".[22] Though strictly limited, exchange of personal data is still possible, and this inevitably trig-

[22] Article 20, para. 4. See also para. 5, which states that: "Any exchange of information under paragraph 1, which provides a third country with information that could be used to identify persons or groups of persons whose request for access to international protection is under examination or who are under a serious risk of being subjected to torture, inhuman and degrading treatment or punishment or any other violation of fundamental rights, shall be prohibited."

gers the question of the fate of those data, once in the hands of TC institutions. The threat of function creep is there. As presented above, the DPD, in its Article 6, paragraph 1, states that personal data have to be "collected for specified, explicit and legitimate purpose and not further processed in a way incompatible with those purposes" (Directive 95/46/EC 1995). This should secure that the data are not misused, that is, it prevents the function creep of the data. However, if within the EUROSUR system the exchange of data across national authorities (border management, law enforcement, etc.) is enabled, the question of the effectiveness of the data protection provisions remains: how can the institutions and the states monitor the respect of data provisions, after they have exchanged data with, for example, Libyan authorities?

Another challenge for privacy and perhaps other more fundamental rights, such as the right to life, is represented by extended surveillance itself and its implications. We will not develop here the answers to the questions on the policy shift(s) enabled by the deployment of drone technology for border surveillance, which seem to go in the direction of externalisation of border surveillance, and not in the direction of humanitarian operations; however, the surveillance of specific areas, outside the EU borders, will probably deter (migrants') vessels from using a specific route and perhaps make them use another, more dangerous route. Drones are able to follow the routes of picked vessels, take images of what the crew is doing or even prevent a cell phone from receiving a signal. Another problem here is that drones fly in an unnoticed and unheard manner, and they are also valuable because of these properties. Therefore, should people entering a shadowed area generally be notified that they are entering a surveillance area (according to Article 18 of the DPD)? This is the case already in our cities when we enter a closed-circuit television camera (CCTV) area. Of course, in practice, this cannot be done at sea. All in all, even if the EUROSUR Regulation embeds border surveillance within the data protection framework, one should be aware that large-scale surveillance in the Mediterranean can also cause a "chilling effect"[23] or self-disciplining effect or even erode the society's expectation of privacy, therefore leading to normalisation of previously unacceptable levels of surveillance (Finn and Wright 2012).

The Challenges for Human Rights Created by EU's Cooperation with Third Countries in the Area of Border Management

Under the concept of integrated border management, great emphasis is attributed to cooperation with TCs. In the aftermath of the Lampedusa tragedy of 3 October 2013, where more than 350 persons died close to the Italian island, the EU set up the Mediterranean Task Force in order to boost a strategy for facing the migratory

[23] In the context of law, a chilling effect is the inhibition or discouragement of the legitimate exercise of a constitutional right because of the fear of potential or threatened prosecution or sanction (Wallace and Wild 2010).

phenomenon (European Commission 2013a). As one can read in the latter report, cooperation with TCs is of crucial importance.

In order to more effectively combat unauthorised migration and cross-border crime, Frontex cooperates with TCs on the basis of working arrangements, on information exchange, risk analysis, training, research and development, joint operations and pilot projects (Article 13, Regulation No 1168/2011 2011). Currently, 18 working arrangements have been signed with TCs' authorities.[24] Namely, cooperation with TCs in migration and surveillance raises many concerns because of the poor safeguards for democratic standards and human rights.

This section will assess how border surveillance carried out with drones can challenge fundamental rights of migrants, with special reference to the cooperation of European and TCs' border surveillance agencies.

First, it is questionable that border surveillance, as defined in the SBC, can include the situation of surveillance behind the external border as it is envisaged by the development of the CPIP with EUROSUR: Hence, there seems to be a gap between law and policy objectives. The extraterritorialisation of surveillance also entails surveillance of territorial waters of TCs: this is a sensitive issue as it touches upon the territorial sovereignty of those states. States increasingly turn to border surveillance in order to deter new departures of migrants from North African coasts and to identify, intercept and send back the boats before they reach European territory (Carling and Hernández-Carretero 2011). Aware of the limits that the *Hirsi* judgment has placed on their extraterritorial action, European states increasingly seek the cooperation of North African countries. This raises concerns as many North African states do not sufficiently protect fundamental rights, and as long as the people, mainly refugees, find themselves on their territory, neither member states nor Frontex can prevent the violation of their rights.

Studies indicate how precarious political rights are in those states (Freedom House 2013). Therefore, there are serious concerns that the fundamental rights are not sufficiently protected in these countries or are even breached in larger scale. Moreover, neither the European Convention nor the EU Charter and the DPD are legally binding outside Europe as long as a non-EU state's authority has full and effective control over a person.

An investigation on cross-border crime needs to cover all aspects of the organisational network, involving the countries of both origin and destination. States contracting the ECHR have the obligation to cooperate effectively with other countries to investigate events of cross-border trafficking, even if that event happens outside their territory (Case of Rantsev v. Cyprus and Russia 2010). However, the same cannot be said for countries which are not contracting the ECHR. European states have the obligation to protect individuals against slavery and inhuman treatment

[24] Among those states, we can name Turkey as a departure country of unauthorised migration and cross-border crime in the Mediterranean. Furthermore, Frontex is also planning to expand its cooperation with other North African countries (such as Libya, Morocco, Egypt, Tunisia as well as Mauritania and Senegal), which have been involved in the Hera operation on the basis of bilateral agreements with Spain (Frontex 2014c).

like human trafficking, according to Article 4 of the ECHR. Although there are not sufficient protections for human rights in North African countries, Frontex still sees them as partners in issues of combating unauthorised migration and cross-border crime.

While Frontex does not have any working arrangement with the African countries referred to (except Turkey), there is however cooperation on the basis of bilateral agreements between the member states and the TCs. Member states can cooperate with TCs even without the involvement of the agency (Article 16(3), Regulation EC No 562/2006 2006). As to surveillance, member states can incorporate the information gained from cooperation with TCs into its national situational picture in EUROSUR (Article 9(2h), Regulation No 1052/2013 2013) and exchange the information with TCs according to Article 20 of this regulation. Aware of the risks nested in cooperation with TCs, the Meijers Committee suggested that these bilateral agreements should be disclosed to the EC (Meijers Committee 2012); the final text of the EUROSUR Regulation provides an obligation of member states simply to just notify any such agreement to the Commission (Article 20(2), Regulation No. 1052/2013 2013) to ensure that the important provisions of the agreement comply with the EUROSUR Regulation. The Regulation does not require the Commission to assess compliance with other European and international legal provisions, including those on fundamental rights, but can we deem that the Commission is not bound to do so by the treaties? From the provisions of Article 20 of the EUROSUR Regulation, it is not clear what actually happens when the Commission denies that the agreement complies with the provisions of the EUROSUR Regulation or makes any recommendation to the member state to change some provisions of the agreement. It can therefore be assumed that the Commission only has political influence powers on the agreements between member states and TCs. Will the Commission start an enforcement action against a state, for breach of the EU's fundamental human rights?

Although these agreements must comply with European and international provisions on human rights (Article 20(3), Regulation No 1052/2013 2013), there is no systematic process of examination of whether the agreements really comply with those provisions. Furthermore, the secrecy of some agreements is an additional issue; the lack of transparency this entails undermines the principle of democratic accountability with the danger of possibly harming migrants' rights: the *Hirsi* case concerning an agreement between Italy and Libya that was not made public affords an example.

For effective cooperation with other countries, it is necessary to exchange important data on the situation concerned. It is most probable that the information collected from border surveillance tools is going to be stored in the EUROSUR network. As discussed above (section 'The Limitations for Drone Surveillance from Privacy and Data Protection'), the EUROSUR Regulation handles the exchange of data with TCs in Article 20. Article 20(5) indirectly states that personal data can be exchanged with TCs. This mainly concerns persons who are not refugees or asylum applicants or who are not at risk of being tortured, subjected to inhuman and degrading treatment or punishment in that TC. Alongside this, the DPD also handles the transfer of data to TCs in Article 25 ff. (Directive 95/46/EC 1995). The DPD has an extraterritorial reach. The Directive states that "the transfer of personal data to a [TC] which does not ensure an adequate level of protection must be prohibited" (Recital 57, Directive 95/46/EC 1995). Personal information exchange via EUROSUR also has to comply with those provisions (Article 20(4), Regulation No 1052/2013 2013). Nevertheless, the Regulation 1052/2013 establishing EUROSUR (EUROSUR Regulation) provides an open-ended list of datasets and agencies that can be included in the information network (Hayes and Vermeulen 2012) and although the regulation states that exchange of personal information should be limited to the extent that is absolutely necessary (Article 20(4)), it is nowhere clearly defined in what situations this absolute necessity arises. This, combined with the unsound legal system of TCs and the sanctions imposed by them for irregularly leaving the country, causes serious concerns on the consequences and effects of externalising border controls to TCs for fundamental rights of migrants. This poses serious risks to persons that may be in need of international protection as their data can be freely collected and transmitted without extensive examination of their migratory and personal status. Moreover, neither the member states nor Frontex can effectively control what is happening with that data or for what it is used after the data are transmitted to TCs. This can jeopardise fundamental rights of migrants as people can be identified on the basis of provided data and thereby face risks of being tortured, imprisoned, etc. (Kenk et al. 2013). Drones deployed outside EU borders could therefore have considerable impacts on asylum seekers because of threats to their lives and human dignity in TCs. These persons, with the help of drones operating in the pre-frontier area (also including the territory of TCs), can be notified and intercepted before they actually reach the high seas without being extensively examined regarding their migratory status. This would then infringe the principle of non-refoulement and provisions on prohibition of collective expulsion. Moreover, even if people are not refugees, drone surveillance would inhibit their right to emigrate and, if they are intercepted before they reach Europe, they can be subjected to considerable threats of being imprisoned or fined for irregularly leaving the country. To sum up, surveillance beyond the external borders of the EU (mainly in the territorial waters of TCs) complemented by cooperation with North African TCs is a threat to the fundamental rights of migrants, does not guarantee that member states will improve their search and rescue capacities, and is not proportionate to the risk these people represent to security for the state, whereas these measures do not improve the security of migrants' lives.

Concluding Observations

This chapter has dealt with border surveillance in the EU, focussing especially on its most sensitive field, that is, the Southern European border (i.e. the Mediterranean Sea). After an introduction presenting the problem and the theoretical framework, it has presented the actors involved, at both the single state and the European level, and discussed border surveillance and its latest developments, namely the deployment of drone technology and the construction of a surveillance infrastructure connecting member states' surveillance systems, EUROSUR. In spite of increased surveillance, a consequence of the security-oriented approach taken by Europe, several critical issues remain: for example, the death toll of migrants in the Mediterranean is still very high, and, second, controversies remain among Southern member states on SAR obligations.

Within this framework, the past few years have witnessed research on and investments in the direction of the deployment of drone technology in border surveillance. Frontex has shown interest in drones and participated actively in their development through research projects; some states, such as Italy, have already deployed them, for example, in the operation *Mare Nostrum*. At the same time, limitations exist as to the cost-effectiveness of this technology, and the American experience should also be studied in Europe in order to avoid copying the mistakes too. Alongside this, there is little information and transparency on the costs of the deployment of this technology and the whole process is technocratic, rather than democratic.

The chapter has shown that the deployment of drones raises several issues: first, as a fact in itself and, second, for its impact on border surveillance. Regarding the first aspect, the chapter stresses the importance of an adequate aviation legal framework being in place, before the deployment of drones in the civil airspace; the regulatory framework should aim at guaranteeing safety in the accommodation of these aircraft into the airspace. Alongside this, using drones in border surveillance also means moving the borders of Europe outwards, in the direction of TCs; hence, border surveillance acquires a preventive character and postulates the cooperation of TCs. In this perspective, drone technology becomes crucial for triggering a policy change, toward the externalisation of border surveillance.

Assessing the impact of surveillance by drones on fundamental rights, the chapter has demonstrated that drones can be very invasive for individuals' privacy; however, the EU and the member states are bound to a strict legal framework for data protection, which also applies in the context of border surveillance. Though EUROSUR's primary aim is not collecting personal data, it provides for a CPIP to be fed with information received by TCs, which is a problematic issue, in light of the level of rule of law and protection of rights of migrants in those states. Once the data are transmitted to the TCs, there is a risk of function creep that can hardly be controlled from Europe. So, even if constrained by fundamental rights, provisions on data protection, increased surveillance of borders and the possibilities for cooperation with TCAs challenge the safeguards provided by the European legislations.

The chapter therefore suggests that, while the deployment of drones for civilian purposes, such as border surveillance, seems to be one of the innovations that cannot be stopped, it is necessary to monitor the effects and the implications of the use of drones in many public policies.

References

Balzacq, T. (2008). The policy tools and securitization: Information exchange, EU foreign and interior policies. *Journal of Common Market Studies, 46*(1), 75–100.

Balzan, J. (7 August 2013). *NGOs welcome Italy's decision to accept migrants*. Accessed 10 Sept 2014, from Malta Today: http://www.maltatoday.com.mt/news/national/28946/ngos-welcome-italy-s-decision-to-accept-migrants-20130807#.U38X5Ch_TKA.

Balzan, J., & Dalli, M. (6 August 2013). *Manuel Mallia: Blocking Salamis is 'point of principle' for Maltese government*. Accessed 10 Sept 2014, from Malta Today: http://www.maltatoday.com.mt/news/national/28925/home-affairs-minister-press-conference-20130806#.U38Q1Ch_TKA.

Besters, M., & Brom, F. (2010). "Greedy" information technology: The digitalization of the European Policy. *European Journal of Migration and Law, 12*(4), 455–470.

Beugels, E. (2011). *UAVs for European Border Surveillance. Frontex, Head of Research and Development Unit*. Warsaw: Frontex.

Bigo, D. (2001). Migration and security. In V. Guiraudon & C. Joppke (Eds.), *Controlling a new migration world* (pp. 121–149). London and New York: Routledge.

Boswell, C., & Geddes, A. (2011). *Migration and mobility in the European Union. The European Union Series*. Basingstoke: Palgrave Macmillan.

Carling, J., & Hernández-Carretero, M. (2011). Protecting Europe and protecting migrants? Strategies for managing unauthorised migration from Africa. *The British Journal of Politics and International Relations, 13*(1), 42–58.

Case of Hirsi Jamaa and Others v. Italy, 27765/09 (European Court of Human Rights February 23, 2012).

Case of Peck v. The United Kingdom, 44647/98 (European Court of Human Rights January 28, 2003).

Case of Rantsev v Cyprus and Russia, 25965/04 (European Court of Human Rights January 7, 2010).

Collett, E. (2011). *Emerging transatlantic security dilemmas in border management*. Washington, D.C.: Migration Policy Institute.

Council Framework Decision (2008/977/JHA of 27 November on the protection of personal data processed in the framework of police and judicial cooperation in criminal matters). (2008, 12 30). *OJ L, 350*, 60–71.

Czaika, M., & De Haas, H. (2013). The effectiveness of immigration policies. *Population and Development Review, 39*(3), 487–508.

Day, M. (14 October 2013). *Italy will use drones and tripled sea patrols in bid to halt deadly migrant boat wrecks*. Accessed 10 May 2014, from The Independent: http://www.independent.co.uk/news/world/europe/italy-will-use-drones-and-tripled-sea-patrols-in-bid-to-halt-deadly-migrant-boat-wrecks-8879844.html.

Dictionary Central. (2012). Media analysis definitions. Accessed 23 May 2014, from Dictionary Central: http://www.dictionarycentral.com/definition/media-analysis.html.

Dijstelbloem, H., & Meijer, A. (Eds.). (2011). *Migration and the new technological borders of Europe*. Basingstoke: Palgrave MacMillan.

Directive 95/46/EC of the European Parliament and of the Council of 24 October 1995 on the protection of individuals with regard to the processing of personal data and on the free movement of such data. (1995, October 24). *OJ L 281/31*, 31–50.

Ellioniki Radiophonia Tileorassi Anonimi Etairia (ERT AE) and Panellia Omospondia Syllegon Prossopikou ERT v Dimotiki Etairia Pliroforissis and Sotirios Kouvelas and Nicolaos Avdellas and others, C-260/89 (European Court of Justice June 18, 1991).

Ente Nazionale per l'Aviazione Civile (ENAC). *Regolamento mezzi aerei a pilotaggio remoto* (1st ed., 16.12.2013, 2nd ed., 16.07.2015). https://www.enac.gov.it/repository/ContentManagement/information/N122671512/Reg_APR_Ed%202_2.pdf. Accessed 29 Sept 2015.

European Commission. (2013a). Communication from the Commission to the European Parliament and the Council on the work of the Task Force Mediterranean (COM (2013) 869). Brussels.
European Commission. (29 November 2013b). EUROSUR: Protecting the Schengen external borders—Protecting migrant lives. Accessed 10 Sept 2014, from Press Release Database: http://europa.eu/rapid/press-release_MEMO-13-1070_en.htm.
European Commission. (7 October 2014a). Press release: Frontex Joint Operation Triton—Concerted efforts to manage migration in the Central Mediterranean. Accessed 13 Nov 2014, from http://europa.eu/rapid/press-release_MEMO-14-566_en.htm.
European Commission. (6 May 2014b). Remotely Piloted Aircraft Systems (RPAS). Accessed 15 May 2014, from European Commission—Enterprise and Industry: http://ec.europa.eu/enterprise/sectors/aerospace/uas/.
European Commission. (8 April 2014c). Communication from the Commission to the European Parliament and the Council. A new era for aviation. Opening the aviation market to the civil use of remotely piloted aircraft systems in a safe and sustainable manner (COM (2014) 0207). Brussels.
European Union Agency for Fundamental Rights. (2013). *Fundamental Rights at the Europe's Southern Sea Borders*. Luxembourg: Publication Office of the European Union.
Fink, M. (2012). Frontex working arrangements: Legitimacy and Human Rights concerns regarding "Technical Relationships". *Merkourios, 28*(75), 20–35.
Finn, R. L., & Wright, D. (2012). Unmanned aircraft systems: Surveillance, ethics and privacy in civil applications. *Computer Law & Security Review, 28*, 184–194.
Fotiadis, A., & Ciobanu, C. (9 January 2013). *Closing Europe's borders becomes big business/ people pay for research against migrants*. Accessed 10 Sept 2014, from Hans & Tamar Oppenheimer Chair in Public and International Law: http://oppenheimer.mcgill.ca/Closing-Europe-s-Borders-Becomes?lang=En-US.
Freedom House. (2013). Map of freedom. Accessed 19 May 2014, from Freedom House: http://www.freedomhouse.org/sites/default/files/Map%20of%20Freedom%202013%2C%20final.pdf.
Frontex. (2 March 2012). RPAS border surveillance workshop—invitation. Accessed 10 Sept 2014, from Frontex: http://frontex.europa.eu/news/rpas-border-surveillance-workshop-2012-invitation-u4Ibe1.
Frontex. (2013a). *General Report 2012*. Warsaw: Frontex.
Frontex. (2 October 2013b). To the rescue…. Accessed 10 Sept 2014, from Frontex: http://frontex.europa.eu/feature-stories/to-the-rescue–ILWGXf.
Frontex. (2014a). Border surveillance. Accessed 23 April 2014, from Frontex: http://frontex.europa.eu/research/border-surveillance.
Frontex. (2014b). Invitation for Border Surveillance Workshop 2014. Accessed 10 Sept 2014, from Frontex: http://frontex.europa.eu/news/invitation-for-border-surveillance-workshop-2014-OfPDf6.
Frontex. (2014c). Third countries. Accessed 10 Sept 2014, from Frontex: http://frontex.europa.eu/partners/third-countries.
Frontex. (28 August 2014d). Concept of reinforced joint operation tackling the migratory flows towards Italy: JO EPN-Triton. Operations Division of Frontex. Accessed 1 Jan 2015, from Avvenire.it: http://www.avvenire.it/Cronaca/Documents/JOU%20Concept%20on%20EPN-TRITON%20%282%29.PDF.
Ghelli, F. (17 October 2013). *Das Flüchtlingselend kommt Italiens Rüstrungsindustrie gelegen*. Accessed 13 May 2014, from Zeit Online: http://www.zeit.de/politik/ausland/2013-10/italien-fluechtlinge-mare-nostrum.
Google Spain SL and Google Inc. v Agencia Espanola de Protección de Datos (AEPD) and Mario Costeja González, C-131/12 (European Court of Justice May 13, 2014).
Hayes, B. (2006). *Arming big brother. The EU's Security Research Programme*. (W. van der Zeijden, Ed.) http://www.tni.org/sites/www.tni.org/files/download/bigbrother_0.pdf.
Hayes, B., & Vermeulen, M. (2012). *Borderline. EU border surveillance initiatives. An assessment of the costs and its impact on fundamental rights*. Berlin: Heinrich-Böll-Stiftung.

Hayes, B., Jones, C., & Töpfer, E. (2014). *Eurodrones Inc*. Amsterdam: Transnational Institute & Statewatch.

Heijer, M. d. (2011). *Europe and extraterritorial asylum*. Leiden: Universiteit Leiden.

Howell, J. L. (25 April 2014). *Spain: Authorities working on future unmanned aerial vehicle (UAV) regulation*. Accessed 10 Sept 2014, from Bird & Bird: http://www.twobirds.com/en/news/articles/2014/spain/spain-authorities-working-on-future-unmanned-aerial-vehicle-regulation.

ICAO. (2014). About ICAO. Accessed 15 May 2014, from International Civil Aviation Organization (ICAO): http://www.icao.int/about-icao/Pages/default.aspx.

International Maritime Organization. (1985). *International convention on Maritime search and rescue*. Chile: International Maritime Organization.

Kenk, V. S., Križaj, J., Štruc, V., & Dobrišek, S. (2013). Smart surveillance technologies in border control. *European Journal of Law and Technology, 4*(2), 1–12.

Kolev, Z. (2012). *RPAS for European border surveillance. Challenges of introducing RPAS technology in an operational context*. Presentation, Frontex, Research and Development Unit, Paris.

Laitinen, I. (March 2013). Border control in practice—The role of technology. (G. Watson (Ed.)) *Government Gazette*, 49. London: Centre for Parliamentary Studies.

Léonard, S. (2010). Frontex and the securitization of migrants through practices. *European Security, 19*(2), 231–254.

Mamakouka, I.-A. (12 April 2014). *Greece to use drones for immigration control*. Accessed 18 June 2014, from Greek Reporter: http://greece.greekreporter.com/2014/04/12/greece-to-use-drones-for-immigration-control/.

Marin, L. (2011). Policing EU's external borders: A challenge for the rule of law and fundamental rights in the area of freedom, security and justice? An analysis of Frontex joint operations at the Southern Maritime Border. *Journal of Contemporary European Research, 7*(4), 468–487.

Marin, L. (2014). Effective and legitimate? Learning from the lessons of 10 years of practice with the European arrest warrant. *New Journal of European Criminal Law, 5*, 327–348.

Marin, L. (2015). The 'metamorphosis' of drone: Challenges arising from the deployment of drone technology in border surveillance. An exploratory study. In D. Bowman, A. Rip, & E. Stokes (Eds.), *Embedding and governing new technologies: A regulatory, ethical & societal perspective*. Singapore: Pan Stanford Publishing (Head Office).

Meijers Committee. (2012). Note on the proposal for a regulation establishing the European Border Surveillance System (COM (2011) 0873). Utrecht, The Netherlands.

Moreno-Lax, V. (2014). *Seeking asylum in Europe: Border controls and refugee rights under EU law*. Oxford: OUP.

Mungianu, R. (2013). Frontex: Towards a common policy on external border control. *European Journal of Migration and Law, 15*(4), 359–385.

New Europe Online. (11 April 2014). Greece to use drones to control immigration. Accessed 14 May 2014, from New Europe Online: http://www.neurope.eu/article/greece-use-drones-control-immigration.

Nielsen, N. (14 January 2013a). *Frontex chief looks beyond EU borders*. Accessed 14 May 2014, from EUobserver: http://euobserver.com/fortress-eu/118471.

Nielsen, N. (19 March 2013b). *EU border agency keen to expand Mediterranean Sea surveillance*. Accessed 10 Sept 2014, from EUobserver: http://euobserver.com/justice/119473.

Nielsen, N. (3 September 2014). *EU migrant mission will not replace Mare Nostrum*. Accessed 11 Jan 2015, from EUobserver: https://euobserver.com/justice/125456.

Nolin, P. C. (2012). *Unmanned aerial vehicles: Opportunities and challenges for the alliance. Special Report*. Canada: NATO Parliamentary Assembly.

Oxford University Press. (2014). Data mining. Accessed 23 May 2014, from Oxford Dictionaries: http://www.oxforddictionaries.com/definition/english/data-mining?q=data+mining.

Plunkett, J. (3 October 2013). *Protecting Spanish borders with drones*. Accessed 10 Sept 2014, from Euro Weekly News: https://www.euroweeklynews.com/news/costa-de-almeria/item/115945-protecting-spanish-borders-with-drones.

Purvis, C. (19 August 2011). *Spain considering UAVs to supplement its maritime border security lineup*. Accessed 10 Sept 2014, from Security Management: http://www.securitymanagement.com/news/spain-considering-uavs-supplement-its-maritime-border-security-lineup-008909.

Regulation (EC) No 45/2001 of the European Parliament and of the Council of 18 December 2000 on the protection of individuals with regard to the processing of personal data by the Community institutions and bodies and on the free movement of such data. (2001, December 18). *OJ L 8/1*.

Regulation (EC) No 562/2006 of the European Parliament and the Council of 15 March 2006 establishing a Community Code on the rules governing the movement of persons across borders (Schengen Borders Code). (2006, April 13). *OJ L 105*, 1–32.

Regulation (EU) No 1168/2011 of the European Parliament and of the Council of 25 October 2011 amending Council Regulation (EC) No 2007/2004. (2011, October 22). *OJ L 304*, 1–17.

Regulation (EU) No 1052/2013 of the European Parliament and of the Council of 22 October 2013 establishing the European Border Surveillance System (EUROSUR). (2013, November 6). *OJ L 295*, 11–26.

Sauter, W. (2013). Proportionality in EU law: A balancing act? In C. Barnard, A. Albors-Llorenc, M. Gehring, & R. Schütze (Eds.), *Cambridge yearbook of European legal studies 2012–2013* (pp. 439–466). Oxford: Hart Publishing.

Scherer, S. (17 June 2014). *Italy to push EU to lead sea mission to rescue migrants*. Accessed 18 June 2014, from Swissinfo.ch: http://www.swissinfo.ch/eng/news/international/Italy_to_push_EU_to_lead_sea_mission_to_rescue_migrants.html?cid=38806102.

Souliotis, J. (20 June 2014). *Αγορά μη επανδρωμένων αεροσκαφών από ΕΛ. ΑΣ.* Accessed 24 Aug 2014, from Kathimerini Greece: http://www.kathimerini.gr/772876/article/epikairothta/ellada/agora-mh-epandrwmenwn-aeroskafwn-apo-elas.

Sternstein, A. (12 June 2012). *Freeze border security drones purchases, IG says*. Accessed 30 July 2014, from Nextgov.com: http://www.nextgov.com/defense/2012/06/freeze-border-security-drone-purchases-ig-says/56212/.

UK Civil Aviation Authority: Directorate of Airspace Policy. (2014). Environmental information sheet—number 12 + 13—Unmanned Aircraft. Accessed 25 May 2014, from Civil Aviation Authority: http://www.caa.co.uk/docs/7/EIS%2012.pdf.

UNHCR. (31 January 2012). Mediterranean takes record as most deadly stretch of water for refugees and migrants in 2011. (S. Wilkes (Ed.). Accessed 10 Sept 2014 from UNHRC. The Refugees Agency: http://www.unhcr.org/4f27e01f9.html.

Wallace, J., & Wild, S. E. (2010). *Webster's new world law dictionary*. Hoboken: Wiley Publishing Inc.

Weaver, O. (1995). Securitization and desecuritization. In R. Lipschutz (Ed.), *On security* (pp. 46–86). New York: Columbia University Press.

Luisa Marin is an assistant professor at the Department of Law of the School of Law, Economics and Governance, University of Utrecht, the Netherlands. Previously, she has served as an assistant professor of European Union (EU) law at the University of Twente, and as a post-doc researcher at the University of Verona, where she also defended her PhD on the principle of mutual recognition in criminal matters. She has been a visiting researcher at the European University Institute (EUI, Fiesole) and the University of Helsinki, and regularly presents her research at European (and American) universities and research institutions. Luisa's research interests cover several policies of the area of freedom, security and justice, from judicial cooperation in criminal matters to border surveillance, the governance of irregular migration and crimmigration. She has widely published on the European Arrest Warrant, and its impact on fundamental rights, and, more recently, on border surveillance, Frontex and on technological developments (such as drone technology) within this context and its ethical, policy and legal implications. She is also a member of the Meijers Standing Committee, a network of independent experts in the areas of international law, migration law and criminal law.

Kamila Krajčíková BA, BSc currently works as a consular assistant and is a master's degree candidate in social sciences at the Heinrich Heine University Düsseldorf. Previously, Kamila defended her double bachelor's diploma in public administration at the University of Twente and the University of Münster. Her bachelor thesis engaged with the topic of border surveillance by Frontex with the use of drone technology and its impact on fundamental rights and civil liberties within the context of EU law.

Part III
Drones, the "War on Terror" and Public International Law

Chapter 7
Death from the Sky: International Legal and Practical Issues on the Use of Armed Drones

Mélanie De Groof

Introduction

> [T]he present sense of uncertainty as to the applicable legal standards, the rapid development and proliferation of drone and robotic technology, and the perceived lack of transparency and accountability of current policies have the potential of polarizing the international community, undermining the rule of law and, ultimately, of destabilizing the international security environment as a whole. (Melzer 2013, p. 1)

Early September 2014, the leader of Al-Shabaab was killed in Somalia. According to the United States (US) Department of Defense, Ahmed Abdi Godane's death was the result of a "US military targeted airstrike" and "reflects years of painstaking work by our intelligence, military and law enforcement professionals" (The White House 2014). The Pentagon confirmed that the attack was carried out by Special Operations forces using both manned and unmanned launching missiles and precision bombs on a gathering of Al-Shabaab commanders (International New York Times 2014b). Late September 2014, US drones were spotted over Islamic State (IS)–controlled territories in Syria and Iraq (International New York Times 2014c; Turkish Weekly 2014). Different sources have reported that American fighter jets and armed Predator and Reaper drones, flying alongside warplanes from Arab and European allies, struck a broad array of IS targets, including IS militants (International New York Times 2014a). These are but two examples of recent extraterritorial drone missile strikes which resulted in the death of individuals.

On an almost weekly basis, the international media makes mention of targeted killings of supposed terrorists in, among others, Yemen, Somalia, Pakistan, Iraq,

This chapter further elaborates on points first made in a note published at GRIP in May 2014 ("Utilisation des drones armés: considerations juridiques et pratiques")

M. De Groof (✉)
The Group for Research and Information on Peace and Security (GRIP), Brussels, Belgium
e-mail: m.degroof@grip.org

A. Završnik (ed.), *Drones and Unmanned Aerial Systems*,
DOI 10.1007/978-3-319-23760-2_7

Afghanistan, and the Palestinian territories by drones deployed by distant States.[1] States turn to armed drones because of their free-ranging capacity to eliminate human targets anywhere on the planet (EUI Seminar 2013). The detailed analysis of actual State practice shows that armed drones kill individuals both in situations which do and which do not reach the armed conflict threshold. The result is that the world has become an open battlefield without borders.

Besides important ethical questions, the diverse uses of armed drones raise numerous legal issues, which to this day remain underexposed. Targeted killing by means of armed drones triggers questions in the domain of international law, and more specifically in the areas of international humanitarian law, human rights law, and the use of force. Analysis of States' obligations under these different branches of international law allows for the conclusion that a State can, under certain circumstances, violate its international obligations if it uses armed drones abroad. Hence, when a drone strike is in violation of international law, the question of international State responsibility also arises.

Therefore, the purpose of this chapter is twofold: First, it aims to clarify whether, and under what circumstances, a State is authorized to use armed drones beyond its own borders for the sake of killing, by establishing a clear dividing line between the legal and illegal use of unmanned combat aerial vehicles (UCAV).[2] Second, it considers some practical implications and legal consequences of an illegal use of armed drones. Actual State practice and doctrinal debate are the red lines across this analysis, which departs from some specific case studies. Interviews with—and views raised at conferences by—experts, scholars, and policymakers at both the national and international level complement this research.[3]

This chapter first focuses on the different types of drones and their different characteristics (section "Different Types of Drones"). Second, it offers a panorama of the different targets of drone strikes (section "Different Targets of Drone Attacks") and the various contexts in which armed drones are used (section "Drone Attacks in Different Contexts"). Third, this chapter analyzes the legality of drone strikes (section "The (Il)Legality of the Use of Armed Drones"). The permissibility of drone strikes is analyzed in view of international human rights law, international humanitarian law, and international law governing the use of force. Thereafter, this chapter examines if, and under what conditions, States may be held internationally responsible for illegal drone strikes (section "Death from the Sky and State Responsibility"). Finally, this chapter concludes with some recommendations for enhancing the respect of international law and promotion of transparency and further accountability in

[1] Among others, on 14 November 2012, the head of the military wing of Hamas was killed by an Israeli military drone strike in Gaza City (Vilmer 2013, p. 121; The Telegraph 2012); a US drone strike in south Yemen killed at least 30 al-Qaeda suspects on 20 April 2014 (Al Arabiya 2012; Le Figaro 2014). On 28 October 2013, the main explosive expert for Al-Shabaab was killed in a suspected US drone strike in Somalia (*The Times* 2013; Le Monde 2013b). On 24 October 2013, a 68-year-old woman was killed in a US drone strike in Pakistan (Amnesty International 2014).

[2] This analysis is limited to the most important instruments of international law which delineate the legality of the use of armed drones.

[3] During most conferences and interviews, the Chatham House Rules applied.

the field of drone strikes and articulates future challenges both jurists and political leaders will have to face (section "Concluding Observations").

Different Types of Drones

General

A drone or an unmanned aerial vehicle (UAV) is an aircraft with no pilot on board, which is either remote controlled or preprogrammed and used in diverse missions. Although there exists no universally accepted definition of drones, the following elements appear in most descriptions: (a) it is a mobile engine; (b) without a crew on board; (c) remotely piloted or guided entirely autonomously through programming; (d) equipped with systems to collect information and/or weapon systems; (e) used for military or civil purposes.[4] A drone forms part of a complex system consisting of both an aircraft and a land engine from which individuals can program and control the aircraft's flight mission (Mekdour 2011, p. 15). For example, in Afghanistan and Iraq, British and American ground crews launch Reaper and Predator drones from the conflict zone itself, but the engines are subsequently manipulated by individuals that are based in the Nevada desert (Drone Wars UK web site). The absence of a human presence on board allows drones to undertake missions of longer duration and/or with higher risks than vehicles with a pilot on board (the UAV web site). In principle, the absence of risks to the individual piloting the engine allows for a more balanced decision-making process. Another advantage of drones is that they are much cheaper than traditional airplanes.

Drones are ranked on the basis of several criteria, such as their speed, their endurance, their range, the altitude they can reach, their takeoff mode, and their use. While there are many different types of drones, according to their use and mission, two broad categories of drones may be distinguished; that is, on the one hand, the drones used for *civil* purposes and, on the other, the drones used for *military* purposes (Table. 7.1).

Civil Drones

Civil drones are used in order to capture and collect aerial data. Drones provide multiple opportunities for civil usage, such as the surveillance of illegal migration, border traffic, sporting events, and electoral processes; the search for victims after natural disasters; inspections in agricultural, industrial, and artistic sectors; the

[4] This definition is inspired by the writings of several authors, among others, Boothby 2012, p. 231; Commission spécialisée de terminologie et de néologie du ministère français de la Défense 2011, p. 1496; Hoppe 2008, p. 45; US Department of Defense 2010.

Table 7.1 Types of drones

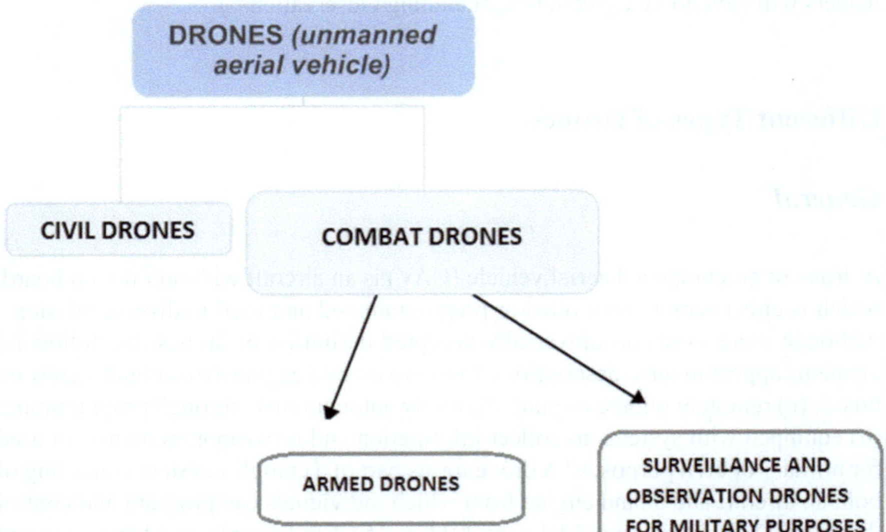

prevention and evaluation of fires and other disasters; and so forth. (Mekdour 2011, p. 11). Civil drones can also be used for commercial purposes (Forbes 2014; The Washington Post 2014). Civil drones are being used by both State and non-State actors and are increasingly popular as demonstrated, for example, by the recent announcement of DHL, Amazon, and Google to launch drone delivery services.

Combat Drones

Combat drones or unmanned combat aerial vehicles are used for military engagement purposes and are equipped with weaponry or information-gathering systems (CICDE 2012; Nogrix and Bergé-Lavigne 2006). Combat drones may thus be used for armed purposes, but this is not necessarily the case. Within the category of combat drones, one may distinguish between those that are armed and those that are not armed; the latter are also known as "surveillance and observation drones for military purposes" (Nogrix and Bergé-Lavigne 2006). The real-time surveillance of the Taliban, for example, was achieved through the operation of these nonarmed combat drones. Contrary to armed drones, to which only a few States would have recourse to today, surveillance and observation drones are currently used by many States. Moreover, there is indeed clear evidence that the technology of combat drones has also fallen into the hands of non-State groups such as Hezbollah or Hamas (France 24 2012; Aljazeera 2014; The Jerusalem Post 2014; IKV Pax Christi 2011, p. 12). The fact that States have lost the monopoly on combat drones is a source of serious concern as it entails that over time terrorist and extremist groups will also be able to scrutinize developments abroad, or even engage in attacks, without their physical presence on foreign territory. This physical absence on the ground will most likely make non-State actors harder to catch and lead them to conduct more risky military operations than those that require a human presence.

Armed Drones

Armed drones are armed military aircraft that are manipulated at a distance from the opponent (Nogrix and M. Bergé-Lavigne 2006; Vilmer 2013, p. 119). Contrary to civil drones or military surveillance and observation drones, armed drones are in a position to conduct air strikes by delivering missiles, bombs, or other explosives. For the remote pilot to obtain all the necessary information for the mission, armed drones are also equipped with other specific devices such as multispectral, infrared optoelectronic or electromagnetic sensors, video recorders, and a camera (Sartre 2013, p. 440–441). While the flight can be almost automatic, combat functions—on the other hand—require human control in real time (Sartre 2013, p. 440). The fact that the combat functions are closely controlled by a human distinguishes the drone from those vehicles called "autonomous lethal robots" that are being developed, which, once activated, can select and attack targets without further intervention on the part of a human operator (Kiergaard 2013; Human Rights Watch 2012; UN News Centre 2013). However, as was explained by Patrice Sartre, to date, "technology is not yet able to confer upon them [the drones] an autonomous capacity to find and recognize the target, and to then judge the opportunity of the shooting, namely with regard to the risks of fratricide and collateral damage" (Sartre 2013, p. 441).[5]

In recent years, there has been a steep increase in the number of attacks by means of armed drones. According to the experts, drone strikes have increased sixfold under the presidency of Barack Obama, and armed drones have now become an essential element in the fight against terrorism led by the Pentagon (Grand 2013; Wall Street Journal 2011; Le Monde 2013a; The Bureau of Investigative Journalism web site). Experts also agree that the technology of armed drones is equally proliferating exponentially worldwide. Although to date armed drones are used only by a few States, between 40 and 76 States are leaning towards developing combat drone systems (EUI Seminar 2013).

This chapter focuses on the latter category of drones—armed drones—and aims specifically to highlight the legal issues that arise following the choice by a State to use an armed unmanned aerial vehicle in an extraterritorial context. Before establishing the contours of the legal use of armed drones, three observations must be made. First, a drone is not a weapon in the strict sense; a drone is first and foremost a system of weapon transmission. It is therefore not the system of drones itself that poses a threat to the proper and legal application of international human rights law,

[5] [Translation]

international humanitarian law, and the principles of nonuse of force and nonintervention under public international law. It is the incorrect use or the abuse of this new technology that can generate illegal situations (Grand 2013; EUI Seminar 2013). It is therefore important to distinguish between technology on one hand and how this technology is used on the other. As J.J. Vilmer rightly stated, "[i]f the drone is today the 'aerial figure of evil', it is largely due to *its use* by the CIA in Waziristan, Yemen and Somalia" (Vilmer 2013, p. 119; Henrotin 2013, p. 50–52).

Second, when used correctly, armed drones can present certain advantages. Among others, they reduce the risk to the military, which is no longer sent to the battlefield. Thanks to the advanced technology embedded in the drones (sensor accuracy, long flight range, surveillance capacity, cool and calm shooting, etc.), they also allow for—in principle—strikes that are more discriminatory, proportionate, and respectful of the precautionary principle than more conventional weapons such as aircraft, missiles, helicopters, machine guns, etc. (Vilmer 2013, p. 119; Henrotin 2013, pp. 50–52). By its nature and assigned characteristics, an armed drone is therefore not an illegal weapon system (CAVV 2013, p. 2; Vilmer 2013, p. 119; EUI Seminar 2013). Indeed, armed drones are perfectly able to meet the demands of the general principles of international humanitarian law, namely necessity, distinction, proportionality, precaution, and the prohibition of causing unnecessary suffering (Vilmer 2013, p. 351–372).

Third, this observation is linked to the previous two; practice shows that these potential advantages of the use of armed drones are not always exploited. Indeed, the civilian casualties and human suffering—physical as much as psychological—caused by drone attacks are precisely the main triggers of a growing reluctance to the use of these devices (Human Rights Watch 2009, 2013; Amnesty International 2013; CNN 2012). According to various sources, between 2500 and 4000 people—many of them civilians—have been killed by drone attacks conducted by the USA in the last decade (The Telegraph 2013; The Huffington Post 2014; The Bureau of Investigative Journalism 2014a; Taking Note 2014). However, because of the lack of transparency at the political level, it is at this moment in time not possible to state with certainty that the number of civilian casualties is massively disproportionate. The differences in proportions mentioned by various reports vary from 15 to 90% of victims as being civilians (Vilmer 2013, p. 119; International Human Rights and Conflict Resolution Clinic and Global Justice Clinic. 2012).

Different Targets of Drone Attacks

Depending on the target of the drone attack, literature differentiates between "signature strikes" and "personality strikes," a distinction that is mainly drawn to indicate the different *modus operandi* of drone strikes carried out by the USA (Columbia Law School Human Rights Clinic 2012, p. 8). With personality strikes the firing drone targets one specific person, whereas this is not at all the case in signature strikes.

"Personality strikes" or "targeted attacks on a person" target individuals whose identity is known with great certainty (Columbia Law School Human Rights Clinic 2012, p. 8). According to officials of the American government, the person who makes the final decision to attack must have a "high degree of certainty" that the identified and searched for individual is situated where the drone will strike (Columbia Law School Human Rights Clinic 2012, p. 8). It is in this way, for example, that the USA killed Ahmed Abdi Godane in Somalia (2014) and Al-Qaeda's number two, Abu Yahya al-Libi, in North Waziristan in Pakistan (2012; International New York Times 2012; Le Monde 2012). American officials told the New York Times that Abu Yahya al-Libi was indeed the target of the attack, which nonetheless took the lives of five suspected insurgents (Le Monde 2012).

In contrast, during "signature strikes," attacks are launched against individuals—often together in groups—who, even if their precise identity is unknown, are presumed to be linked to a terrorist group following various behavioral patterns (travel, place of meeting, social circles and various contacts, transport of weapons, etc.) or visible characteristics (gender, age, religious symbols, etc.; CAVV 2013, p. 5; Human Rights Watch 2013, p. 86; EUI Seminar 2013). Strikes against individuals whose signature would be linked to terrorist activity or association constitute a significant proportion of the drone strikes conducted by the USA (Dommage civils web site; Wall Street Journal 2011; Columbia Law School Human Rights Clinic 2012, p. 9). The legality of signature strikes is very controversial, especially because this practice, first, does not require any specific knowledge concerning the participation of the targeted individuals in hostilities and would therefore violate the presumption of civilian status and, second, does not require the presence of an established and imminent threat (Human rights and Civil Rights Groups 2013; Heller 2013; Amnesty International 2013, p. 18; EUI Seminar 2013). The reluctance of the USA to disclose specific information when these strikes occur may be indicative of the discomfort of this State with regard to the legality of such a practice which, according to some sources, causes—sometimes innocent—casualties that are far greater than those caused by personality strikes States (Columbia Law School Human Rights Clinic 2012, p. 9; Human rights Watch 2013). The practice of signature strikes has a tendency to damage the relationship between the State using the drones and the State on which the attack is conducted, especially when civilians are affected (Wall Street Jounal 2011). Thus, there have been several instances of high tension between Pakistan and the USA as a result of the drone strikes that the USA was conducting in Pakistani territory in the context of the fight against terrorism (The Guardian 2012; France 24 2013). In Afghanistan, in November 2013, a drone attack in Helmand province caused civilian casualties, prompting President Karzai to refuse once again to sign a long-term security agreement with Washington as long as this shooting continued, despite an apology from the commander of American forces and of the North Atlantic Treaty Organization (NATO) in Afghanistan (International New York Times 2013).

Drone Attacks in Different Contexts

Drones have been used in situations of conventional armed conflict for decades. In the Vietnam War, for example, drones were sent out by American forces during the 1960s. Israel was using drones during the armed conflict against Lebanon in the 1980s. However, their use in these cases was limited to monitoring and observational purposes. It was only in late 2001, after the 9/11 attacks, that the USA decided to arm the drones. A few months later, the first combat mission using armed drones took place in Afghanistan (The Nation 2012; IKV Pax Christi 2011). Since then, armed drones have been used in several wars, both noninternational (Afghanistan, Libya, Mali, Syria, etc.) and international (Afghanistan, Iraq, etc.). To illustrate, between 2008 and 2012, the Central Intelligence Agency (CIA) and British crews have carried out almost 1200 drone strikes in Afghanistan, Libya, and Iraq (The Bureau of Investigative Journalism 2012). The use of drones in armed conflicts is not particularly worrying and gives place to the same legal concerns (for example, the distinction between combatants and civilians, proportionality and precautions in attack, etc.) that arise when other means of warfare are used, including air attacks using inhabited systems (Grand 2013).

On the other hand, armed drones also kill individuals that are considered enemies without there being any declaration of war, or without there being an armed conflict clearly limited both in time and space. The current debate—both in the political and in the legal doctrinal arena—focuses mainly on the use of combat drones in such situations. According to many experts, it is outside the context of any armed conflict that the majority of drone attacks take place and that the vast majority of casualties are reported (EUI Seminar 2013). To illustrate, the CIA led, according to the Bureau of Investigative Journalism, 392 drone strikes in Pakistan between 2004 and 2014, causing the death of 2354–3806 people (including 416–957 civilians) and injuring 1104–1662 people; between 65 and 77 confirmed drone attacks in Yemen from 2002 to 2014, killing 339–494 people (including 64–83 civilians) and injuring between 92 and 125 people; between 6 and 9 drone attacks in Somalia from 2007 to 2014, killing 16–30 people (including 0–1 civilian; The Bureau of Investigative Journalism 2014b). Of interest, there is no armed conflict between any of these States and the USA. An important question that arises is on what legal grounds are these drone attacks based? Washington is of the opinion that it is the worldwide "war against terrorism" that justifies the use of armed drones in these territories. This argument is rejected by most experts in the field of public international law, as discussed next.

The (Il)Legality of the Use of Armed Drones

Given the many benefits for the user State, the logic of using drones is overwhelming and seems to have reached a point of no return (The Bureau of Investigative Journalism 2014b). The real challenge today is thus not so much to delegitimize the use of drones but rather to regulate their use. Under international law, although armed drones may be used for legal purposes, they are too often used illegally (EUI Seminar 2013). As will be explained in greater detail below, in particular in a war context, targeted killings involving drones may certainly be justifiable, but many drone attacks are carried out in violation of the law regulating the use of force, the law of armed conflict, and/or human rights law. It is therefore quite surprising that, while the use of drones for targeted killings is often criticized and questioned by the media and advocacy groups, it is not, or barely, criticized or condemned by other States. For example, not a single Member State of the European Union (EU) has officially and openly condemned the armed drone attacks of the USA and the UK in foreign States, not even when there was sufficient evidence that it was mostly civilians who were killed. As has been argued by Ben Emmerson, United Nations (UN) Special Rapporteur on the promotion and protection of human rights and fundamental freedoms while countering terrorism, this silence may be interpreted as acquiescence (Emmerson 2013). It is therefore important to articulate the limits of the use of armed drones and to prevent the materialization of such an overall acceptance of the unlawful use of force.

When a State makes use or considers making use of combat drones, it must meet its obligations under international law. This body of law establishes binding rules regarding the use of force, human rights, and humanitarian law—three separate legal spheres of interest in the case of armed drone use (Melzer 2013, p. 21). Where international law governing the use of force determines the legality of strikes according to the territorial State, international humanitarian law and human rights law determine the legality of the use of drones according to citizens as potential victims of weapons dropped by drones (Melzer 2013, p. 21). Thus, when establishing the legal implications of the use of drones, these three different sets of norms must be taken into account. The aim of this section is to analyze the main international instruments that define the legality of using armed drones. In doing so, it reveals some gaps in the legal and political reasoning of the USA, the State which now has the largest fleet of combat drones and which, often erroneously, bases its drone strikes on the concepts of consent of the territorial State, legitimate self-defense, and the war against terrorism (Table 7.2).

It should be noted that the legal analysis below exceeds the strict regime of rights and obligations that must be taken into account by State "users" of drone technology. Indeed, a well-defined theoretical framework for the legality of targeted attacks also allows State "producers" of armed drones (or of its components) including several EU Member States—to best respond to import requests from the States using these drones. Finally, the legal framework also defines the rights and obligations of territorial States that—before agreeing to any possible strikes by third countries—are obliged to assess the legality of such consent under international law.

International Law on the Use of Force: State Sovereignty and the Principles of Nonintervention and Nonuse of Force

The use of armed drones raises questions about the legality under the international law governing the use of force and the relations between States. It has sometimes

Table 7.2 Evaluation of the legality of the use of armed drones. (authors: Mélanie De Groof & Emmanuel De Groof)

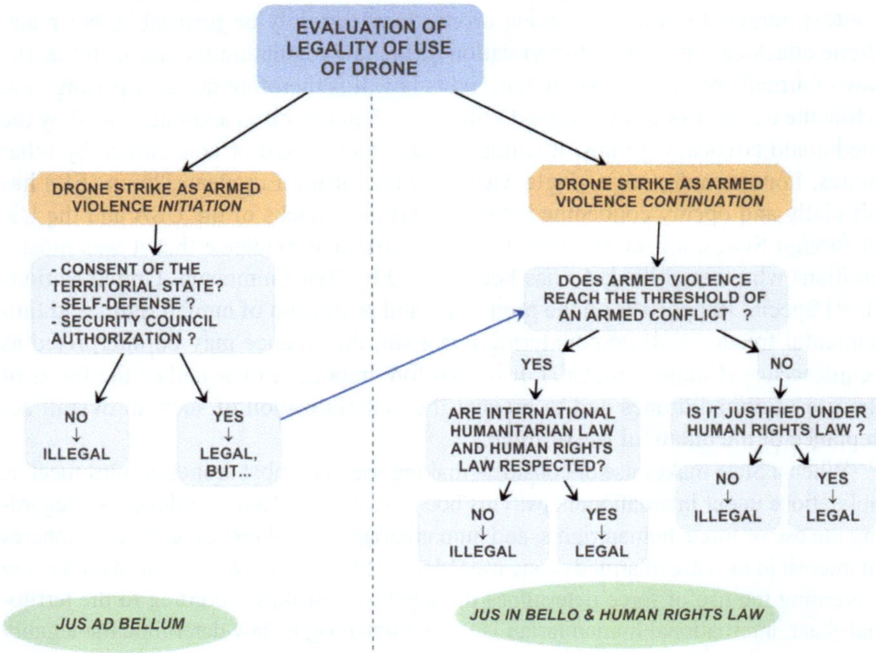

been argued that the most serious objection that one may make to the use of armed drones is precisely that it lowers the threshold for the use of force *(jus ad bellum)* to a degree that renders the criterion of last resort inapplicable (Vilmer 2013, p. 126; Brunstetter and Braun 2011, p. 346).

Article 2.4 of the Charter of the United Nations sets out the main rule regarding the use of force in international relations, where it declares: "All Members shall refrain in their international relations from the threat or use of force against the territorial integrity or political independence of any state, or in any other manner inconsistent with the Purposes of the United Nations." Do the principles of non-intervention and nonuse of force articulated in the Charter then limit the right of third States to use armed drones in foreign territories? In international law, there are several circumstances in which the use of force, including through air strikes, is legitimate. This section will briefly analyze (1) the question of whether the consent of the government of the State on whose territory the attacks are carried out, could render drone strikes legitimate, (2) the concept of self-defence, and (3) the notion of authorization of the Security Council of the UN.

Intervention upon Request by the Government The first question that arises is whether the State on whose territory the attacks are carried out has expressly consented to the drone strikes. In practice, the answer varies from one case to another. For example, in September 2012, Yemeni President Mansour Hadi publicly acknowledged to have given consent to deadly strikes perpetrated by American drones on Yemeni territory (PressTV 2012). By contrast, the drone attacks conducted by the USA in Pakistan may have not all been approved by the government (The Guardian 2013). The situation differs once more in Somalia, where for many years there was no effective government that was even in a position to allow drone attacks to be perpetrated on its territory.

If the user State has received approval from the territorial State, the principles of nonintervention and nonuse of force are not necessarily violated. Under international law, the government of one State may request assistance of other States in order to preserve law and order or to defend its borders against an outside attack.[6] For both the purposes, it may validly request the presence of foreign military forces on its own territory and consent to this third State using force on its territory (Jennings and Watts 1992, p. 437; Gray, pp. 84–85). However, such an intervention based on the request or on the consent of the territorial State is only allowed in cases where several conditions are met.[7] First, the government of the State in which the intervention would occur must have given its consent, either prior to any intervention or on an ad hoc basis. Second, consent must be given freely and may not be the result of pressure or influence exerted by the intervening State. Third, the request must come from the legitimate government of the State, which represents the population and controls the territory (Kunig 2013, p. 7). These are therefore factual elements that must be considered in light of the particular situation, an exercise that is sometimes convenient but often difficult. What about the examples of drone attacks in Libya, Syria, or Somalia, where the official government has no control over large parts of its territory? In sum, under the law governing the use of force, the legality of drone attacks in cases where there has been consent varies from case to case. Moreover, even if the user State has effectively received valid approval from the territorial State, this does not mean that the drone attack is necessarily lawful under public international law, because besides the law regulating the use of force, every drone attack must also pass the legality test under international humanitarian law—at least if it is applicable—and under human rights law, two branches of international law discussed below (sections "International Humanitarian Law" and "International Human Rights Law").

On the other hand, when the user State has not received the approval of the territorial State, a drone attack will only be legal if the measure meets the requirements established, first by the other *jus ad bellum* rules and second by the *jus in bello* which sets out rules for the conduct of military operations. Moreover, the legality of the use of drones will also be conditioned by international human rights law (EUI

[6] See, for example, Security Council Resolution 387, S/RES/387, 31 March 1976 (it is 'the inherent and lawful right of every State, in the exercise of its sovereignty, to request assistance from any other State or group of States'); International Court of Justice, *Nicaragua v. United States of America,* 27 June 1986, § 246 (intervention is 'allowable at the request of the government of a State'); International Court of Justice, *Democratic Republic of the Congo v. Uganda*, 19 December 2005, § 42 and 53.

[7] International Court of Justice, *Nicaragua v. United States of America,* 27 June 1986, § 246. See also Kunig (2013), "Prohibition of Intervention", Max Planck Encyclopedia of Public International Law, p. 6.

Seminar 2013). The next section thus addresses the following question: under what circumstances may a drone attack constitute a case of self-defence or may an attack be permitted by the Security UN Council *(jus ad bellum)*.

Authorization of the Security Council Under the UN Charter, the authorization of the Security Council may also justify intervention in the sphere of sovereignty of another State (Chapter VII, Charter of the United Nations). Therefore, depending on the terms used in the resolutions of the Security Council, a drone intervention may be legitimate. For example, if, following a genocide taking place in State A, the Security Council authorizes States B and C to resort to force in order to put an end to the genocide, these intervening States may resort to the use of drones to carry out this mandate. It is important to underline that the Security Council need not explicitly authorize the use of drones because when the Security Council authorizes the use of force, the resolutions do not specify the type of weapons that may be used (CAVV 2013, p. 4).

Self-Defense Under international law, self-defence is a recognized justification for intervention in the sphere of sovereignty of another State (Article 51, Charter of the United Nations). Therefore, an armed aggression by a State can justify the use of force against this State, including through the use of armed drones, so long as this action of self-defense meets the requirements of self-defense. An action in self-defense is only permitted in situations in which there is an imminent armed attack and if it meets the criteria of necessity, proportionality, and immediacy as established by customary international law.[8] Also, the armed drone attack must be limited to rebutting the armed attack, and the action in self-defense is permitted so long as the Security Council has not taken the measures necessary to maintain international peace and security (Article 51, Charter of the United Nations).

The USA regularly invokes the right of self-defense to justify drone strikes conducted in foreign jurisdictions such as in Pakistan and Yemen. The act of killing people through drones, the USA argues, fits perfectly into the context and logic of the right to self-defense (EUI Seminar 2013). However, this argument raises a number of legal questions, such as (1) may an armed attack be accomplished by a terrorist group; (2) is there now a right to self-defense against non-State actors; (3) what is the gravity threshold for an armed attack to be classified as aggression; (4) what does an "imminent" armed attack mean; and (5) does a right of anticipatory self-defense exist? (EUI Seminar 2013) All these questions lie at the core of a doctrinal and political debate that is constantly progressing.[9] The majority of experts and politicians agree that the concepts of "armed attack" and "self-defense" have evolved and must continue to evolve in order to address current challenges such as the emergence of "private terrorism" (Gery 2012, p. 16). However, there are major

[8] Authoritative case law of the International Court of Justice reaffirms this. See, for example, International Court of Justice, *Nicaragua v. United States of America,* 27 June 1986, § 194 and International Court of Justice, *Democratic Republic of the Congo v. Uganda*, 19 December 2005, § 147. On this topic, see also Gery 2012, p. 14–15; CAVV 2013, p. 4 and Gray 2008, p. 148–156.

[9] For an interesting overview and analysis of these questions, see Ruys 2013.

differences—at the academic, political, and diplomatic level—regarding the precise contours of these two concepts, which leaves room for different interpretations. In turn, this creates significant uncertainty about the legality of certain drone strikes, which must be assessed on a case-by-case basis. The scope of this chapter does not allow a thorough analysis of the concepts of self-defense or armed attack. It is important to recall, however, that the meaning and scope of the concepts of armed aggression and self-defense do not allow for the conclusion that each drone attack against terrorists (or suspected as such) is justified under the law governing the use of force. For example, it is highly uncertain that the drone attack in Somalia in January 2012 killing Bilaal al-Barjawi, the mastermind of the bombing in Kampala 2 years earlier, would pass the self-defense test. This, the argument that there would exist an imminent danger of armed attack by this individual or his terrorist group that would allow for an act of legitimate self-defense under Article 51 of the UN Charter, is not very tenable.

International Humanitarian Law If there is no agreement by the territorial state, both the *jus contra bellum* and the *jus in bello* should be fulfilled! This is the law as it stands and we should adhere to this! (EUI Seminar 2013)

If the *jus ad bellum* authorizes drone strikes, the *jus in bello* may further constrain the way in which the attacks were implemented (Van Schaack 2011, p. 281). This, of course, depends on whether or not international humanitarian law applies to the attacks. The US government often relies on concepts of international humanitarian law, such as the notion of "combatant," to justify the use of armed drones for killing purposes (Milanovic 2010). The question that thus arises is whether international humanitarian law, also known as the law of armed conflict, broadens or limits the right of States to use drones abroad. In some cases, international humanitarian law provides a legal basis for strikes perpetrated on foreign territory by a government, at least if that government is engaged in an armed conflict. In these cases, it does not matter whether such an attack was carried out through an inhabited system or not; in both cases, the same legal rules apply. For a targeted attack to be legal, the following conditions must all be met: (a) there exists an international or noninternational armed conflict; (b) between identified parties to the armed conflict; (c) the targeted person is related to a party to the conflict; (d) the targeted person performs military duties or is a civilian, but is nonetheless taking part in the hostilities; (e) the targeted person is killed in a location that can be described as , or has a substantial connection with, the current theater of war; (f) the attack is in conformity with the principles of necessity, distinction, proportionality, precaution, and the prohibition of causing unnecessary suffering (Scheinin 2013, p. 37; CAVV 2013, p .5).

As stated by the Advisory Committee on Issues of Public International Law (CAVV) of the Netherlands, the fact that a target has no opportunity to defend himself against a drone does not influence in any way the legality of the attack, as international humanitarian law does not require that opponents have equal chances on the battlefield (CAVV 2013, pp. 4 and 20). Still, considering the criteria listed above, international humanitarian law sets out very strict conditions that must be met for a drone attack to be allowed. Practice shows that these cumulative conditions are often not met in instances of drone attacks. In the following paragraphs, this argument is briefly explained.

Existence of an Armed Conflict In the context of armed conflict, the armed forces of the parties to the conflict are allowed to conduct drone strikes. By contrast, international humanitarian law does not justify a drone in a foreign country in situations where there is no armed conflict, as this body of law only applies in armed conflict. In other words, if an attack using drones was not requested or consented to by the territorial government, and if the attack is also not justified under the legal systems of self-defense or the authorization of the use of force by the Security Council (see section "International Law on the Use of Force: State Sovereignty and the Principles of Nonintervention and Nonuse of Force"), it then makes sense to analyze whether this attack fits well within the context of an armed conflict.

According to the USA, drone attacks perpetrated abroad—such as those in Yemen, Somalia, and Pakistan—are legal to the extent that the USA is involved in a new type of armed conflict, namely the worldwide war on terrorism (Emmerson 2013). To the extent that the conflict does not take place between two States but rather between a State and a non-State armed group, Washington believes that its drone attacks are perpetrated as part of a "transnational noninternational armed conflict" against Al-Qaeda. According to Washington, this gives them the right to target any fighter with ties to Al-Qaeda, in any part of the world.[10]

This argument, which is highly controversial, raises important questions. First, can we accept an armed conflict in any place and at any time? Second, who are the targets and do they know that they are at war with the USA? Third, can there be a conflict between the government of a State and a non-State actor given that this actor is abroad, in addition to being dispersed? In summary, the argument of the USA—if it is accepted—would require that international humanitarian law be entirely revised (Emmerson 2013). Indeed, the American concept of a war against terrorism tends to extend the applicability of international humanitarian law to all operations against terrorism, regardless of whether the legally required thresholds for armed conflict are met (Melzer 2013, p. 20). The following paragraphs develop this argument.

The International Criminal Tribunal for the former Yugoslavia (ICTY) in the Tadic case (1995) stated that "an armed conflict exists whenever there is resort to armed force between States, or protracted armed violence between governmental authorities and organized armed groups, or between such groups within a State".[11] This definition, which is accepted by international practice and doctrine, distinguishes between (1) a conflict between States, that is to say an international armed conflict, and (2) a conflict within one State, meaning a noninternational armed conflict. Indeed, current international humanitarian law only recognizes the existence

[10] Several authors have supported this argument. See, for example, Vilmer (2013), p. 126.

[11] International Criminal Tribunal for the former Yugoslavia, *Dusco Tadic case*, Interlocutory Appeal on Jurisdiction, 2 October 1995, § 70.

of two types of armed conflict, i.e. international armed conflicts and noninternational armed conflicts.

The concept of international armed conflict refers to armed hostilities between two sovereign States. This concept also includes all cases of partial or total military occupation of one State by another (Article 2 common to the four Geneva Conventions (1949)), as well as wars of national liberation (Article 1(4) of Additional Protocol I to the Geneva Conventions (1977)). Thus, when a drone attack targets a non-State actor abroad, it is difficult to argue that this is within the context of an international armed conflict.

The next question is whether the drone attacks on non-State actors abroad are covered by the law on noninternational armed conflicts. Noninternational armed conflict may exist (1) between governmental authorities and organized groups, or (2) between such groups within a State.[12] There is a general consensus within both international practice and legal doctrine that for armed violence to amount to an "armed conflict" and for the application of international humanitarian law to be therefore triggered, two essential conditions must be met: (1) the armed violence must reach a certain level of intensity; and (2) the groups involved in the conflict must attain a certain level of organization.[13] The authoritative jurisprudence on the matter concludes that "the determination of the intensity of a conflict and the organisation of the parties are factual matters which need to be decided in light of the particular evidence and on a case-by-case basis."[14]

It is difficult to establish whether the variety of groups affiliated to or inspired by Al-Qaeda—that is, the principal targets of US drone strikes—are sufficiently organized. The text of Article 3 common to the four Geneva Conventions (1949), which deals with "conflicts not of an international character,"[15] does not specify the required level of organization of fighting groups for armed clashes to amount to "armed conflict." International practice remains vague as well: The ICTY merely refers to "organized groups"; the Inter-American Commission on Human Rights (IACHR) stipulates that the armed confrontations must involve "relatively organized armed forces or groups"; the International Criminal Tribunal for Rwanda (ICTR) and the International Committee for the Red Cross (ICRC) refer to armed

---

[12] Ibid.

[13] See among others, International Criminal Tribunal for the former Yugoslavia, *Dusco Tadic case*, Interlocutory Appeal on Jurisdiction, 2 Octobre 1995, § 70; ICTY, *Ljibe Boskoski case,* Trial Judgment, 10 July 2008, § 177; International Criminal Tribunal for the former Yugoslavia, *Fatmir Limaj case,* Trial Judgement, 30 November 2005, § 84; International Criminal Court, *Prosecutor v. Thomas Lubanga Dyilo,* Trial Judgment, 14 March 2012, § 506. See also Moir (2002), at p. 43 and ICRC 2008.

[14] See, for example, International Criminal Tribunal for the former Yugoslavia, *Fatmir Limaj case,* Trial Judgement, 30 November 2005, § 90; International Criminal Tribunal for Rwanda, *George Rutaganda case,* Trial Judgment, 6 December 1999, § 93; International Criminal Tribunal for the former Yugoslavia, *Haradinaj, Balaj and Brahimaj case*, Trial Judgment, 3 April, 2008, § 63–100.

[15] Article 3 common to Geneva Conventions I–IV states: "In the case of armed conflict not of an international character occurring in the territory of one of the High Contracting Parties, each Party to the conflict shall be bound to apply, as a minimum, the following provisions: (…)."

forces which are "organized to a greater or lesser degree/extent."[16] Legal scholars also reflect haziness when it comes to assessing the organization requirement. Some scholars and national judges have questioned the application of Common Article 3 to conflicts involving unorganized groups (Geiss 2005, p. 491; Cahin 2007, p. 198) or have formulated a narrow interpretation of the "organization" requirement and argued that parties to the conflict must have a minimum military organization, responsible command, capacity to fulfil the minimum humanitarian standards, and territorial control (ICRC 1998, p. 9). If the assessment of the existence of an internal armed conflict depends on the latter interpretation, numerous armed hostilities that take place in Pakistan, Yemen, and Somalia would fall below the threshold of Common Article 3 for many armed factions operating in these territories do not comply with these stringent conditions.

The next question to be answered is whether the intensity criterion is met. International tribunals, in determining the intensity of a conflict, have taken into account factors such as the severity of the attacks, the spread of confrontations in a territory over a given period, the intensification of arming, etc. In practice, we can see that these elements are interpreted in a very strict manner. International tribunals, international and regional organizations, States, and even the ICRC recognize that there is an armed conflict only in cases where the legal criteria that define a noninternational armed conflict are unquestionably met. To illustrate, it was only in July 2012 that the ICRC for the first time labeled the Syrian conflict as a "noninternational armed conflict" (American Red Cross Blog 2012; BBC 2012; Aljazeera 2012). Thereafter, the UN and the most important regional organizations soon accepted the qualification as an armed confict, with all its implications in terms of legal rights and duties. Therefore, in light of the current intensity of the fighting in Pakistan, or in Yemen, it is uncertain whether the US drone attacks on the territory may be considered to take place within the context of an armed conflict (Amnesty International 2013, p. 9).

Finally, international law as it stands does not recognize the concept of armed conflict unlimited in time and space to which international humanitarian law would apply. In contrast, both noninternational and internal armed conflicts are delimitated by geographical and temporal boundaries.

In short, the argument that the USA is in a constant and global armed conflict with non-State actors abroad is not consistent with international humanitarian law in its current form. The meaning and scope of the concepts of both international and noninternational armed conflict clearly do not allow for each drone attack against suspected terrorists abroad to fall within the scope of international humanitarian law.

When a drone attack cannot be justified in this way under the law of war, the last question is whether international human rights law can be invoked to legally justify drone attacks (see section "International Human Rights Law"). Conversely, when a drone attack does take place within the context of an armed conflict, one will have to proceed to the analysis of other conditions also listed because, even in

[16] On this topic, *see* doctoral thesis of the author to be published in 2016.

times of war, there are rules that govern the conduct of military operations.[17] This brings us to the important analysis of the *jus in bello* when assessing the legality of drone attacks.

Jus in Bello A drone attack, even if it falls within the context of an armed conflict, is only legal if the rules governing the conduct of hostilities are complied with. This section is limited to pointing out some fundamental principles which—based on the information included in the reports and the press regarding drone strikes—are regularly violated by States that make use of armed drones.

First, the principle of distinction requires every drone pilot to distinguish between civilians and combatants. Civilians cannot be targeted by drone attacks, except when participating in hostilities and then only for the duration of this participation (Henckaerts and Doswald-Beck 2005, pp. 2–8 and 3743; Fleck 2008, p. 614). The second fundamental principle of international humanitarian law prohibits the means and methods of warfare that cause unnecessary suffering. Unlike, for example, cluster bombs and chemical weapons, the use of drones is not a priori problematic in this regard. Inherently related to the second principle is the principle of necessity and proportionality.[18] According to the former, acts of war must be guided by the requirements of military necessity and the obligation to minimize the extent of the use of force. The principle of proportionality imposes a comparative assessment of the military benefits and the expected harm (Henckaerts and Doswald-Beck 2005, Rule 14). Finally, the precautionary principle requires that "[i]n the conduct of military operations, constant care shall be taken to spare the civilian population, civilians, and civilian objects" (Article 57 of Additional Protocol I to the Geneva Conventions (1977); Henckaerts and Doswald-Beck 2005, Rule 15).

In light of these principles, it can be concluded that air strikes by means of drones are regularly—and especially in the case of "signature strikes" and when civilians are killed—being carried out in flagrant violation of international humanitarian law.

International Human Rights Law

> There is very little space for targeted killing in international human rights law! (EUI Seminar 2013)

Several international and regional treaties recognize the right to life and, consequently, establish that no one shall be deprived of his life "intentionally" or "arbitrarily."[19] The International Court of Justice (ICJ) has determined that the pro-

[17] In this chapter, only few of these rules will be examined.

[18] See, for example, International Court of Justice, *Nicaragua v. United States of America,* Judgment, 27 June 1986, § 194; International Court of Justice, *Democratic Republic of the Congo v. Uganda*, Judgment, 19 December 2005, § 147; International Court of Justice, *Advisory Opinion on the Legality of the Threat or Use of Nuclear Weapons,* 8 July 1996, § 41–46.

[19] See, for example, Article 2(1) of the European Convention on Human Rights (1948) and Article 4(1) of the American Convention on Human Rights (1969).

visions of Article 3 common to the four Geneva Conventions, which articulates the prohibition of extrajudicial executions of individuals that are not involved in military hostilities, are rules which reflect "elementary considerations of humanity."[20] These elementary considerations of humanity, the Court made clear, are "even more exacting in peace than in war."[21] Accordingly, under the protection afforded by international human rights and fundamental freedoms, drone attacks are also problematic. To what extent may a State conduct drone strikes abroad without this constituting a violation of its obligations under this body of fundamental rules? The essence of the legal reasoning in the field of human rights is explained in the following paragraphs.

First and foremost, it should be highlighted that the tripartite typology of human rights obligations distinguishes between the obligation to respect, protect, and fulfil human rights (De Schutter 2010, pp. 264–256). The obligation to respect human rights means that States must abstain from destroying human rights and freedoms (Gondek 2008, p. 60; Skogly 2006, p. 67). When analyzing the legality of drone strikes, it is important to note that this negative duty, not to interfere with rights and freedoms, exists both in a State's own territory and outside the State's territory. The extraterritorial application of human rights treaties thus requires a State to not violate civil and political rights abroad, a negative duty that is widely acknowledged by both international practice and legal doctrine (Gondek 2008). In an extraterritorial context, two scenarios are possible. First, in cases where a State is in control of territory abroad (territorial control), it is responsible for human rights violations committed in this territory. For example, as Russia controls a part of the territory of Ukraine, the Russian State must respect and ensure respect for fundamental rights and freedoms in the region it controls. Furthermore, given that Israel controls the Palestinian territories, Israel is compelled to guarantee all human rights in these territories. Second, if a State only has control over individuals abroad (personal control), the State is required to meet its human rights obligations with regard to the people that it controls. Several States have therefore been held responsible for having intentionally killed or violated other human rights beyond its borders. A perfect example of such an extraterritorial violation is the case of a public official belonging to State A, who goes to State B to capture and torture a person for information-gathering purposes. In this case, State A is violating the right to liberty and security, and the right not to be tortured, of the captured individual.[22] In doing so, State A incurs international responsibility even if these acts are committed outside its own borders. Although there is no international case law to this day relating to the application and enforcement of human rights when there are drone attacks, this reasoning on extraterritorial human rights obligations equally applies to targeted

[20] International Court of Justice, *Nicaragua v. United States of America,* Judgment, 27 June 1986, § 215.

[21] International Court of Justice, *United Kingdom of Great Britain and Northern Ireland v. Albania*, Judgment, 9 April 1949, § 22. *See* also N. Melzer (2013), p. 19.

[22] See, for example, European Court of Human Rights, *Öcalan v. Turkey,* Preliminary Objections and Merits, Judgment, 12 May 2005.

drone killings by one State in the territory of another State. By controlling a drone, an agent of the State is exercising personal control over his target—it can even kill that person—and is therefore in principle (see below) obligated to respect the right to life and all other rights of the individual abroad.

Second, one must keep in mind that international human rights law applies at all times and therefore also in situations of armed conflict. Therefore, when there is an armed conflict (which allows a State to take measures derogating from its human rights obligations if the situation requires so and only for a given period of time), the legality of the use of drones for targeted attacks is still governed by international humanitarian law and international human rights law (Melzer 2013, p. 4; Amnesty International 2013, p. 8). In summary, even when accepting the argument for a "global war against terrorism," drone strikes must be conducted in respect of human rights law in order to be legal (Melzer 2013, p. 4; Amnesty International 2013, p. 8).

Third, outside the context of an armed conflict, targeted killings are in principle prohibited (CAVV 2013, p. 6). The use of armed drones for killing will only be permitted in exceptional circumstances and under strict conditions (CAVV 2013, p. 6). First, in times of public emergency threatening the security and independence of a State, the State may take measures derogating from its human rights obligations if the situation requires so and only for a given period of time. Second, the loss of lives may be permitted within the context of individual self-defense. Third, the loss of human lives may be permitted when there is an established imminent threat to the life of other individuals (Scheinin 2013, p. 37; CAVV 2013, p. 6). In all three cases, the use of lethal force is only allowed as a last resort and this force must be both necessary and proportionate. In summary, as indicated by the CAVV, "[t]he deployment of an armed drone in a law enforcement situation will hardly ever constitute a legal use of force. The principle of proportionality as it applies within international human rights law is considerably stricter than under IHL [international humanitarian law], particularly to prevent innocent people falling victim to such attacks" (CAVV 2013, p. 6).

In daily practice, drone attacks are regularly carried out outside the context of war and of public emergency as defined above. Indeed, drone attacks often strike people who are not involved in military hostilities or who do not constitute a real threat to the government that either uses or consents to the use of armed drones. As mentioned above, according to various sources, drone attacks do not only strike fighters or terrorists; quite the contrary, many of the victims of drone attacks appear to be civilians. In summary, drone attacks are generally not compatible with human rights law in their current form.

Death from the Sky and State Responsibility

State responsibility is a legal regime that arises "whenever a duty established by any rule of international law has been breached by act or omission" and is thus a corollary of an international obligation (Jimenez de Arechaga and Tanza 1991, p. 347).

As each State has obligations under international human rights law, international humanitarian law, and the international law governing the use of force, in principle it can be held accountable for violations of its legal obligations under these bodies of law. A State is compelled to make reparation for the harmful consequences that arise from internationally wrongful conduct for which the State is held internationally responsible. Accordingly, the last issue that will be addressed in this chapter is whether States using armed drones can be held accountable.

For State responsibility to arise, two main elements must be fulfilled: the act or omission must be attributable to the State and the action or omission must be a breach of an international obligation of the State (Article 2 ILC's Articles on the Responsibility of States for Internationally Wrongful Acts 2001).[23] When the two conditions of attribution and breach are cumulatively fulfilled, one may conclude that there exists an internationally wrongful act entailing the international responsibility of that State. The second element, that of a breach, was extensively analyzed in the above sections discussing the substantive obligations of States using armed drones. As for the first element, that of attribution, a State can be responsible only if a certain conduct can be ascribed to the State. With regard to targeted killings by means of armed drones, the attribution criterion does not, in principle, raise problems. This is because when States make use of armed drones, they will always call on State agents or State organs, often military personnel, to pilot the engine and control its combat functions. The general rule is that the conduct of organs of the State, or of others who have acted under the direction, instigation, or control of those organs, that is, as agents of the State, is attributable to the State at the international level (Articles 4, 5, 6, and 8 ILC's Articles on the Responsibility of Internationally Wrongful Acts 2001).

Different types of behavior can give rise to international responsibility. International law prohibits a State, through its organs or agents, to itself commit the wrongful conduct. As such, it is forbidden for a State to give the order to strike, even abroad, if these drone strikes will most likely breach international law standards. International law also prohibits the facilitation, contribution, or support to international wrongful conduct. If a State knowingly aids or assists the government of another State in the commission of illegal acts, it may be internationally responsible as well. This is explicitly acknowledged by Article 16 of the ILC's Articles on State responsibility.[24] Thus, when a State makes available (components of) armed drones at the request of another State, knowing that these aerial vehicles may be deployed for the commission of international law violations (extrajudicial killings, armed attack, etc.), it may be held accountable.

[23] Article 2 is a positive reflection of pre-existing customary rules. See among others International Court of Justice, *United States of America v. Iran*, Judgment, 24 May 1980, § 56.

[24] Article 16 ILC's Articles only concerns relations between States and is not applicable inasmuch as the aid or assistance concerns entities that have no international status. See, for example, International Court of Justice, *Bosnia and Herzegovina v. Serbia and Montenegro*, Judgment, 26 February 2007, § 420.

Concluding Observations

An analysis of State obligations under the international law governing the use of force, human rights law, and humanitarian law leads to the conclusion that drone attacks are only permitted under very limited circumstances. In other words, the current use of armed drones by States is often in violation with international law as many drone attacks are carried out in situations that may not be legally justified. Thus, although drones may offer many benefits, the way in which several States are using this technology to kill people on foreign territory, often difficult to access, is reprehensible. That is why some recommendations that could contribute to an improved protection of the fundamental values protected by international law can be made.

First, it is essential for the EU and other regional and international organizations, but also for the individual nation States, to define some common guidelines for legal use of drones, especially when they are used in targeted attacks. Given that the use of armed drones must respect both the rights of the individuals and the rights of the territorial States, such debates should link on the one hand legal experts on human rights and humanitarian law and on the other hand specialists of the principles relating to the use of armed force (Melzer 2013, pp. 21–22). The EU and the UN have shown a growing interest in the question of the legality of drone strikes in recent months. In February 2014, for example, the European Parliament adopted a resolution on the use of armed drones (European Parliament 2014), and several UN special rapporteurs have examined various aspects of this practice and pointed at the responsibility of the user States (Emmerson Report 2014; Alston Report 2010). It is essential that these international actors continue to look into the legal implications of drone strikes and condemn any use that violates international law.

Second, States that use armed drones must be made aware of the risks that may result from unlawful use of armed drones. Among others, it is most likely that dictatorial States and non-State armed groups will equip themselves not only with the drone technology but also with the legal reasoning that current States users base themselves on to justify their drone strikes in foreign territory. This is why user States need to rethink their arguments, especially the argument brought forward by the USA that there exists a legal *right* to target the enemy anywhere at any time.

Third, in view of the exponential development and use of armed drones, it is urgent to promote transparency, the international responsibility of user States, and the rule of law (Melzer 2013, p. 4). When a State launches a drone attack, it should a posteriori make this attack publicly known and report to the world the justifications and legal basis for such an attack. Ideally, an independent international body should be assigned for the assessment of the legality of drone attacks. If it turns out that the drone strike cannot be justified under international law, the user State should incur international responsibility therefore and victims should be compensated.

Fourth, there is an urgent need to determine the real impact and effectiveness of drone attacks. How many terrorists were killed? What is the fighter/civilian ratio? What is the impact of drone attacks on the territory? To what extent are the target populations psychologically scarred after the drone attacks?

Finally, it is necessary that international law reflects current realities and evolves accordingly. To the extent that the use of armed drones is now an official policy of States and seems irreversible, there can therefore be no question of prohibiting each use of armed drones. Yet, even if this does not serve the direct interests of the user States, the latter must respect the most basic rules of international law incorporated in the law governing the use of force, human rights law, and humanitarian law and abstain from drone strikes when these strikes cannot meet the legality test requirements. Otherwise, States are lapsing into arbitrariness, and arbitrariness will never serve the cause of international peace, security, or justice.

References

Advisory Committee on Issues of Public International Law (CAVV). (2013). Advisory report on armed Drones. *Advisory Report*. http://cms.webbeat.net/ContentSuite/upload/cav/doc/CAVV_advisory_report_on_armed_drones_(English_translation_-_final)_(2).pdf. Accessed 23 July 2013.

Arabiya, A. (2012). Tribe: Drone attack kills 30 'Qaeda' suspects in Yemen. http://english.alarabiya.net/en/News/middle-east/2014/04/20/Drone-strike-in-Yemen-kills-5-Qaeda-militants.html. Accessed 20 April 2014.

Aljazeera. (2012). Red cross declares Syria conflict a civil war. http://www.aljazeera.com/news/middleeast/2012/07/2012716231913738.html. Accessed 16 July 2012.

Aljazeera. (2014). Israel fears future Drone attacks. http://www.aljazeera.com/news/middleeast/2014/03/israel-fears-future-drone-attacks-2014310145427998548.html. Accessed 10 March 2014.

Alston, P. (2010). Report of the UN Special Rapporteur Philip Alston on extrajudicial, summary or arbitrary executions, UN Doc. A/HRC/14/24/Add. 6. Accessed 28 May 2010.

American Red Cross Blog. (2012). ICRC confirms international humanitarian law applies to conflict in Syria. http://redcrosschat.org/2012/07/15/icrc-confirms-international-humanitarian-law-applies-to-conflict-in-syria/#sthash.aQVrCS8G.dpbs. Accessed 15 July 2012.

Amnesty International (2013). Serai-je la prochaine? http://www.amnesty.org/fr/library/info/ASA33/013/2013/fr. Accessed Oct 2013.

Amnesty International (2014). Pakistan. Pourquoi Mamana Bibi a-t-elle été tuée par une attaque de Drone américain? http://www.amnestyinternational.be/doc/s-informer/notre-magazine-le-fil/le-fil-en-ligne/le-fil-24-janvier-fevrier-2014/article/pakistan-pourquoi-mamana-bibi-a-t. Accessed 11 March 2014.

BBC (2012). Syria in civil war, Red Cross says. http://www.bbc.com/news/world-middle-east-18849362. Accessed 15 July 2012.

Boothby, W. H. (2012). *Weapons and the law of armed conflict*. Oxford: Oxford University Press.

Brunstetter, D., & Braun, M. (2011). The implications of Drones on the just war tradition. *Ethics and International Affairs, 25*(3), 337–358.

Cahin, G. (2007). L'Etat défaillant en droit international: quel régime pour quelle notion? In *Mélanges offerts à Jean Salmon: Droit du pouvoir, pouvoir du droit*. Brussels: Bruylant.

CICDE. (2012). Emploi des systèmes de Drones aériens, *Réflexion doctrinale interarmées, 136*. http://www.cicde.defense.gouv.fr/IMG/pdf/20120606_np_cicde_rdia-2012-010-esda.pdf. Accessed 6 June 2012.

CNN. (2012). Drone strikes kill, maim and traumatize too many civilians, U.S. study says. http://edition.cnn.com/2012/09/25/world/asia/pakistan-us-drone-strikes/. Accessed 26 Sept 2012.

Columbia Law School Human Rights Clinic and Center for Civilians in Conflict. (2012). The civilian impact of Drones. Joint Report. institute/files/The%20Civilian%20Impact%20of%20Drones.pdf. Accessed Sept 2012.

Commission spécialisée de terminologie et de néologie du ministère français de la Défense. (2011). *Official Journal,* Nr. *0141*, 19 June 2011.
De Schutter, O. (2010). *International human rights law*. Cambridge: Cambridge University Press.
Dommage civils. website. http://dommagescivils.wordpress.com/2012/12/25/signature-strike/.
Drone Wars UK. website. http://dronewars.net/aboutdrone/.
Emmerson, B. (2013). EU Parliamentary Update Nr. 27. DROI and SEDE Subcommittees. Brussels. 25 April 2013.
Emmerson, B. (2014). Report of the UN Special Rapporteur Ben Emmerson on the promotion and protection of human rights and fundamental freedoms while countering terrorism, UN Doc. A/HRC/25/59, 28 February 2014.
EUI High-Level Policy Seminar on Targeted Killing, Unmanned Aerial Vehicles and EU Policy. (2013). EUI, Florence, 22 February. http://cadmus.eui.eu/handle/1814/28500.
European Parliament (2014). Resolution on the use of armed Drones. Resolution 2014/2567. http://www.europarl.europa.eu/sides/getDoc.do?pubRef=-//EP//TEXT+MOTION+P7-RC-2014-0201+0+DOC+XML+V0//EN&language=en. Accessed 25 Feb 2014.
Fleck, D. (2008). The law of non-international armed conflict. In D. Fleck (Ed.), *The handbook of international humanitarian law*. Oxford: Oxford University Press.
Forbes. (2014). Under orders from Congress, FAA makes way for commercial Drones. http://www.forbes.com/sites/robertbowman/2014/07/22/under-orders-from-congress-faa-makes-way-for-commercial-drones/. Accessed 22 July 2014.
France 24. (2012). Drone abattu en Israël: le Hezbollah et Téhéran défient Tel-Aviv. http://www.forbes.com/sites/robertbowman/2014/07/22/under-orders-from-congress-faa-makes-way-for-commercial-drones/. Accessed 12 Oct 2012.
France 24. (2013). Infographie: la guerre des Drones américains au Yémen et au Pakistan. http://www.forbes.com/sites/robertbowman/2014/07/22/under-orders-from-congress-faa-makes-way-for-commercial-drones/. Accessed 29 Oct 2013.
Geiss, R. (2005). Failed states: Legal aspects and security implications. *German Yearbook of International Law, 8,* 457–501.
Gery, A. (2012). "Licéité de l'opération américaine au Pakistan contre Ben Laden", Research Report. http://www.lepetitjuriste.fr/wp-content/uploads/2012/09/Lice%CC%81ite%CC%81%20de%20l%27ope%CC%81ration%20ame%CC%81ricaine%20au%20pakistan%20contre%20Ben%20Laden.pdf?aa0226.
Gondek, M. (2008). *The reach of human rights in a globalising world: Extraterritorial application of human rights treaties*. Antwerp: Intersentia.
Grand, C. (2013). EU Parliamentary Update Nr. 27. DROI and SEDE Subcommittees. Brussels. 25 April 2013.
Gray, C. (2008). *International law and the use of force*. Oxford: Oxford University Press.
Heller, K. J. (2013). One hell of a killing machine: Signature strikes and international law. *Journal of International Criminal Justice, 11*(1), 89–119.
Henckaerts, J. M., & Doswald-Beck, L. (Eds.)(2005). *Customary international humanitarian law*. Cambridge: Cambridge University Press.
Henrotin, J. (2013). Le Drone, figure aérienne du mal? *DSI*, 30 (hors série), June–July 2013.
Hoppe, L. (2008). *Le statut juridique des Drones aeronefs non habités*. Aix-en-Provence: Presses universitaires d'Aix-Marseille.
Human Rights and Civil Rights Groups. (2013). Joint letter to President Obama on US Drone strikes and targeted killings. http://www.hrw.org/sites/default/files/related_material/4-11-13_US_LetterToPresidentObamaOnTargetedKillings.pdf. Accessed 11 April 2013.
Human Rights Watch. (2009). Precisely wrong. http://www.hrw.org/sites/default/files/reports/iopt0609webwcover_0.pdf. Accessed June 2009.
Human Rights Watch. (2012). Losing Humanity. http://www.hrw.org/sites/default/files/reports/arms1112_ForUpload.pdf. Accessed 19 Nov 2012.
Human Rights Watch. (2013). Between a Drone and Al-Qaeda. http://www.hrw.org/sites/default/files/reports/yemen1013_ForUpload.pdf. Accessed Oct 2013.

ICRC. (1998). Armed conflicts linked to the disintegration of state structures. *ICRC First meeting held at Geneva, Switzerland,* in 19–23 January 1998.

ICRC. (2008). How is the term 'Armed Conflict' defined in International Humanitarian Law? ICRC Opinion Paper. https://www.icrc.org/eng/assets/files/other/opinion-paper-armed-conflict.pdf. Accessed March 2008.

IKV Pax Christi. (2011). Does unmanned make unacceptable? May 2011.

International Court of Justice. (1986). *Nicaragua v. United States of America,* 27 June, § 246.

International Human Rights and Conflict Resolution Clinic and Global Justice Clinic. (2012). Living under Drones. http://www.livingunderdrones.org/wp-content/uploads/2013/10/Stanford-NYU-Living-Under-Drones.pdf. Accessed Sept 2012.

International New York Times. (2012). Qaïda deputy targeted in Drone strike in Pakistan. http://www.nytimes.com/2012/06/05/world/asia/drone-strike-kills-14-in-pakistan.html?_r=0. Accessed 4 June 2012.

International New York Times. (2013). Recent Drone strikes strain U.S. ties with Afghanistan and Pakistan. http://www.hrw.org/sites/default/files/reports/yemen1013_ForUpload.pdf. Accessed 29 Nov 2013.

International New York Times. (2014a). U.S. jets and Drones attack militants in Iraq, hoping to Stop Advance. http://www.nytimes.com/2014/08/09/world/middleeast/iraq.html. Accessed 8 Aug 2014.

International New York Times. (2014b). Airstrike in Somalia by U.S. Forces Targets Shabab Leader. http://www.nytimes.com/2014/09/03/world/africa/us-airstrike-in-somalia-targets-shabab-leader.html. Accessed 2 Sept 2014.

International New York Times. (2014c). Airstrikes by U.S. and Allies hit ISIS targets in Syria. http://www.nytimes.com/2014/09/23/world/middleeast/us-and-allies-hit-isis-targets-in-syria.html. Accessed 22 Sept 2014.

Jennings, R. Y., & Watts, A. D. (1992). *Oppenheim's international law*. London: Longman.

Jimenez De Arechaga, A. E., & Tanza, A. (1991). International state responsibility. In M. Bedjaoui. (Ed.), *International law: Achievements and prospects*. Paris: UNESCO/Martinus Nijhoff Publishers.

Kiergaard. (2013). Élements sur les robots létaux autonomes et la tendance à l'automatisation des Drones. *Points de vue alternatifs*. http://www.points-de-vue-alternatifs.com/2013/10/%C3%89l%C3%A9ments-sur-les-robots-l%C3%A9taux-autonomes-et-la-tendance-%C3%A0-l-automatisation-des-drones.html. Accessed 9 Oct 2013.

Kunig, K. (2013). Prohibition of intervention. Max Planck Encyclopedia of Public International Law.

Le Figaro (2014). Yémen: 30 morts dans l'attaque de Drone. http://www.lefigaro.fr/flash-actu/2014/04/20/97001-20140420FILWWW00116-yemen-30-morts-dans-l-attaque-de-drone.php. Accessed 20 April 2014.

Le Monde (2012). Le numéro 2 d'Al-Qaida tué par un Drone au Pakistan. http://abonnes.lemonde.fr/international/article/2012/06/05/le-numero-2-d-al-qaida-vise-par-un-drone_1712624_3210.html. Accessed 5 June 2012.

Le Monde (2013a). Barack Obama, président des Drones. http://abonnes.lemonde.fr/ameriques/article/2013/06/18/barack-obama-president-des-drones_3432234_3222.html. Accessed 18 June 2013.

Le Monde (2013b). Somalie: un chef chabab tué par un tir de Drone. http://abonnes.lemonde.fr/afrique/article/2013/10/28/somalie-des-chefs-shebabs-pourraient-avoir-ete-tues-par-une-frappe-de-drone-americain_3504511_3212.html. Accessed 29 Oct 2013.

Mekdour, M. (2011). Les Drones: succès commercial d'un outil controversé. GRIP Analysis Note. http://abonnes.lemonde.fr/afrique/article/2013/10/28/somalie-des-chefs-shebabs-pourraient-avoir-ete-tues-par-une-frappe-de-drone-americain_3504511_3212.html. Accessed 2 Feb 2010.

Melzer, N. (2013). Human Rights Implications of the Usage of Drones and Unmanned Robots in Warfare. Directorate-General for External Policies of the Union Policy Department Study. http://www.europarl.europa.eu/RegData/etudes/etudes/join/2013/410220/EXPO-DROI_ET(2013)410220_EN.pdf. Accessed May 2013.

Milanovic, M. (2010). Drones and targeted killings: Can self-defence preclude their wrongfulness? EJIL:Talk! http://www.ejiltalk.org/drones-and-targeted-killings-can-self-defense-preclude-their-wrongfulness/. Accessed 10 Jan 2010.

Moir, L. (2002). *The law of internal armed conflict*. Cambridge: Cambridge University Press.

Nogrix, P., & Bergé-Lavigne, M. (2006), Le rôle des Drones dans les armées. French Senat Information report nr. 215. http://www.senat.fr/rap/r05-215/r05-215_mono.html#toc8.

Press TV. (2012). Yemeni president says he approves US assassination Drone strikes. http://www.senat.fr/rap/r05-215/r05-215_mono.html#toc8. Accessed 30 Sept 2012.

Ruys, T. (2013). 'Armed Attack' and Article 51 of the UN Charter. Evolutions in Customary Law and Practice, Cambridge University Press.

Sartre, P. (2013). Drone de guerre. *Etudes, 11,* 439–448.

Scheinin, M. (2013). The lawful scope for targeted killings by governments, and resulting considerations about transparency and accountability, Policy Paper of the High-Level Policy Seminar on Targeted Killing, Unmanned Aerial Vehicles and EU Policy, EUI, Florence. http://cadmus.eui.eu/handle/1814/28500. Accessed 22 Feb 2013.

Security Council Resolution 387. (1976). S/RES/387, 31 March 1976.

Skogly, S. (2006). *Beyond national borders: states' human rights obligations in international co-operation*. Antwerp: Intersentia.

Taking Note (2014). How many non-combatants are killed by Drones? http://takingnote.blogs.nytimes.com/2014/02/04/how-many-non-combatants-are-killed-by-drones/. Accessed 4 Feb 2014.

The Bureau of Investigative Journalism. (2012). Revealed: US and Britain launched 1200 Drone strikes in recent wars. http://www.thebureauinvestigates.com/2012/12/04/revealed-us-and-britain-launched-1200-drone-strikes-in-recent-wars/. Accessed 4 Dec 2012.

The Bureau of Investigative Journalism. (2014a). August 2014 update: US covert actions in Pakistan, Yemen and Somalia. http://www.thebureauinvestigates.com/2014/11/03/october-2014-update-us-covert-actions-in-pakistan-yemen-and-somalia/. Accessed 31 Aug 2014.

The Bureau of Investigative Journalism. (2014b). Covert Drone war. http://www.thebureauinvestigates.com/category/projects/drones/drones-graphs/.

The Bureau of Investigative Journalism website. http://www.thebureauinvestigates.com/category/projects/drones/drones-graphs/.

The Guardian. (2012). Drone blitz on Pakistan enters third straight day. http://www.theguardian.com/world/2012/jun/04/15-dead-drone-pakistan. Accessed 4 June 2012.

The Guardian. (2013). US Drone strikes in Pakistan 'carried out without government consent. http://www.theguardian.com/world/2013/mar/15/us-drone-strikes-pakistan. Accessed 15 March 2013.

The Huffington Post. (2014). The toll of 5 years of Drone strikes: 2400 dead. http://www.huffingtonpost.com/2014/01/23/obama-drone-program-anniversary_n_4654825.html. Accessed 23 Jan 2014.

The Jerusalem Post. (2014). Hamas military wing releases pictures of 'captured' IDF Drone. http://www.jpost.com/Defense/Hamas-military-wing-releases-pictures-of-captured-IDF-drone-345024. Accessed 11 March 2014.

The Nation. (2012). A brief history of Drones. http://www.thenation.com/article/166124/brief-history-drones. Accessed 27 Feb 2012.

The Telegraph. (2012). How Israel killed Ahmed Jabari, its toughest enemy in Gaza. http://www.telegraph.co.uk/news/worldnews/middleeast/israel/9685598/How-Israel-killed-Ahmed-Jabari-its-toughest-enemy-in-Gaza.html. Accessed 17 Nov 2012.

The Telegraph. (2013). UN inquiry finds more civilians killed by US Drone strikes than reported. http://www.telegraph.co.uk/news/worldnews/asia/pakistan/10387810/UN-inquiry-finds-more-civilians-killed-by-US-drone-strikes-than-reported.html. Accessed 18 Oct 2013.

The Times. (2013). Al-Shabaab explosives chief 'killed by US Drone'le Monde. http://www.thetimes.co.uk/tto/news/world/africa/article3907118.ece. Accessed 29 Oct 2013.

The UAV. http://www.theuav.com/index.html#.

The Washington Post. (2014). As Drones evolve from military to civilian uses, venture capitalists move in. http://www.washingtonpost.com/business/as-drones-evolve-from-military-to-civilian-uses-venture-capitalists-move-in/2013/10/31/592ca862-419e-11e3-8b74-d89d714ca4dd_story.html. Accessed 1 Nov 2013.
The White House. (2014). Statement by the press secretary on the death of Ahmed Godane. http://www.whitehouse.gov/the-press-office/2014/09/05/statement-press-secretary-death-ahmed-godane. Accessed 5 Sept 2014.
Turkish Weekly. (2014). IS fighters besiege Kurdish city as Drone reportedly belonging to US spotted over Syria. http://www.turkishweekly.net/news/172253/is-fighters-besiege-kurdish-city-as-drone-reportedly-belonging-to-us-spotted-over-syria.html. Accessed 18 Sept 2014.
UN News Centre. (2013). Un expert de l'ONU appelle à un moratoire sur les 'robots létaux autonomes. http://www.un.org/apps/newsFr/storyF.asp?NewsID=30428&Cr=drones&Cr1=#.U1jIp_1_uFF. Accessed 30 May 2013.
US Department of Defense. (2010). Joint publication 1–02 Dictionary of Military and Associated Terms. http://www.dtic.mil/doctrine/new_pubs/jp1_02.pdf. Accessed 15 Dec 2014.
Van Schaack, B. (2011). The killing of Osama Bin Laden and Anwar Al-Aulaqi: Uncharted legal territory. *Yearbook of International Humanitarian Law, 14,* 255–325.
Vilmer, J. J. (2013). Légalité et légitimité des Drones armés, *Revue politique étrangère, 3*, 119–132.
Wall Street Journal. (2011). U.S. tightens Drone rules. http://www.wsj.com/news/articles/SB10001424052970204621904577013982672973836. Accessed 4 Nov 2011.

Mélanie De Groof holds a master's degree in law (Antwerp University) and an LLM in European and International law (Maastricht University). In February 2013, she became a researcher at GRIP. Within the Small Arms and Light Weapons division she conducts research on issues related to arms transfers controls as well as on legal aspects relating to the use and threat of armed violence, peace, and security. She is currently finalizing a doctoral research at Maastricht University. Her study addresses the international legal challenges resulting from State collapse. She is also an assistant at the KU Leuven. Prior to joining GRIP, Mélanie de Groof worked as a researcher at Maastricht University (2005–2010), Human Rights Watch (2011), and Handicap International (2011). She conducted several field missions in Ivory Coast, Ethiopia, Haiti, Colombia, and Kenya. She is a member of the editorial board of the Belgian UN Association Review Wereldbeeld.

Chapter 8
The Predators' Rule of Terror

Vasja Badalič

Introduction

One of the best-known anecdotes about the renowned military strategist Sun Tzu, the author of *The Art of War*, describes how Ho Lu, King of Wu, once challenged Sun Tzu to put his theory of war into practice (Sun Tzu 1971, pp. 57–59). Ho Lu wanted to see whether Sun Tzu could transform a group of 180 concubines into a well-disciplined military unit. Sun Tzu accepted the challenge. After dividing the concubines into two companies, and appointing two concubines as company commanders, Sun Tzu explained to the girls the meaning of specific orders that soldiers have to follow while marching. But when Sun Tzu gave his first order, the girls just burst into fits of laughter. After that first failed attempt, Sun Tzu, convinced that he had not been clear enough in explaining his orders, said the responsibility lay with the general, in that case himself, and so he re-explained the meaning of the orders to the girls. Despite his effort, the concubines again giggled when he gave his orders. Following that second failed attempt, Sun Tzu concluded that his orders had been clear, and, therefore, the responsibility for failing to adhere to them lay with the company commanders. Consequently, Sun Tzu ordered the beheading of both company commanders and replaced the commanders with the next concubines then standing at the head of each company. When Sun Tzu again started giving orders, the concubines finally obeyed him. Both companies quickly morphed into a disciplined, well-trained military unit that adhered to the general's orders.

Sun Tzu's method is based on selective executions that serve as a threat to convince a group of individuals—those who are not executed—that it makes no sense to challenge the executioner's authority. In other words, the aim of this approach is to use arbitrary decapitations as a means to "educate" or "train" subjected individuals and turn them into docile, disciplined individuals who always obey the dominant class. Perhaps it was Sun Tzu's supposed efficiency in establishing disciplined sub-

V. Badalič (✉)
Institute of Criminology at the Faculty of Law, Poljanski nasip 2, 1000 Ljubljana, Slovenia
e-mail: vasja.badalic@gmail.com

A. Završnik (ed.), *Drones and Unmanned Aerial Systems*,
DOI 10.1007/978-3-319-23760-2_8

jects that prompted some contemporary military strategists to adopt his approach as a model for establishing power relations between an invading army and its enemies. Ullman and Wade, for instance, included this Sun Tzu example in their Rapid Dominance doctrine, envisaged as the new US military doctrine for the twenty-first century (Ullman and Wade 1996). Within the Rapid Dominance doctrine, the Sun Tzu example is presented as one of the nine examples of how the US military could impose *shock and awe* on its adversaries. The aim of imposing a sufficient level of shock and awe is to achieve the principal goal of the doctrine—that is, to affect the will, perception, and understanding of adversaries in order to compel them to accept US strategic objectives (Ullman and Wade 1996, pp. xxiv–xxv). In this context, the Sun Tzu method consists of imposing shock and awe on targeted populations through instant decapitations, through the very selective and utterly ruthless application of force on military and civilian targets, with the aim of breaking the will of the majority of the people living in the targeted area and convince them that resistance to the USA is futile (Ullman and Wade 1996, p 27).

When Ullman and Wade compiled a list of military hardware that could be used to impose shock and awe on targeted populations, they included drones. In their vision of future wars, drones would be deployed in large numbers for surveillance, target selection, and carrying out strikes in order to create their own shock-and-awe impact on adversaries (Ullman and Wade 1996, pp. 17). Over the past decade or so, that vision has become, in large part, a reality. In 1996, when Ullman and Wade published the paper in which they laid out the characteristics of the Rapid Dominance doctrine, drone warfare was still in its early stages, but today it is already one of the central pillars of the US "war on terror." Drone warfare has become, according to former Central Intelligence Agency (CIA) chief Leon Panetta, the only game in town in terms of confronting al-Qaeda (AQ) and its associated forces (Shachtman 2009). By virtue of their capability to launch "surgically precise" strikes, armed drones—the Predator and Reaper drones (Rodgers and Hill 2014, pp. 33–39; Singer 2009, pp. 33–35)—emerged during the "war on terror" as the weapon of choice for imposing a Sun Tzuian shock-and-awe campaign through decapitations of individuals identified by US counterterrorism officials as "terrorists." On the one hand, the US administration has implemented its drone program in declared theaters of war, such as Afghanistan and Iraq, where drone operations are conducted by the US Army Special Forces under the command of the US Joint Special Operations Command (JSOC) (Scahill 2013). On the other hand, the US administration also implemented a secret targeted killings program in undeclared combat zones, for example, in some areas in Pakistan, Yemen, and Somalia, where drone strikes are carried out by the CIA, which in some cases hires private military contractors such as Blackwater (or Xe Services or Academi) for a variety of tasks, including gathering intelligence on possible targets, maintaining and loading the drones' missiles, and even flying the drones (Alston 2010, p. 7; Emmerson 2013, pp. 7–10; Scahill 2013; Mayer 2009). In addition, targeted areas expanded to new battlefields in Libya and Mali (Emmerson 2013, pp. 9–10; Schmitt and Saraye 2013, p. A4; Whitlock 2013).

Strategic defense documents prepared by the Obama administration indicate that drones, along with Special Operations Forces, are set to become one of the key military tools for achieving the goals of US military strategy in the twenty-first century (Shaw 2013, pp. 5–7). In January 2012, the US Department of Defense released a strategic document entitled *Sustaining U.S. Global Leadership: Priorities for 21st Century Defense*, in which it articulated a major shift in US military strategy, a shift from large ground wars and counterinsurgency operations—such as those conducted in Iraq and Afghanistan—to much more limited, selective operations against "terrorists" and their affiliates (Department of Defense of the United States of America 2012, p. 6). In accordance with this reorientation of its military policy, the US administration plans to reduce the size of its armed forces designed to conduct large-scale "stability" operations and create a more flexible and technologically advanced force (Department of Defense of the United States of America 2012, p. 6). As it unveiled its strategic document, which serves as a blueprint for the development of US forces until 2020, Leon Panetta, the then US Secretary of State, said that the US administration plans to protect and in some cases increase its investments in, among other things, new technologies such as unmanned systems (Pilkington 2012). Therefore, although the USA plans to reduce its large-scale ground presence in conflict areas, the US unmanned aerial presence, and with it the Sun Tzu method, will expand across selected targeted areas.

The aim of this chapter is to provide an analysis of the main characteristics of the targeted killings program that has been implemented by the US administration through its deployment of armed drones across selected conflict zones.[1] The starting point of this chapter is the hypothesis that drones are not merely a new weapon to combat "terrorism," a weapon used solely to eliminate "terrorists," but, more importantly, a tool that enables the US administration to establish power relations in which entire civilian populations living in targeted areas are subjected to the US rule. In order to test this hypothesis, I will, in a general manner, adopt Jabri's definition of two modes of power in contemporary armed conflicts—first, the power to kill or the direct use of violence that aims at annihilating the adversary, and second, disciplinary power that aims at establishing passive, subordinated communities (Jabri 2010, p. 55). Drawing on these two intertwined modes of power, both of which were adopted by Jabri from Foucault's analytics of power, I will first focus on the power to kill, on the US administration's decision to endow itself with the right to conduct arbitrary killings with drones. In this context, I will, in the first section of the chapter, try to show that the power to kill—the power to launch lethal drone attacks—is exercised within a system of disciplinary mechanisms, the "means of correct training," which are supposed to ensure "surgically precise" strikes and, consequently, a shock-and-awe effect that is supposed to convince the targeted population to submit to US rule. In the second section of the chapter, I will proceed to argue that those disciplinary mechanisms cannot guarantee precise strikes. I will explain which factors cause drone strikes to be inherently inaccurate and how these

[1] Although other states run similar drone programs (e.g., the UK army conducts drone strikes in Afghanistan, the Israeli army uses drones in the Gaza Strip), in this chapter I will focus exclusively on the US drone program. The US drone program is currently the most extensive one and, consequently, most of the available literature on drones is about US drones. However, many of the points made in this chapter about the US drone program can be generalized and applied to other drone programs.

factors undermine the drones' disciplinary effect vis-à-vis the targeted population. In the third section of the chapter, I will focus on the drones' constant panoptic surveillance of the civilian population living in targeted areas and on the disciplinary power relations being established through this constant surveillance. In the fourth and final section of the chapter, I will, drawing in part on Foucault's account of disciplinary power, briefly explain the characteristics of power relations established by drones in areas where they operate.

The Power to Kill/"The Right of the Sword"

According to Dillon, the legitimacy of the use of force in armed conflicts depends on the specific political regime that at the time of a particular conflict decides which use of force is legitimate (Dillon 1998, pp. 543–569). In this context, the primary aim of deliberations over the legitimacy of the use of violence is to decide which belligerent side (e.g., a national army, a rebel group, etc.) is allowed to legitimately use violence in an armed conflict. In the process of making legitimate a specific use of violence, the dominant political–military class, the class that has the power to proclaim itself as the ultimate arbiter of the legitimacy of the use of violence, bestows upon itself the right to the legitimate use of violence. In addition, the dominant class also decides which kind of violence is illegitimate, that is, violence perpetrated by its enemies, the subordinated class. Thus, the legitimacy of the use of violence at any particular moment in history depends on power relations that permit the dominant class to grant itself the right to decide whose force is legitimate, and it is precisely on the basis of this right that the dominant class grants itself the right to use violence, the right that, in the most extreme cases, manifests itself as the power to kill. On the other hand, the dominant class refuses to give legitimacy to the use of force by the dominated class. The dominant class tries to delegitimize its enemies' use of violence by criminalizing it, by defining it as something barbaric, and at the same time it also identifies its enemies as legitimate targets of violence perpetrated by the armed forces of the dominant class. To put this differently, it is the dominant class that sets up the framework that defines who is allowed to kill and who is allowed to be killed.

In the post-9/11 era, the Bush and Obama administration acted as the supreme arbiter with the right to decide on the legitimacy of the use of force in the "war on terror". From their perspective, imperial violence, unleashed against "terror" groups, is categorized as legitimate violence, while violence perpetrated by "terrorists" is portrayed as illegitimate. This distinction also applies in drone warfare, where targeted killings, in which individuals identified by the US administration as "terrorists" are being eliminated, are interpreted as legitimate violence. One of the key characteristics of targeted killings is the US President's right to arbitrarily decide where and when, and against whom, drone strikes will be carried out. Although in the majority of drone strikes, CIA officials decide when and where to go ahead with a strike, the US Commander-in-Chief, the *executioner-in-chief*, acts as the supreme authority for green-lighting strikes in which, as the *New York Times* reported, CIA officials do not have "near certainty" that there will be zero civilian casualties (Becker and Shane 2012, p. A1). Therefore, if there is an opportunity to kill an individual allegedly involved in "terrorist" activities, and that individual is, at the time of the strike, surrounded by civilians (e.g., his family members, friends, etc.), it is the *executioner-in-chief* who has the authority to make the final decision whether to launch the strike or not.

By granting the US President the authority to make the final decision in targeted killings carried out by drones, the US administration introduced into its drone policy a new version of what had been defined in classical theory of sovereignty, in the seventeenth and eighteenth centuries, as the sovereign's right to decide on life and death, the right of the sovereign to let his subjects live or to put them to death (Foucault 2003, pp. 240–241). At the beginning of the twenty-first century, that right—the sovereign's "right of the sword," as Foucault called it—materialized again through the articulation of the US sovereign's authority to decide who is allowed to live and who has to die in drone strikes. In this contemporary articulation of the sovereign's "right of the sword," the highest representative of the US executive branch became the supreme authority that decides which individuals surveilled by drones have to be eliminated and, consequently, which individuals are allowed to live. This authority is not limited by national boundaries. In drone warfare, the imperial sovereign has been granted the right to authorize drone attacks on citizens of both his own country and those of any other country providing safe haven to alleged "terrorists" (Scahill 2013).

How are these strikes executed? Which procedures for selecting targets were instituted to enable the *executioner-in-chief*, or his counterterrorism officials, to make the decision when and where to launch a strike? In order to create the illusion that drone strikes are precise and effective, the US administration introduced a target selection procedure that, generally speaking, consists of three disciplinary techniques, the three "means of correct training," that is, hierarchical observation, examination, and normalizing judgment (Foucault 1995, pp. 170–194). First, in the target selection procedure, the aim of *hierarchical observation* is to collect intelligence about suspicious individuals who may pose a threat to US interests. To be effective, such observation needs to fulfil three requirements: first, it has to be complete, which means that it has to form an uninterrupted network of surveillance that covers the entire targeted area without omitting any unmonitored spots; second, it has to be discreet, which means that those who supervise must not be visible to the surveilled population while they perform their duty, and third, it has to be constant, continuous, it has to be performed all the time (Foucault 1995, p. 174). In drone warfare, hierarchical observation is carried out through the drones' video cameras and local informers deployed in targeted areas. The drones' high-tech multiple video cameras, including infrared devices that enable drones to produce live video at night, serve as a means to conduct constant surveillance of the targeted population. The drones' video cameras penetrate the entire public area below them, they follow the targeted individuals' movements and monitor their daily activities, they collect all information that surveillants/executioners could use in identifying "terrorists." In addition to video surveillance, the monitoring of possible targets is also being carried out by local informers stationed in targeted areas. For example, in Pakistan's Federally Administered Tribal Areas (FATA), the CIA runs a paramilitary group of about 3000 local tribesmen who, besides conducting operations against insurgents, collect intelligence that supposedly helps identify "terrorist" targets. In Pakistan's tribal areas, the US forces also rely on members of the Pakistani Army stationed there (Zenko 2013, p. 7). As a result, the central panoptic gaze, the drones' gaze, multiplies on the ground into thousands of gazes that monitor the population not from above, but from within, from a privileged position among the surveilled subjects.

Second, data gathered through the drones' video surveillance and on-the-ground intelligence operatives enable the implementation of the next disciplinary technique, *examination*, a process in which suspicious individuals selected from the surveilled population are evaluated and classified. In the examination process, selected individuals are turned into "cases" that constitute a "hold," an object of knowledge, for those in power who have the authority to evaluate the information on these selected individuals and make decisions on which category they fit into (Foucault 1995, pp. 184–194). In the context of drone warfare, those in power, those with access to intelligence gathered through hierarchical observation, compare various pieces of evidence on suspicious individuals—they scrutinize the individuals' activities and personal connections—in order to figure out which one of them could be identified as what in contemporary imperial newspeak is called a "terrorist". The entire process consists of four main stages: identification, vetting, validation, and nomination (Shaw and Akhter 2014, p. 226). The decision on who is categorized as a "terrorist" is made in what has been described on one occasion as a "bureaucratic ritual" within the US national security apparatus (Becker and Shane 2012, p. A1). That ritual takes place every week or so when more than 100 members of the US security apparatus assemble, by secure video teleconference, to examine suspects' biographies and recommend to the president who should be the next to die (Becker and Shane 2012, p. A1). Bureaucrats working in the security apparatus evaluate specific "terrorist" threats, they try to figure out which individual could be included in the kill list, and by submitting their nominations for "terrorists" to the commander-in-chief they indicate their preferences regarding possible targets.

Third, hierarchical observation and the examination process pave the way for the introduction of the next disciplinary technique, *normalizing judgment*. Generally speaking, one of the key principles of normalizing judgment is punishing individuals based on a definition of the limit that separates "normal" from "abnormal" behavior (Foucault 1995, p. 183). In defining the distinction between "normal" and "abnormal" behavior, "abnormal"/unacceptable behavior is interpreted as behavior that does not comply with the interests of the dominant class. And it is this "abnormal" behavior that is then neutralized or removed by the dominant class through a system of punishments. In drone warfare, the US administration has defined as "abnormal" all insurgent activities conducted against US interests, and punishment for such "abnormal" behavior is execution carried out by drones. Under the Obama administration, the authority to make the final decision on which individual falls into the "abnormal" category—in other words, the final decision on the inclusion of enemy combatants in the kill list—lies with the commander-in-chief. It is the commander-in-chief, who insisted on approving every new name included on the ever-expanding kill list, that makes the final decision on who has to be eliminated (Becker and Shane 2012, p. A1). The primary kill list, also termed the "disposition matrix," was introduced to amalgamate the separate but, in some cases, overlapping kill lists of the CIA and JSOC into a single, continually evolving database that includes the suspects' biographies, their locations, their known associates, and affiliated organizations (Miller 2012). In addition to that, the "disposition matrix" also includes various strategies for taking targets down, for example, extradition requests, kill-or-capture operations, and drone missions (Miller 2012). It is based on this kill list that punishments—targeted killings—for "abnormal" behavior are carried out.

At first sight, speaking strictly from a Foucaultian perspective, it is perhaps too hasty to claim that targeted killings of alleged "terrorists" establish disciplinary power relations. According to Foucault, disciplinary power relations are defined as relations where a dominant class *governs* subjected individuals and not as relations where a dominant class *exterminates* the subjected individuals (Foucault 2001, pp. 340–341). If disciplinary power relations are understood as relations in which subjected individuals are being governed, then these individuals have to be kept alive. But, despite the fact that in drone warfare some subjected individuals are exterminated, drones nevertheless establish disciplinary power relations, that is, power relations that do not aim at subjecting the targeted individuals but the entire population inhabiting the areas where strikes are being conducted. By carrying out targeted killings, the US forces create power relations in which the targeted area's entire population is included against its will. US forces establish themselves as the dominant entity in relation to the targeted population, with the aim of forcefully "educating" the targeted population about the distinction between what the US forces define as "normal," acceptable behavior (docile behavior which complies with US interests), and "abnormal," unacceptable behavior (rebellion against the US forces). Thus, targeted killings become a source of terror, a military means to inflict shock and awe on the targeted population in order to "normalize" the population, to discipline members of the population under the US rule. In other words, targeted killings carried out by drones have exactly the same role as executions in accordance with the Sun Tzu method.

Random Strikes

The secrecy of the drone program and the limited access of independent researchers to areas where drones operate cause many insurmountable obstacles in the process of evaluating the exact impact of such strikes on civilians (International Crisis Group 2013, pp. 7–11). It is impossible to obtain exact figures on civilian casualties—estimates of the number of civilians killed or injured in drone strikes cited by independent nongovernmental organizations (e.g., The Investigative Bureau of

Journalism, The New America Foundation) considerably contrast with the numbers provided by US administration officials (International Human Rights and Conflict Resolution Clinic and Global Justice Clinic 2012, pp. 43–54; International Crisis Group 2013, pp. 7–11). However, despite the claims of drone proponents as to strikes only being carried out against those combatants who supposedly pose an imminent threat to US security, it is evident, based on available reports, that civilians who have never participated in any "terrorist" organization are often among the victims of strikes (Amnesty International 2013; Human Rights Clinic 2012; International Human Rights and Conflict Resolution Clinic and Global Justice Clinic 2012). Despite strikes being based on disciplinary mechanisms that should guarantee accuracy in target selection, a critical analysis of the outcomes of strikes reveals that "surgical precision" is just an illusion that aims at convincing the general public in the USA and other countries that these strikes exclusively target "terrorists."

Analyses of the outcomes of drone strikes reveal that in many strikes those killed or injured were either exclusively civilians or the majority of them were civilians. Numerous times drone strikes have targeted different public and privately owned areas (e.g., a bazar, a road, a cemetery, a house, a field) where civilians were carrying out their daily activities at the time of the strikes. In order to gain at least a general overview of such strikes, it is worth taking a closer look at some of them. Here are some examples. On 30 October 2006, US drones targeted a *madrasa*, an Islamic school, located in the village of Chenagai in Bajaur, a Pashtun tribal area in Pakistan. In that attack, which killed 83 people in total, the majority of the victims were minor students, some of them as young as 9-years-old, who had never participated in "terrorist" activities (Ali 2006). On 23 June 2009, the CIA launched a drone strike on a funeral procession in the city of Makeen in South Waziristan, a tribal area in Pakistan. About 5000 people attended the funeral of Khwaz Wali Mehsud, a mid-ranking Pakistani Taliban commander, who himself had been killed in a previous drone strike. The CIA planned to use Wali Mehsud's funeral as bait to hook Baitullah Mehsud, then leader of the Pakistani Taliban. The missiles fired at the funeral killed up to 83 people. Among those killed, there were as many as 45 civilians, including ten children and four tribal leaders (Woods and Lamb 2012). On 31 December 2009, a US drone fired a missile at Karim Khan's house, located in the village of Machikhel in North Waziristan, a tribal area in Pakistan. The drone's missile leveled the house, thus killing Khan's brother Asif Iqbal, employed as a schoolteacher in a nearby village; Khan's son Zaeenullah, who worked as a guard at the same school as his uncle Asif; and Khaliq Dad, a stone mason who worked at the local mosque and was staying at Khan's house as a guest (Benjamin 2012, pp. 109–112). On 21 February 2010, a US drone struck a car convoy of civilians driving through Oruzghan province, Afghanistan. About two dozen people were traveling in three vehicles, all of them were civilians—shopkeepers going to a nearby city to buy supplies, students going back to school, patients going to the city to seek medical treatment, and families with children planning to go visit their relatives. By the US military's count, 15 or 16 civilians were killed in the strike, while local tribal elders claimed that 23 civilians had been killed, including two boys, 3-year-old Daoud and 4-year-old Murtaza (Cloud 2011). On 17 March 2011, drones bombed a *jirga*, a gathering of tribal elders, organized at a bus depot in Datta Khel village in North Waziristan. Evidence gathered through interviews with local witnesses indicated that at least 42 people were killed in the attack, 38 of them civilians. Only four of the participants at the jirga were allegedly members of the Pakistani Taliban (International Human Rights and Conflict Resolution Clinic and Global Justice Clinic 2012, pp. 57–62; Abbot 2012). On 6 July 2012, a drone strike hit a group of workers from Zowi Sigdi village in North Waziristan. The strike killed ten workers who had gathered at a tent after a day's work. A few minutes after the first strike, drones fired another series of missiles targeting rescuers who had come to the scene of the attack to help the victims of the first strike. In total, 18 people, all of them civilians, were killed in both attacks, and at least 22 were injured, according to witnesses and relatives of victims (Amnesty International 2013, pp. 24–27). On the afternoon of 24 October 2012, a US drone circling above Ghundi Kala village in North Waziristan fired two Hellfire missiles on 68-year-old Mamana Bibi while she was working on one of her family's fields, collecting ocra to cook that evening. She was killed instantly. At the time of the strike, she was with three granddaughters, aged 5–8 years, who had been helping her on the field. Two of them sustained shrapnel injuries (Amnesty International 2013, pp. 18–23; Devereaux 2013). On 4 June 2013, a drone operated by members of the International Security Assistance Force (ISAF) fired two missiles at a humanitarian deminer as he was working with 80 deminers clearing mines at a site in Panjwayi District, Kandahar Province, Afghanistan. The deminer, who was working on the Emirates Mine Clearance Project Afghanistan (EMCPA), died instantly, but none of his colleagues present with him at the demining site sustained injuries (United Nation Assistance Mission in Afghanistan 2013, p. 41). On 12 December 2013, a US drone launched its Hellfire missiles on a car convoy on the outskirts of the city of Rad'a in Yemen. After the attack, in which at least 12 people were killed and at least 15 were injured, it turned out that the convoy was a wedding procession that was bringing the bride and her family members to the groom's hometown (Human Rights Watch 2014, pp. 1–2).

On the basis of these accounts of drone strikes that targeted public and private areas where at the time of the strike civilians were conducting all sorts of mundane activities—young students studying at a religious school, a female farmer working on a field, a deminer clearing mines at a demining site, a group of people driving from their villages to a city, families living in their houses, laborers resting after finishing their shift, tribal elders attending a public gathering, mourners attending a funeral, and a group of people driving in a car convoy to a wedding—it is clear that, in their hunt for "terrorists," strikes are often imprecise. Among the victims, there are civilians of all ages, of both sexes, civilians of different professions who at the time of the strikes were carrying out various activities.

From the perspective of the people living in drone-surveilled areas, where they can observe firsthand the outcomes of drone warfare, drone strikes appear to be random strikes in which anyone can be hit at any time. Due to the strikes' unpredictability, it is impossible for the civilian population to discern a clear pattern of these strikes, a pattern that would help them to adapt their daily activities and movements in such a way that they would be able to completely avoid the danger of being hit.

There are five main reasons that cause imprecise strikes.[2] First, imprecise attacks are a consequence of a too broad, flexible definition of what constitutes a "terrorist." Within the drone warfare framework introduced by the Bush and Obama administrations, "terrorists" are not only defined as individuals who directly participate in hostilities and pose an imminent threat to the USA but also as individuals who behave *as if* they are members of some "terrorist" organization. Within this flexible framework, "terrorists" are broadly defined as individuals that "bear the characteristics of al-Qaeda or Taliban leaders on the run" (Schmitt and Sanger 2008) or as individuals that act "in a manner consistent with AQ-linked militants" (Abbot 2012). Therefore, in the process of identifying a "terrorist" target, it is not necessary for US counterterrorism officials to prove that a particular individual is actually a combatant engaged in hostilities, but merely that the individual is acting as if he were a "terrorist." In this identification process, US officials try to capture the individual's behavioral pattern, his "pattern of life," which supposedly discloses whether the individual is a "terrorist" or not. US counterterrorism officials try to analyze the suspected individual's social environment by assessing the locations frequented by the individual and by evaluating his relationships with other people (Shaw and Akhter 2014, p. 227). And what exactly does it mean—for the US administration—to act as a "terrorist"? In their quest of characteristics indicating who could be identified as a "terrorist," US officials went so far as providing a definition in which "terrorists" are defined as all military-age males in a strike zone (Becker and Shane 2012, p. A1). According to this definition, all military-age men, all males between the age of 20 and 40 (McKelvey 2012), who find themselves at the site of a drone strike that targeted an alleged "terrorist" are automatically identified as "terrorists." The disturbing logic behind such definition is that all military-age men killed or wounded in a drone attack are confirmed as "terrorists" just because at the wrong time they found themselves at the wrong place, in the vicinity of at least one alleged "terrorist." Being at the site of a drone strike, without actually participating in any hostility, already confirms the "pattern of life" of a "terrorist." US counterterrorism officials support such definition by arguing that "people in an area of known terrorist activity, or found with a top Qaeda operative, are probably up to no good" (Becker and Shane 2012, p. A1). This kind of logic does not take into consideration that innocent civilians may find themselves in the vicinity of insurgents against their will or that civilians may come across insurgents while carrying out their everyday tasks (e.g., local tribal elders negotiating with insurgents, shopkeepers selling something to insurgents, journalists interviewing insurgents). In addition to identifying all military-age individuals in a strike zone as "terrorists," the US administration also expanded its definition of a "terrorist" by including in it drug lords who have supposedly supported members of the Taliban movement by contributing money to

[2] Due to the covert nature of the drone program and the incessant attempts by US authorities to prevent any kind of independent evaluation of the program, it is very difficult to critically examine its characteristics. However, by delving into a number of studies and newspaper articles that in recent years have disclosed specific aspects of the program, it is possible to provide a detailed, genuine picture of the program's main flaws.

their cause. A report by the US Senate Foreign Relations Committee disclosed that the "joint integrated prioritized target list"—the kill list—was expanded to include about 50 Afghan drug traffickers who were suspected of helping the Taliban insurgency (US Senate Foreign Relations Committee 2009, pp. 15–16; Mayer 2009; Risen 2009). Although these individuals did not directly participate in hostilities and did not pose an imminent threat to the USA, the US administration nevertheless branded them as "terrorists" and placed them on the kill list.

Second, the inaccuracy of drone attacks stems from the fact that current drone surveillance technology provides images of low quality. Consequently, drone crews are unable to clearly identify suspicious individuals and discern possibly incriminating objects (e.g., weapons) that they see on their screens. According to Linebaugh, an ex-US service member who worked on the drone program, the pixelated video provided by a drone is not, even in perfect weather conditions, usually clear enough to detect someone carrying a weapon (Linebaugh 2013). Due to the low quality of the pixelated images, there is much more room for mistakes in the process of identifying targets. The identification process relies, to a certain extent, on the drone operators' guesses as regards what the blurred images on the screens actually depict. As became evident after the abovementioned drone strike carried out in Oruzghan Province, Afghanistan, on 21 February 2010, drone operators completely misjudged the situation on the ground because they could not clearly identify their targets. The results of a US military investigation into the Oruzghan strike revealed that the images provided by the drone were not clear enough for the drone operators to discern whether the individuals seen on the screens were adult males or women and children and whether the grainy objects on the screens were weapons or not. Moreover, drone operators could not even figure out how many people were traveling in the targeted convoy (Radio Transmissions and Intercom 2010; Cloud 2011; Benjamin 2012, pp. 92–94).

Third, the low quality of images provided by drones paves the way for the third factor influencing the precision of the strikes, that is, the drone operators' biased interpretations of these images. In the process of interpreting ambiguous images, there is always a possibility that drone operators will make mistakes when trying to figure out whether the images indicate some kind of "terrorist" activity. But, in addition to making mistakes, there is also the possibility that drone operators "creatively" interpret these images with the aim to deliberately use false interpretations as "evidence" of "terrorist" activity. Drone operators, as Walls and Monahan pointed out, may reduce a wide range of data (e.g., unclear images of the movements and activities of people) to preestablished functional categories (e.g., "terrorist" activity) that conform to their needs and biases (Wall and Monahan 2011, p. 240). Such interpretation of images where the goal is not to find out what the images actually represent, but rather to satisfy the interpreters' biases, can distort the reality on the ground in ways that may cause fatal consequences. The transcription of the discussion of the prestrike planning that took place among the members of the drone crew and other military personnel involved in the abovementioned strike on a car convoy of civilians in Oruzghan Province, Afghanistan, provides us with unique insight into the process of constructing a web of "creative" interpretations of images, which in that case served as "evidence" for green-lighting the strike (Radio Transmissions and Intercom 2010). In the transcription we can notice two distinct ways of distorting the reality on the ground. First, the drone operators dismissed or relativized each image indicating that not all individuals in the convoy were military-age men (which in their world meant "terrorists"). For example, when images emerged indicating that children were present in the convoy, the drone operators either completely dismissed them—"I don't think they have kids out at this hour"—or they relativized them by claiming that those were probably teenagers, "something more towards adolescents," who are just as dangerous as adults—"But like I said, 12–13 years with a rifle is just as dangerous" (Radio Transmissions and Intercom 2010). And second, many movements that in normal circumstances would have been seen as completely unsuspicious movements were consistently interpreted by the drone operators as movements indicating some sort of tactical "terrorist" manoeuvre. Here are some examples of such interpretations: When a third vehicle joined the two vehicles in the convoy under surveillance, the drone's sensor operator said "That looks like a, uh, grouping of forces" (Radio Transmissions and Intercom 2010), suggesting that the "terrorists" were grouping in order to launch an ambush on US ground forces stationed in that area. When a scuffle erupted between the surveilled individuals, the drone operators did not interpret it as just a scuffle but rather as "terrorists" forcing innocent civilians into the vehicle in order to use them as human shields during the supposed ambush on US ground forces. When the vehicles on the screen took a road that meandered away from the location where the US troops were stationed, the drone operators did not interpret that movement as a sign indicating that the surveilled vehicles were actually not a Taliban convoy planning to attack US troops. The drone operators interpreted that movement as "evidence" indicating that the "terrorists" wanted to flank the US forces and attack them from another location. It can be concluded that the drone operators in that strike consistently dismissed or falsely interpreted all images that could have potentially undermined their preestablished belief that what they saw below them was a convoy of "terrorists." In that strike, the reality on the ground became almost completely irrelevant—what really counted in the preparation for the strike were the "creative" interpretations provided by drone operators involved in the strike. The decision to launch the strike was not based on what the images on the screen actually represented but rather on what the drone operators claimed they saw in them.

Fourth, another factor that causes imprecise strikes is faulty intelligence provided by local informants who serve the US forces as confirming witnesses for selecting individuals for targeting (Mayer 2009; Alston 2009, pp. 10–12; Alston 2010, p. 25). In Afghanistan and Pakistan, US officials identified three reasons why local informants deliberately provide false information: first, the US military sometimes worked with local tribal chiefs who often falsely identified their own enemies (e.g., their adversaries in local tribal feuds) as members of al-Qaeda or the Taliban, because they wanted to take advantage of the US forces' firepower to eliminate these enemies; second, informants sometimes provided false information because, despite not knowing the insurgents' whereabouts, they wanted to curry favor with their US paymasters, they wanted to convince their paymasters that they possessed valuable information about insurgents; and third, informants in some cases made "intelligence" up because they simply wanted to earn some money (Mayer 2009)—in Pakistan's tribal areas, for example, informants get paid US$100 for providing information about insurgents and the locations of their training camps (Benjamin 2012, p. 117).

Fifth, the strikes' imprecision is also a consequence of the tactics applied during the execution of the strikes. There are two types of strikes—"signature" strikes and "double tap" strikes—which considerably increase the possibility of hitting civilians. On the basis of the US administration's flexible definition of a "terrorist," "signature" strikes, also termed "crowd killings," target groups of anonymous people who, in the view of US counterterrorism officials, behave in a manner consistent with that of a "terrorist." In contrast to "personality" strikes, in which the precise identity of the targeted individual is known, "signature" strikes aim at unidentified individuals who bear certain "signatures," or characteristics that the US administration believes are associated with "terrorist" activities. Introduced by the Bush administration in 2008 and later adopted and expanded by the Obama administration, "signature" strikes target all individuals who have the "pattern of life" of a "terrorist," which usually means all military-age men who find themselves in the vicinity of an alleged "terrorist" (Zenko 2013, p. 12; Human Rights Clinic 2012, p. 8; Entous et al. 2012). The problem with this type of strike is that the vague definition of a "terrorist" increases the danger of killing and injuring innocent civilians, as happened in the abovementioned "crowd killing" carried out on 17 March 2011 in North Waziristan, which killed 38 civilians, most of them pro-government tribal elders and members of the local police force, and only four alleged Pakistani Taliban fighters (Abbot 2012). On the other hand, "double tap" attacks consist of a sequence of two drone strikes in which the second strike hits, after a short period of time, the site of the first strike, thus killing or harming rescuers who rush to the site of the strike to help the victims of the first strike (International Human Rights and Conflict Resolution Clinic and Global Justice Clinic 2012, pp. 74–76; Woods and Lamb 2012). The short period of time between the two consecutive strikes—the second strike may be launched just some minutes, or half an hour, after the first strike—indicates that at least in the second strike members of the drone crew do not bother to identify their targets in order to make sure that civilians are not harmed. It is impossible in such a short period of time for drone operators to determine whether the people helping the victims of the first strike are civilians or "terrorists," which means that drone operators do not care who gets hit in the second strike. Thus, it is not surprising that, according to reports from the tribal areas in Pakistan, many civilians, first responders, who come to the rescue of those injured in the first strike, are killed in follow-up strikes (Woods and Lamb 2012; Woods and Yusufzai 2013).

Although these five factors significantly differ from each other (the first one refers to the discursive framework defining a "terrorist," the second one is about the drones' technical capabilities, the third one refers to the drone operators' biased interpretations of images, the fourth one concerns unreliable intelligence, while the fifth factor refers to the tactics used while carrying out strikes), we can nevertheless put them into two categories. The first category consists of two factors—the flexible definition of a "terrorist" and the low quality of real-time video feed provided by drones—which constantly influences the target selection procedure and the execution of the strikes. In the process of preparing and conducting drone attacks, these two factors do not appear sporadically as some sort of isolated, unpredictable instances, but—on the contrary—they are immanently integrated within this process. To put this differently, these two factors are systemic factors that are continuously present in drone strikes, and, therefore, they continuously influence the execution of the strike in a way that may cause civilian casualties. On the other hand, the other three factors—"signature" strikes and "double tap" strikes, biased interpretations of images, and faulty intelligence—appear sporadically in the process of conducting strikes, and, therefore, occasionally cause "mistakes" in the target selection procedure.

Despite the target selection procedure put into place by the US administration to identify "terrorists" among the surveilled population and prevent any harm to civilians, it is evident that, based on the five abovementioned factors, drone strikes are conducted, to a siginificant extent, indiscriminately. And the fact that all those involved in the drone program—US policy makers, counterterrorism officials, and drone operators—continue to carry out strikes despite being aware of the systemic factors causing the inaccuracy of the strikes, shows us that they deliberately conduct strikes that, in many cases, they know to be indiscriminate.

This argument about the strikes' indiscriminate nature goes against the "surgical precision" argument advanced by drone proponents, for example, Strawser (Strawser 2010). The main weakness of Strawser's argument is its reliance on two narrow premises, both of which distort the reality of the drone program in a way that conceals some of its characteristics that heavily influence the precision of such strikes. First, Strawser's argument narrowly focuses only on the technical aspect of drone warfare, on the drones' technical capabilities, thus ignoring broader issues of the drone program. According to Strawser, the drones' cutting-edge technology (e.g., the drones' capability to launch missiles in a *fire, observe, and update mode*, which allows the pilot to steer the missile off its trajectory if he notices mid-course that the selected target is a civilian) increases the pilot's ability to discriminate between combatants and noncombatants (Strawser 2010, pp. 351–352). The main problem with Strawser's narrow focus on the technical capabilities of the drones is that he avoids taking into consideration many other factors of the drone program (e.g., the flexible definition of a "terrorist," faulty intelligence, biased interpretations of images, and the periodic launch of "signature" strikes and "double tap" strikes) which cause indiscriminate strikes. By dismissing these factors, which have, based on available reports, caused many "mistakes" in target selection, Strawser outlines a very twisted picture of how drones operate. But, in addition to that, even if we focus solely on the technical aspect of drone warfare, there is a crucial factor—the low quality of the live video feed provided by the drone—which constantly influences the drone operators' ability as regards noncombatant discrimination. As has been explained above, the pixelated images are of such low quality that drone operators are unable to clearly identify persons and objects under their surveillance (Linebaugh 2013; Cloud 2011).

Second, Strawser's argument about the precision of drone strikes is built on a comparison with the precision of manned aircraft strikes—it is on the basis of such comparison that Strawser concludes that drones increase the pilot's ability to discriminate between civilian and military targets (Strawser 2010, p. 352). But the suggestion that drones may be more precise in targeting "terrorists" than traditional manned aircraft does not mean that drones cannot launch indiscriminate strikes. Even if drones increase the pilot's ability to both identify a "terrorist" and avoid significant "collateral damage", as Strawser claims, the drone program in general is still so flawed that it leads to a considerable number of indiscriminate attacks. If Strawser is right in his comparison of both types of aircraft, drones may cause fewer "mistakes" in target selection and less indiscriminate strikes than manned aircraft, but this still does not mean that indiscriminate strikes do not happen in drone warfare.

By taking into account the five factors causing the inaccuracy of such strikes, I argue that the drone program in general is inherently inaccurate. And it is this inaccuracy that causes indiscriminate, random strikes and, consequently, introduces a new power regime in drone-surveilled areas. The abovementioned drones' disciplinary power regime, implemented through the application of the "means of correct training", aims at eliminating exclusively "terrorists" in order to compel the population in targeted areas to stop supporting "terrorist" networks. The key premise of the drones' disciplinary power regime is that targeted killings of "terrorists" can deter potential recruits from joining "terrorist" groups only if exclusively "terrorists" are killed in drone strikes. Only accurate strikes can show the targeted population that it is better for them not to join or support "terrorists" and their associates. On the other hand, the application of indiscriminate strikes introduces a different power regime, a power regime in which anyone, "terrorist" or civilian, can be targeted, which means that such strikes cannot serve as a warning about what can happen to those who want to join "terrorist" groups. When anyone can get killed in drone strikes, there is no disciplinary effect on the targeted population because the strikes do not indicate what kind of behavior is being punished. With indiscriminate attacks taking place, it is impossible for the targeted population to discern what kind of behavior is being punished, and, therefore, it is impossible for them to discern the distinction between "normal" and "abnormal" behavior, between acceptable (docile) and unacceptable (rebellious) behavior. Moreover, random strikes do not provide any guarantee to members of the surveilled population that they will not be targeted in the event they decide not to participate as "terrorists" in hostilities or in any manner support a "terror" group. When members of the local population see that anyone can be eliminated in a drone strike, they realize that even if they decide not to oppose the US forces, they may still become victims of attacks. For civilians who decide not to rebel, there is no reward in terms of having guarantees that they will be allowed to remain alive.

Indiscriminate drone strikes indicate that the US military is conducting a drone war in which it deliberately aims at terrorizing the entire population living in its sphere of influence. But although indiscriminate strikes cannot, in a Sun Tzuian way, "educate" or discipline the terrorized population, they nevertheless consolidate the position of the dominant entity, the US military, in relation to the subjected entity, the targeted population. The dominant entity's ability to conduct, with impunity, arbitrary indiscriminate strikes is perhaps the most extreme example indicating the radical schism that divides both sides involved in the conflict. The civilian population is completely subjected to the whims of the dominant military entity, without having the possibility to understand the logic that drives the strikes perpetrated by that entity.

Permanent Panoptic Surveillance

As has been explained in the first section of this chapter, the drone's panoptic gaze is a mechanism of surveillance employed in the process of evaluating and selecting targets. In that context, the drone's panoptic gaze serves as a means to collect actionable intelligence based on which acts of terror—targeted killings—are being carried out. But, as will be explained in this section of the chapter, the panoptic gaze is also a mechanism that directly, on its own, inflicts terror, a constant wave of shock and awe, on the surveilled population.

Perhaps the best approach to explaining how the drone's panoptic gaze generates terror is by comparing this kind of terror with the effects of power generated within panoptic institutions. Such an approach, I think, will show us that the terror inflicted by the drones' panoptic gaze is generated in a similar way as effects of power are being produced within the panopticon's specific distribution of visibility and invisibility. The two key principles of the distribution of visibility and invisibility which in panoptic institutions produces effects of power on the surveilled subjects are as follows: first, those who exercise power, those who supervise, must never become visible to the surveilled subjects, while the surveilled subjects must be visible to their supervisors at any moment; second, the surveilled subjects must always be able to see a material entity that indicates to them the possibility that those supervising them are constantly present (Bentham 1995; Foucault 1995, pp. 200–201). Therefore, in order to explain how the drone's panoptic gaze generates terror in a similar way as panoptic institutions generate their power effect, I will show how drone terror is generated on the basis of these two key principles of the panoptic distribution of visibility and invisibility.

First, if we focus on the supervisors' constant invisibility and the surveilled subjects' constant visibility in panoptic apparatuses, we notice that drones perfectly emulate the panoptic arrangement in the sense that drone operators are always invisible to the targeted population, while the targeted population is constantly visible to the drone operators. In a similar way as Bentham's architectural apparatus, drones establish a specific division of the visible and invisible in which the dominant entity is never seen by the surveilled population, while the subordinated, surveilled population is permanently visible, always perceptible to their supervisors' eyes.

Second, in terms of creating the awareness in surveilled individuals that their supervisors might be constantly present, drones also dutifully adopted the panoptic arrangement. In Bentham's panopticon, the awareness of the possibility of the supervisors' constant presence is induced in the surveilled individuals through the presence of the central tower, which is positioned in such a way that the surveilled individuals, locked in cells situated in the annular building encircling the central tower, are able to see the tower but cannot see whether their supervisors are actually present in the tower (Bentham 1995). This kind of arrangement, in which the surveilled subjects constantly have the central tower before their eyes but are not able to actually see the supervisors supposedly monitoring them from the tower, ensures that the supervisors are only indirectly visible, through the central tower's presence, to the surveilled subjects. The position of the central tower makes the presence of the supervisors unverifiable by the surveilled subjects—the tower indicates to the surveilled population that the supervisors may be present at any time, but at the same time it prevents the surveilled subjects from knowing whether the supervisors are actually there. In the case of drones, we have a similar arrangement—the only difference is that the awareness of the possibility of the supervisors' constant presence is induced in the surveilled population not only by the drones' constant visibility but also by their sound. On the one hand, members of the surveilled population are able to see drones flying above them; they are able to see the small dots in the sky that mark the presence of drones. David Rohde, a journalist who saw firsthand how drones operate when he spent months in Taliban captivity in the Pakistani tribal areas, explained that drones could be seen with the naked eye as dark specks in the sky (Rohde 2012). On the other hand, the presence of drones is also made felt by the sound they emit. For example, inhabitants of the Pakistani tribal areas confirmed that drones can be heard every time they are in the air (International Human Rights and Conflict Resolution Clinic and Global Justice Clinic 2012, p. 150). Referring to drone presence in South Waziristan, Pakistani journalist Pir Zubair Shah reported that the drones' buzzing was there all the time, especially during the night when it was all someone could hear (Shah 2012). Both the drones' constant visibility and the sound they generate create the impression in the surveilled population that the drone operators are constantly present, while at the same time they prevent the surveilled population from verifying whether the drone operators are actually spying on them. The presence of drones only suggests that drone operators may be monitoring a targeted area, but it gives no clues whether the operators are actually doing the monitoring. Thus, the drone performs the same function as the panopticon's central tower—the drone serves as an object whose presence ensures that the supervisors' presence is indirectly visible and unverifiable by the survellied individuals. The surveilled population is allowed to see the object/drone, but it is not allowed to know whether the supervisors/drone operators are actually spying on them.

The panoptic apparatus, the specific architectural arrangement that makes possible, first, the permanent visibility of surveilled subjects and, second, the indirect visibility and unverifiability of supervisors, ensures the establishment of disciplinary power relations between both groups involved (Foucault 1995, p. 201). The constant visibility of surveilled subjects and the indirect visibility/unverifiability of supervisors create a power effect that coerces the surveilled individuals to alter their behavior in accordance with the supervisors' expectations. Because the surveilled subjects know that at any given moment they may be under surveillance, they start to behave as the supervising authority expects them to behave. According to Foucault, the disciplinary power regime's constraining force passes over to the side of the surveilled subjects—those who find themselves in the field of visibility assume responsibility for the constraints of power, they become the principle of their own subjection (Foucault 1995, pp. 202–203). Even if, at some moment, there is no supervisor in the central tower, the panoptic apparatus ensures the automatic functioning of disciplinary power relations because the subjected individuals, knowing and/or fearing that they are probably constantly spied upon, continue to behave in keeping with the supervisors' instructions.

Drawing on Bentham's work, Foucault argued that the panopticon can be understood as "a generalizable model of functioning," as an abstract political technology which can be applied in any disciplinary institution striving to create docile subjects (Foucault 1995, p. 205). The aim of this kind of political technology has always been, from its inception on, to spread such means of control across various institutions, to cover an ever larger surface in order to establish a disciplined, "normalized" society (Foucault 1995, p. 209; Dreyfus and Rabinow 1983, p. 153). In our time, drones represent the latest incarnation of this political technology. With the advent of drones, the panoptic gaze spread from closed institutions (e.g., prisons, schools, army barracks, factories, etc.) into open areas inhabited by targeted populations. The drones' panoptic gaze is freed from the limits posed by the architectural designs of closed institutions, and, as a result, it does not create a disciplinary power effect only on particular groups (e.g., prisoners, students, soldiers, workers, etc.), but on entire populations living in the areas under its control.

By adopting Foucault's interpretation of how panoptic apparatuses create a power effect on surveilled subjects, we can understand how drones generate their own power effect—their own terror or shock and awe—on the populations kept under surveillance. Even if drone operators do not monitor the population continuously, at every single moment, the drones' mere presence nevertheless constantly inflicts terror on the targeted population (Williams 2011). The drones' presence inflicts terror on the population by indicating to the population the possibility that drone operators are continuously spying on them, and, consequently, by suggesting that a strike may happen at any time. Without ever knowing precisely at which moment drone operators are actually monitoring them, members of the surveilled population remain permanently caught in this all-encompassing terror-generating surveillance. Because it is impossible to determine from the ground who or what is being tracked by the drones, the entire population below the drones lives in fear (Rohde 2012). Therefore, the mere presence of drones creates the shock-and-awe effect that aims at breaking the will of the targeted population in order to turn them into docile subjects. Even when drones just circle above the surveilled population, without actually launching a strike, they already establish power relations in which they terrorize members of the subjected population. In a similar way as panoptic institutions aim at establishing effective disciplinary power relations, drones aim at creating power relations in which the constraining force of power—the force that coerces surveilled individuals into accepting their subjugation and into behaving in keeping with the supervisors' expectations—passes over to the surveilled subjects. In drone-surveilled areas, US forces expect the surveilled population to assume, under the constant gaze of the drone operators, the responsibility for the constraints of power and discipline themselves in accordance with US interests.

However, the awareness of being permanently subjected to the drone operators' gaze, the executioner's gaze, inflicts terror on the surveilled population only because it is complemented—and this is crucial—by awareness of the constant possibility of an arbitrary execution, the constant possibility of death actualized by means of random strikes. As Rohde put it, the buzz generated by the drone's propeller is "a constant reminder of imminent death" (Rohde 2012). Without the sporadic random strikes, without the periodic use of the "right of the sword," the mere presence of drones could not effectively affect the surveilled population. Without knowing that the speck and buzz in the sky, which both indicate the possibility of the drone operators' presence, are linked to the possibility of an unpredictable use of the "right of the sword," the drones' panoptic gaze could not create a potent power effect on the population. In other words, the power effect created within the drones' panoptic arrangement depends, to a large extent, on the constant threat of the use of the "right of the sword" that gives "credibility" to this power regime in general.

The constant presence of drones, combined with the constant threat of imminent extermination, causes significant changes in the ways people in targeted areas, insurgents and civilians, conduct their everyday activities. According to Rohde, the presence of drones in South and North Waziristan changed the way in which insurgents carried out their activities. In order to reduce the possibility of being detected by the drones' pervasive gaze, insurgents started to avoid organizing gatherings of large groups, they started to conduct training of suicide bombers and roadside bomb-makers in smaller groups, while insurgents' commanders decided to change their traveling habits—they started to change their vehicles more frequently and moved around with fewer bodyguards (Rohde 2012).

The drones' terrorizing effect is not limited to insurgents—it encompasses the entire civilian population living in targeted areas, it dramatically affects the civilian population's willingness to engage in a wide variety of daily activities. Because of the drones' constant presence in North Waziristan, Pakistan, terrified civilians changed the ways in which they conduct some of the most mundane activities. First, because of drone strikes on funerals and spaces where mourning families gathered to express their condolences to drone victims' relatives, people started to avoid attending funerals. Second, because of strikes targeting schools, some families decided to stop sending their children to school. In some cases, it was not possible for children to continue their education because teachers and staff decided to not continue their work at schools. Thus, the fear of drone strikes affected the children's right to education. Third, because of drone strikes on social gatherings—for example, on *jirgas*, the Pashtun community-based conflict resolution process—some members of the targeted population decided to avoid attending such events. Furthermore, because of strikes on private houses and *hujras*, drawing rooms where guests are entertained, many people are now afraid to gather in large groups in their or their friends' houses. Fourth, because of numerous "double tap" strikes in which rescuers helping victims of the first strike were bombed in follow-up strikes, professional health-care workers and bystanders who find themselves in the vicinity of a strike are now afraid to assist drone victims immediately after the first strike takes place. As people are terrified of follow-up strikes, they either completely avoid locations hit by drone missiles or they wait a considerable time before going to the site of the strike to help the victims (International Human Rights and Conflict Resolution Clinic and Global Justice Clinic 2012, pp. 73–99). In addition to these changes in daily life that are a consequence of indiscriminate strikes and the constant presence of drones, the drone program also undermined community trust in surveilled areas. The presence of local informers gathering information on "terrorist" activities, information that is then used by the US forces to select targets, increased mistrust among the people living in surveilled areas. Locals suspect that their neighbors, other members of their tribe, and especially outsiders may be working for the USA to collect intelligence on potential targets. Thus, in surveilled communities the drone program has generated a state of permanent mistrust that hampers the normal functioning and development of these communities (International Human Rights and Conflict Resolution Clinic and Global Justice Clinic 2012, pp. 99–101).

The drones' impacts on the civilian population described here are similar to the impacts on civilians reported in other conflict zones, which may point to the conclusion that drones do not represent a new form of terror in the context of war. However, I argue that it is the drones' constant presence and, consequently, the constant terror they create that represents an "upgrade" in modern warfare, a new mode of terrorizing the civilian population. The advent of drones introduced a new era in which populations living in conflict zones are constantly subjected to the executioner's gaze signaling the possibility of imminent death.[3]

Conclusion: The "Ideal" Society

The part of the "war on terror" that is being fought with drones is not merely a war in which the ultimate goal is to seek out and eliminate, with "surgical precision", alleged "terrorists", but rather a war in which the US military systematically aims at subjecting the wider population living in drone-surveilled areas. The goal of drone warfare is not solely to decapitate "terrorists," as pro-drone apologists would like to convince us, but rather to establish pervasive power relations in which the

[3] In Pakistan's tribal areas, it is difficult to evaluate how violence perpetrated by other belligerent sides—the insurgents and members of the repressive apparatus of the Pakistani state—affects the civilian population. It is true that to a certain extent violence perpetrated by these groups terrorizes the local population and forces them to change their behavior. However, based on interviews with locals from North Waziristan, it is clear that the drones' constant presence plays a very significant role in producing changes in the way mundane activities are being carried out in targeted areas.

population living in surveilled areas would be disciplined under the US rule.[4] The drones constantly affect all people living in surveilled areas, insurgents and innocent civilians alike. All the time, sometimes merely through their presence, drones terrorize all members of the population and force them to significantly change their way of life.

In the process of establishing an effective power regime in drone-surveilled areas, the architects of the drone program relied, in large part, on disciplinary power techniques. The newly articulated "right of the sword" has been implemented across declared and undeclared war zones within a framework that consists of three disciplinary techniques, the three "means of correct training," which are supposed to guarantee "surgically precise" strikes. And it is these targeted executions that are supposed to have a disciplinary effect on the targeted population—these strikes try to break the will of the targeted population and force its members to accept US domination. Moreover, the drone program also relies on the panoptic distribution of visibility and invisibility in order to create disciplinary power effects vis-à-vis the surveilled population. If we agree with Foucault's interpretation of panoptic structures as places permeated by disciplinary power relations, it is possible to come to the conclusion that drone-surveilled areas, where we have a similar panoptic distribution of visibility and invisibility, are dominated by disciplinary power relations.

Despite the similarities between the disciplinary power regime and the drones' power regime, there is also a key difference between them. The power regime introduced by drones is not simply a copy of disciplinary power but, to a certain extent, an original invention that differs from disciplinary power. Within the walls of panoptic institutions, disciplinary power strives to create an "ideal society"—a docile society made up of "normalized" individuals placed within a multilayered pyramidal hierarchical system, a docile society in which subjected individuals dutifully carry out tasks assigned to them by the dominant class (e.g., workers carrying out production work in a factory, soldiers conducting operations in keeping with their superiors' orders, students carrying out tasks under their teachers' supervision; Foucault 1995, pp. 135–169). In drone-controlled areas, the "ideal society" is also envisioned as a disciplined society of docile individuals, but the drones' "ideal society" does not consist of productive, laboring individuals placed within a pyramidal hierarchical structure but rather of a mass of terrified individuals who are expected *to not* conduct, or stop conducting, certain activities—that is, all kinds of insurgent activities perceived as a threat by the US administration. This difference is crucial—the primary aim of disciplinary power is to create productive individuals who will carry out specific tasks in accordance with the dominant class's interests, while the aim of the power regime established by drones is not to create productive

4 To a certain extent, I argue, drone warfare also aims to discipline, through the display of untrammeled US power, other populations that do not live under drone surveillance. By displaying their capability to launch drone strikes in many declared and undeclared combat zones, the US projects its power not only on the drone-surveilled population, but on a global audience. Drone strikes serve as a reminder of US power directed at a wider, global audience, as a reminder of what can happen to those who oppose US power. Drones not only terrorize populations kept under their control, but also serve as a means to instill fear in the public at large.

individuals but rather passive, frightened, "morally reformed" individuals who will, by renouncing attacks on US military or civilian targets, stop posing a threat to US strategic interests.

However, in addition to this disciplinary effect on the targeted population, drones also produce what at first glance may seem to be a dramatic "side effect"—strikes killing or injuring civilians. But, by taking into account the various factors that constantly influence the drones' target selection procedure, it is clear that the strikes' imprecision is not merely a "side effect" that happens from time to time but rather a permanent feature of the drone program. And it is precisely this randomness in the strikes' execution that undermines the drones' disciplinary structure. When members of the surveilled population become aware that anyone can be hit by a missile fired from a drone, they consequently realize that it makes no sense for them to subject themselves to the US military rule of terror because they know that such subjection will not bring them the gratification of being allowed to live. When members of the population realize that subjection to US domination will not guarantee them a safer life, they also comprehend that it is not logical for them to not cooperate in an armed struggle against this domination. When indiscriminate strikes become the norm, the only solution for the subjected population is to do whatever it takes to prevent the launch of such strikes. Therefore, the seeds of armed rebellion are already planted by the drone warfare itself.

References

Abbot, S. (26 February 2012). New light on drone war's death toll. *Associated Press*. http://news.yahoo.com/ap-impact-light-drone-wars-death-toll-150321926.html. Accessed 28 April 2014.

Ali, Y. (5 November 2006). Most Bajaur victims were under 20. *The News*. http://www.thenews.com.pk/TodaysPrintDetail.aspx?ID=4043&Cat=13&dt=11/5/2006. Accessed 28 April 2014.

Alston, P. (2009). Report of the special rapporteur on extrajudicial, summary, or arbitrary executions. United Nations Human Rights Council. http://daccess-dds-ny.un.org/doc/UNDOC/GEN/G09/131/17/PDF/G0913117.pdf?OpenElement. Accessed 6 June 2014.

Alston, P. (2010). Report of the special rapporteur on extrajudicial, summary, or arbitrary executions. United Nations Human Rights Council. http://www2.ohchr.org/english/bodies/hrcouncil/docs/14session/A.HRC.14.24.Add6.pdf. Accessed 10 April 2014.

Amnesty International. (2013). "Will I be next?": Drone strikes in Pakistan. http://amnesty.org/en/library/asset/ASA33/013/2013/en/041c08cb-fb54-47b3-b3fe-a72c9169e487/asa330132013en.pdf. Accessed 12 May 2014.

Becker, J., & Shane, S. (29 May 2012). Secret 'kill list' proves a test of Obama's principles and will. *The New York Times*, A1.

Benjamin, M. (2012). *Drone warfare: Killing by remote control*. New York: OR Books.

Bentham, J. (1995). *The panopticon writings*, London: Verso.

Cloud, D. S. (10 April 2011). Anatomy of an Afghan war tragedy. *Los Angeles Times*. http://articles.latimes.com/print/2011/apr/10/world/la-fg-afghanistan-drone-20110410. Accessed 25 March 2014.

Department of Defense of the United States of America. (2012). Sustaining U.S. Global Leadership: Priorities for 21st Century Defense. http://www.defense.gov/news/defense_strategic_guidance.pdf. Accessed 25 March 2014.

Devereaux, R. (27 October 2013). Family of grandmother killed in US drone strike arrive for Congress visit. *The Guardian*. http://www.theguardian.com/world/2013/oct/27/drones-attack-pakistan-family-rehman-congress Accessed 15 May 2014. Accessed 21 May 2014.

Dillon, M. (1998). Criminalizing social and political violence internationally. *Millennium, 27*(3), 543–567.

Dreyfus, H. L., & Rabinow, P. (1983). *Michel foucault: Beyond structuralism and hermeneutics*. Chicago: The University of Chicago Press.

Emmerson, B. (2013). Report of the special rapporteur on the promotion and protection of human rights and fundamental freedoms while countering terrorism. United Nations Human Rights Council. http://justsecurity.org/wp-content/uploads/2013/10/2013EmmersonSpecialRapporteurReportDrones.pdf. Accessed 7 June 2014.

Entous, A., Gorman, S., & Barnes, J. E. (26 April 2012). U.S. Relaxes Drone Rules. *Wall Street Journal*. http://online.wsj.com/article/SB10001424052702304723304577366251852418174.html. Accessed 25 March 2014.

Foucault, M. (1995). *Discipline and punish: The birth of the prison*. New York: Vintage Books.

Foucault, M. (2001). The subject and power. In J. D. Faubion (Ed.), *Power: Essential works of foucault 1954–1984*. New York: The New Press.

Foucault, M. (2003). *"Society must be defended": Lectures at the Collège de France 1975–76*. New York: Picador.

Human Rights Clinic—Columbia Law School. (2012). The civilian impact of drones: Unexamined costs, unanswered questions. http://web.law.columbia.edu/sites/default/files/microsites/human-rights-institute/files/The%20Civilian%20Impact%20of%20Drones.pdf. Accessed 8 June 2014.

Human Rights Watch. (2014). A wedding that became a funeral: US drone attack on marriage procession in Yemen. http://www.hrw.org/sites/default/files/reports/yemen0214_ForUpload_0.pdf. Accessed 12 May 2014.

International Crisis Group. (21 May 2013). Drones: Myths and reality in Pakistan. http://www.crisisgroup.org/en/regions/asia/south-asia/pakistan/247-drones-myths-and-reality-in-pakistan.aspx. Accessed 1 June 2014.

International Human Rights and Conflict Resolution Clinic—Stanford Law School, Global Justice Clinic—NYU School of Law. (2012). Living under drones: Death, injury, and trauma to civilians from US drone practices in Pakistan. http://www.livingunderdrones.org/wp-content/uploads/2013/10/Stanford-NYU-Living-Under-Drones.pdf. Accessed 31 March 2014.

Jabri, V. (2010). *War and the transformation of global politics*. London: Palgrave Macmillan.

Linebaugh, H. (29 December 2013). I worked on the US drone program. The public should know what really goes on. *The Guardian*. http://www.theguardian.com/commentisfree/2013/dec/29/drones-us-military. Accessed 31 March 2014.

Mayer, J. (26 October 2009). The predator war: What are the risks of the CIA's covert drone program? *The New Yorker*. http://www.newyorker.com/reporting/2009/10/26/091026fa_fact_mayer. Accessed 3 June 2014.

McKelvey, T. (20 November 2012). A former ambassador to Pakistan speaks out. *The Daily Beast*. http://www.thedailybeast.com/articles/2012/11/20/a-former-ambassador-to-pakistan-speaks-out.html. Accessed 3 June 2014.

Miller, G. (23 October 2012). Plan for hunting terrorists signals US intends to keep adding names to kill lists. *The Washington Post*. http://www.washingtonpost.com/world/national-security/plan-for-hunting-terrorists-signals-us-intends-to-keep-adding-names-to-kill-lists/2012/10/23/4789b2ae-18b3-11e2-a55c-39408fbe6a4b_story.html. Accessed 6 June 2014.

Pilkington, E. (5 January 2012). Barack Obama sets out plans for leaner military in historic strategy shift. *The Guardian*. http://www.theguardian.com/world/2012/jan/05/barack-obama-plans-leaner-military. Accessed 28 April 2014.

Radio Transmissions and Intercom. (2010). http://www.scribd.com/doc/52825019/Transcripts-of-Drone-Attack. Accessed 14 April 2014.

Risen, J. (9 August 2009). U.S. to hunt down Afghan drug lords tied to Taliban. *The New York Times*, A1.

Rodgers, A., & Hill, J. (2014). *Unmanned: Drone warfare and global security*. London: Pluto Press.

Rohde, D. (26 January 2012). The drone wars. Reuters. http://www.reuters.com/article/2012/01/26/us-david-rohde-drone-wars-idUSTRE80P11I20120126. Accessed 31 March 2014.

Scahill, J. (2013). *Dirty Wars: The world is a battlefield*. New York: Nation Books.

Schmitt, E., & Sanger, D. E. (22 February 2008). Pakistan shift could curtail drone strikes. *New York Times*. http://www.nytimes.com/2008/02/22/washington/22policy.html?_r=0. Accessed 14 April 2014.

Schmitt, E., & Saraye, S. (23 February 2013). U.S. troops at drone base in West Africa. *The New York Times*, A4.

Shachtman, N. (19 May 2009). CIA Chief: Drones 'Only game in town' for stopping al Qaeda. Wired. http://www.wired.com/2009/05/cia-chief-drones-only-game-in-town-for-stopping-al-qaeda/. Accessed 14 April 2014.

Shah, P. Z. (27 February 2012). My drone war. Foreign Policy. http://www.foreignpolicy.com/articles/2012/02/27/my_drone_war. Accessed 2 July 2014.

Shaw, I. (2013). Predator Empire: The geopolitics of US drone warfare. *Geopolitics, 18*(3), 536–559.

Shaw, I., & Akhter, M. (2014). The dronification of state violence. *Critical Asian Studies, 46*(2), 211–234.

Singer, P. W. (2009). *Wired for war: The robotics revolution and conflict in the twenty-first century*. New York: The Penguin Press.

Strawser, B. J. (2010). Moral predators: The duty to employ uninhabited aerial vehicles. *Journal of Military Ethics, 9*(4), 342–368.

Tzu, S. (1971). *The art of war*. Oxford: Oxford University Press.

Ullman, H., & Wade, J. (1996). Shock and awe: Achieving rapid dominance. The National Defense University. http://www.dodccrp.org/files/Ullman_Shock.pdf. Accessed 31 March 2014.

United Nation Assistance Mission in Afghanistan. (2013). Afghanistan mid-year report 2013: Protection of civilians in armed conflict in Afghanistan for 2013. http://www.unama.unmissions.org/LinkClick.aspx?fileticket=EZoxNuqDtps%3d&tabid=12254&language=en-US. Accessed 12 May 2014.

U.S. Senate Foreign Relations Committee. (10 August 2009). Afghanistan's narco war: Breaking the link between drug traffickers and insurgents—A Report to the Senate Committee on Foreign Relations. United States Senate. http://www.voltairenet.org/IMG/pdf/Afghanistan_s_Narco_War.pdf. Accessed 4 June 2014.

Wall, T., & Monahan, T. (2011). Surveillance and violence from afar: The politics of drones and liminal security-scapes. *Theoretical Criminology, 15*(3), 239–254.

Whitlock, C. (20 July 2013). U.S. military drone surveillance is expanding to hot spots beyond declared combat zones. *The Washington Post*. http://www.washingtonpost.com/world/national-security/us-military-drone-surveillance-is-expanding-to-hot-spots-beyond-declared-combat-zones/2013/07/20/0a57fbda-ef1c-11e2-8163-2c7021381a75_story.html. Accessed 25 June 2014.

Williams, A. (2011). Enabling persistent presence? Performing the embodied geopolitics of the unmanned aerial vehicle assemblage. *Political Geography, 30*(7), 381–390.

Woods, C., & Lamb, C. (4 February 2012). CIA tactics in Pakistan include targeting rescuers and funerals. The Bureau of Investigative Journalism. http://www.thebureauinvestigates.com/2012/02/04/obama-terror-drones-cia-tactics-in-pakistan-include-targeting-rescuers-and-funerals/. Accessed 12 May 2014.

Woods, C., & Yusufzai, M. (1 August 2013). Bureau investigation finds fresh evidence of CIA drone strikes on rescuers. The Bureau of Investigative Journalism. http://www.thebureauinvestigates.com/2013/08/01/bureau-investigation-finds-fresh-evidence-of-cia-drone-strikes-on-rescuers/. Accessed 9 June 2014.

Zenko, M. (2013). Reforming U.S. drone strike policies. Council on Foreign Relations. http://www.cfr.org/wars-and-warfare/reforming-us-drone-strike-policies/p29736. Accessed 31 March 2014.

Vasja Badalič is a researcher at the Institute of Criminology at the Faculty of Law in Ljubljana, Slovenia. His primary fields of research are political philosophy, social philosophy, and French structuralism. He combines theory with frequent fieldwork, focusing on political, military, and labor issues in the Middle East (Iraq, Lebanon, and Syria), Central Asia (Afghanistan), South Asia (India, Bangladesh, and Pakistan), and Latin America (Venezuela, Colombia, Bolivia, Honduras, and Nicaragua). He is the author of three single-authored monographs: *The Terror of 'Enduring Freedom': War in Afghanistan and Pakistan*, Krtina Publishing House, Ljubljana 2013 (in Slovenian only); *The Red Machinery: The Beat of Chávez's Revolution*, Študentska založba Publishing House, Ljubljana 2011 (in Slovenian only); and *For 100 Euros a Month: The Production System of Global Capitalism*, Krtina Publishing House, Ljubljana 2010 (in Slovenian only). His work frequently appears in Slovenian newspapers and magazines such as *Delo, Dnevnik, Večer, and Mladina.*

Part IV
Drones and International Air Law

Chapter 9
An Analysis of Unmanned Aircraft Systems Under Air Law

Pablo Mendes de Leon and Benjamyn Ian Scott

Introduction

The history of unmanned aircraft systems (UAS) is older than that of piloted aviation. However, it was World War II that saw pilotless aircraft being utilised on a large scale with the V1 which was a flying bomb that was preprogrammed to crash after a determined duration (Degarmo 2004, p. 1–2). As UAS history is well situated within the military context, it is not surprising that they have predominantly been used for military applications. One of the most pertinent examples of a modern military UAS is the US Predator, which has been used by the US Air Force in armed conflicts over the skies of Iraq, Afghanistan, Bosnia, Kosovo and Korea (Lazarski 2002, p. 75) and non-conflict areas such as the cases involving Pakistan (Amnesty International 2013). The Predator was used by the US Air Force to fly missions whereby it captured live video footage of enemy actions and streamed it back to the operator, used its on-board weapon systems to destroy enemy targets, and aided precession distance weapons by locating targets (Peterson 2006, p. 548). In addition to combat roles, UAS have also been used by governments for policing, border control with the aim of preventing illegal immigration, search and rescue, and remote sensing.

Not only do UAS serve a purely governmental role, they are also used in dual governmental and private activities. Within this area, their use is becoming more relevant and their application is expanding. This is because UAS are beginning to be used to monitor disaster relief actions, such as flooding, forest fires, earthquakes,

The authors are most grateful for the helpful comments and suggestions of Rita Sousa Uva, Legal Adviser at the European Aviation Safety Agency.

P. Mendes de Leon (✉) · B. I. Scott
International Institute of Air and Space Law, Leiden University, Leiden, Netherlands
e-mail: p.m.j.mendesdeleon@law.leidenuniv.nl

B. I. Scott
e-mail: benjamyn.scott@gmail.com

A. Završnik (ed.), *Drones and Unmanned Aerial Systems*,
DOI 10.1007/978-3-319-23760-2_9

volcanic eruptions and chemical clouds. In addition, they are also being used for safety inspections of infrastructure, such as rail tracks, dams, dykes and power grids. The above examples have been utilised by both private entities and governments, but UAS also have a purely private application; for example, private entities are currently developing micro-UAS which may be able to fix gas pipe leaks or imitate bees by pollinating plants. Furthermore, they are being used in cartographic photography, cinematography and sporting events. Finally, they are also developing rapidly within the private non-commercial sector as individuals are partaking in amateur building and flying of UAS for recreational purpose.

As a matter of fact, the civil use of UAS is growing, and thus it is becoming more pertinent and necessary to regulate this expanding industry. In the world, there are currently approximately around 500 UAS manufactures, whereby Europe is hosting over 30 % (European Commission 2014b, p. 2). The European Commission (Commission) has predicted that this number will increase and this expansion will have the effect of creating up to 150,000 European jobs by 2050 (European Commission 2014a, p. 4). Within Europe, there have been some legal steps taken to help facilitate this developing area, as national authorities have granted around 1000 operating licences to operators. The European growth is nowhere more evident than in France which currently has the largest market share: in December 2012, the French authorities approved 86 operators. However, with the rapid European growth in this area, this number has since risen, as in February 2014, it totalled 431 approved operators. The Commission has stated in a memo that it is predicted, due to the positive growth trend in civil UAS, that within the next 10 years, this industry could amount to €15 billion per year which would be equivalent to 10 % of the world's aviation market (European Commission 2014, p. 2).

The emerging civil UAS sector within Europe has a wide application and will have a significant impact on jobs, the economy, how business is carried out and the aviation industry. Consequently, regulators should consider this in order to ensure that the market can develop appropriately and that the market is properly regulated.

There is, however, currently an absence of a clear legal framework governing UAS; therefore, steps are being taken by the European Union (EU), national authorities and international organisations, such as the International Civil Aviation Organization (ICAO), to address this issue. This work will address the current legal situation regarding the regulation of UAS giving particular attention to ICAO and the EU. In addition to this, attention will also be given to the regulatory proposals coming out of these two entities.

Terminology

Throughout their history, UAS have been referred to under many different names, and because of this, it can lead to some confusion and the nuances between the different terms being overlooked.

Probably, the most well-known and frequently used term is 'drone' which was often officially used by governments in pre-Gulf War times. However, as it is becoming politically unpopular, governmental and non-governmental entities now often avoid using this term (Peterson 2006, p. 528). The term 'drone' is not recognised or used, for example, by the US Federal Aviation Administration (FAA). With the move away from the term drone, 'unmanned aerial vehicle' (UAV) is often popularly used, especially among the media.[1] The Chicago Convention (1944; CC44) on International Civil Aviation (see below under the section 'Pilotless Aircraft Under the Chicago Convention') speaks of 'pilotless aircraft'. The FAA and the National Aeronautics and Space Administration (NASA) have also in the past referred to them as 'remotely operated aircraft' (ROA) (Peterson 2006, p. 528).

There has been a significant terminological shift that has recently occurred which has seen the term 'UAS' being used. For example, the FAA has moved away from ROA and has adopted UAS (US FAA 2014). In addition, the EU, the European Aviation Safety Agency (EASA) and ICAO also widely use UAS.[2]

The above-mentioned terms are often used interchangeably to refer to either just the unmanned aircraft or the unmanned aircraft plus the whole system required to fly it, such as the communication links, controls and operator. Due to the general consensus among the international community, this work will utilise the term 'UAS' in a way to include both the unmanned aircraft and its system.

Within the scope of UAS, there is a distinction made which needs to be appreciated as it provides a nuanced difference within the term UAS.[3] The first distinction that is made is called remotely piloted aircraft systems (RPAS). RPAS are UAS whereby the vehicle is controlled by a human pilot from a remote location. In its (draft) *Manual on Remotely Piloted Aircraft Systems,* ICAO has defined RPAS as a

> remotely-piloted aircraft, its associated remote pilot station(s), the required command and control links and any other components as specified in the type design. (ICAO 2012, p. 22)

The second subdivision is called 'unmanned drones', and these are UAS that do not have a remotely located pilot as they are autonomously controlled. These are said to not yet be authorised for use by ICAO or under EU rules (European Commission 2014b, p. 1); however, this could change with the introduction of new rules.

International Law

The CC44 with its 191 signatory states (per July 2015) provides an overarching and underpinning legal framework to international civil aviation and has been referred to as the Magna Carta of international aviation. Due to this, it must be examined within the scope of UAS as it may contain relevant provisions that affect UAS.

---

[1] The term 'UAV' is, however, losing popularity and has been described by some as an 'obsolete term' (Steer Davies Gleave 2014, p. ii.).

[2] See EASA (2009). See also ICAO (2011).

[3] See European Commission (2014b).

Application to International Aviation Only

The CC44 regulates *international* aviation, that is, the operation of services carried out by civil aircraft passing through the airspace of more than one contracting state.[4] Domestic applications of civil aircraft do not fall under the regime set forth by the CC44 and its ICAO Annexes. Therefore, civil UAS may fall under the scope of the CC44 as long as they satisfy the international criterion.

Within the military context, the *international* element is often satisfied, as for example, Reaper UAS can have a range of 5900 km, a maximum airspeed of 250 knots and can ascend to 15,300 m with missions capable of totalling 18 h (The Guardian 2013), thus providing them with the capacity to conduct cross-border activities. However, within the civilian context, the satisfaction of the international criterion needed for the invocation of the CC44 is not so apparent as most operations will take place within the territory of one state.

The majority of UAS activities in Europe involve small UAS used for civil purposes and these, where regulation is provided, often have to be flown with visual line of sight (VLOS).[5] Therefore, there is limited exposure to international borders for the majority of UAS activities because their range would be markedly limited. Nevertheless, the CC44 may still be applicable and should not be disregarded as states share land borders, and this may introduce an international element. This is even relevant within the context of small UAS in the EU, as most states share borders, and this could have significant legal implications for those that border with non-EU states, such as Russia, Belarus, Ukraine and Turkey.

The Notion of Aircraft

The CC44 applies to *aircraft,* and in the Annexes to the CC44, ICAO has produced a standard definition of the term 'aircraft' which is to mean:

> Any machine that can derive support in the atmosphere from the reactions of the air other than the reactions of the air against the earth's surface. (CC44, Annex 2, 6 & 8)

This, therefore, includes any winged or helicopter vehicles capable of achieving lift. The majority of UAS utilise winged and helicopter technologies in order to fly, so operating in the same manner as manned aircraft. Consequently, it follows that these UAS satisfy ICAO's definition.[6] This conclusion is widely accepted, as for example, ICAO has adopted this definition within the context of UAS and has

---

[4] '"International air service" means an air service which passes through the air space over the territory of more than one state' (Chicago Convention 1944, Article 96(b)).

[5] The UK, for example, necessitates VLOS for small unmanned aircraft. 'The person in charge of a small unmanned aircraft must maintain direct, unaided visual contact with the aircraft sufficient to monitor its flight path in relation to other aircraft, persons, vehicles, vessels and structures for the purpose of avoiding collisions' (CAP393 2014, Article 166(3)).

[6] It may be possible with the advancement of technology that not all forms of UAS may fall under the definition of 'aircraft', so caution must be given when categorising these developing vehicles.

begun to take active steps within its aviation-based competencies to regulate this developing activity. Thus, Amendment 43 of ICAO Annex 2 specifically modifies the Annex to cover UAS (Amendment 43 2014), of which EASA is currently working with stakeholders to implement Amendment 43 of Annex 2 into the EU system (EASA 2014b). Therefore, the criterion of what determines an aircraft is not dependent on the pilot being situated within the vehicle, so UAS are aircraft for the purpose of the CC44.

In addition to traditional UAS, those which operate in the same way as manned airplanes and helicopters, ICAO's definition also provides other kinds of 'aircraft', such as balloons and gliders, and each of these can have unmanned versions (ICAO 2011, p. 4). Therefore, the term 'aircraft' has a wide scope, and this will have an impact on these other UAS technologies as they are also to be considered as aircraft.

Whilst UAS are aircraft as prescribed under the CC44's Annexes, there is, however, a restriction to this inclusive definition. ICAO has noted that:

> In the broadest sense, the introduction of UAS does not change any existing distinctions between model aircraft and aircraft. (ICAO 2011, p. 3)

The term 'model aircraft' refers to those aircraft generally recognised as intended for only recreational purposes. Therefore, their use determines their label. This distinction has also been acknowledged by the EU and is shaping the future Regulations within EASA (EASA 2014b, p. 13). Due to this distinction, those aircraft categorised as 'models', which will constitute a significant portion of the world's UAS, will fall outside the scope of the CC44 and will, therefore, then be subjected to different regulatory standards.

An additional category which is made at an EU level is 'toy aircraft'. These fall under the scope of Directive 2009/48/EC (2009) which

> shall apply to products designed or intended, whether or not exclusively, for use in play by children under 14 years of age. (Directive 2009/48/EC 2009, Article 1)

Some of the toys under this Directive are capable of flight and are thus aircraft. There are some legal obligations prescribed as follows:

> The Directive aims at protecting the 'user' of the toy from any possible hazard (e.g. toxic materials), but it does not cover risks for third parties on the ground or in the air. (EASA 2014b, p. 14)

EASA holds the position that both model and toy aircraft UAS fall under the scope of Regulation 923/2012 (2012; Standardised European Rules of the Air, SERA), and because of this, SERA has to be aligned with Amendment 43 to ICAO Annex 2 in order to include UAS (EASA 2014b, pp. 2.4.5.4.3.). Therefore, under EU law, both model and toy aircraft UAS will have to obey the 'common rules of the air'.

Whilst the definition of 'aircraft' is prima facie inclusive, the exclusion of model and toy aircraft produces a legal patchwork. This may lead to an overlap and contradiction between the rules. For example, an operator may fly a large UAS across an international border, thus seemingly bringing it under the scope of the CC44 and its Annexes, but if it is for only recreational purposes, it would fall outside the scope of ICAO due to definitional purposes. This could lead to legal, safety and security concerns if this is not addressed.

Sovereignty

Prior to the twentieth century, international flight was virtually unregulated, and it was not until the beginning of the last century that significant attempts were made to construct a body of law that matched the advances in technology. Some of the earliest debates can be traced back as far as 1902,[7] but it was not until 1919 that the Aeronautical Commission of the Peace Conference met in Paris to discuss the rudimentary elements of air navigation.[8] The resulting Paris Convention (1919) 'cemented the prevailing customary principle of State sovereignty over air space' (Milde 2008, p. 11). Article 1 of the Paris Convention declares:

> The High Contracting Parties recognise that every Power has complete and exclusive sovereignty over the air space above its territory.

The matter was raised again in 1944 when the US invited ally and neutral states to discuss international civil aviation in Chicago, between 1 November and 7 December 1944 at the International Civil Aviation Conference. The Conference resulted in the CC44. Article 1 of the CC44 reiterated the opening article of the Paris Convention:

> The contracting States recognize that every State has complete and exclusive sovereignty over the airspace above its territory.[9]

Therefore, each state has the 'unilateral and absolute right…to permit or deny entry into its territory and to control all movements therein' (Petras 2010, p. 9). It is clear that civil aircraft have no automatic right to fly over or land in the territory of another state. The CC44 Article 3(c) also affirms this principle within the context of state aircraft as:

> No state aircraft of a contracting State shall fly over the territory of another State or land thereon without authorization by special agreement or otherwise, and in accordance with the terms thereof. (Chicago Convention 1944, Article 3(c))

The sovereignty principle is also applicable for state aircraft, and because of this, UAS and their operators, whether engaged in civil or state activities must respect the sovereignty of foreign states according to international law.

Civil Versus State Aircraft

The CC44 makes a distinction between state aircraft and civil aircraft based on the *use* of the aircraft in question. Article 3(a) and (b) of the CC44 declares:

---

[7] See Reinhardt (2007), p. 70.

[8] Attendance consisted of most European states, as well as the US, Brazil, Cuba and Japan.

[9] The extent of a state's territory, of which it has sovereignty over, is that declared by Article 2. 'For the purposes of this Convention the territory of a State shall be deemed to be the land areas and territorial waters adjacent thereto under the sovereignty, suzerainty, protection or mandate of such State' (Chicago Convention 1944, Article 2).

(a) This Convention shall be applicable only to civil aircraft, and shall not be applicable to state aircraft.
(b) Aircraft used in military, customs and police services shall be deemed to be state aircraft.

There is a clear divide at an international level between civil and state aircraft. Despite this division, Article 3 does not contain a definition of 'state aircraft'.[10] Rather, it lays down a number of services which may be deemed to be those operated by state aircraft. Therefore, the term is somewhat ambiguous and this has given states a wide freedom to qualify aircraft as state aircraft pursuant to their national legislation and international practices.

There is a strong functional approach as the use of the aircraft will mostly determine its label.[11] State practice shows that the examples given by the CC44 are not exhaustive as such practice has widened the scope to include:

- Military
- Police functions
- Border control
- Fire fighting
- Search and rescue
- Pollution control
- Landings on navy ships with the purpose of measuring the environment

The operation of state aircraft is subject to national law and international agreements, including, but not limited to, the Geneva Conventions and Protocols,[12] NATO arrangements[13] and EUROCONTROL requirements.[14]

If an aircraft is not used for public services as mentioned in Article 3(b) of the CC44, then it is deemed to be a civil aircraft. Aircraft are often deemed to be for civil purposes, but not limited to, if they are engaged in:

- Sensing of water (depth; salt or sweet water)
- Registration of small sea animals and plants
- Photography
- Videos for cinema and commercials
- Meteorology
- Surveillance of traffic and pollution (Mendes de Leon 2013)

---

[10] An authoritative definition of 'state aircraft' has been provided by EUROCONTROL as follows: 'Principle 1: For ATM (Air Traffic Management) purposes and with reference to article 3(b) of the Chicago Convention, only aircraft used in military, customs and police services shall qualify as State Aircraft. Accordingly: Aircraft on a military register, or identified as such within a civil register, shall be considered to be used in military service and hence qualify as State Aircraft; Civil registered aircraft used in military, customs and police service shall qualify as State Aircraft; Civil registered aircraft used by a State for other than military, customs and police service shall not qualify as State Aircraft' (Eurocontrol 2001).

[11] However, attention is also given to the registration of the vehicle.

[12] See International Red Cross (2014).

[13] See, for instance, NATO's Standardisation Agreements (STANAG) which are designed to facilitate the use of military aircraft and to enhance the safe transportation of cargo. STANAG 4441 encompasses a manual regarding the carriage of military munitions and explosives.

[14] See EUROCONTROL (2014).

Therefore, UAS may be identified as civil or state aircraft as the determining factor is the *use* of the aircraft and not the location of the pilot. UAS are thus not distinct from manned aircraft as they will be used for numerous tasks, both civil and state. Consequently, depending on the use of the vehicle, different legal regimes may apply to UAS. The civil use will be the focus of this work.

Within the civil category, there is a subcategory between commercial and non-commercial aircraft, which in traditional aviation activities has been regulated differently.[15] It is currently not clear if this distinction will be maintained within the context of UAS or even if it is necessary to make such a distinction for UAS.

Pilotless Aircraft Under the Chicago Convention

The CC44 under Article 8 reflects the sovereignty requirements contained in Article 1 within the scope of 'pilotless aircraft'.

> No aircraft capable of being flown without a pilot shall be flown without a pilot over the territory of a contracting State without special authorization by that State and in accordance with the terms of such authorization. Each contracting State undertakes to insure that the flight of such aircraft without a pilot in regions open to civil aircraft shall be so controlled as to obviate danger to civil aircraft.

Article 8 is titled 'Pilotless Aircraft' and makes reference specifically to aircraft that are 'capable of being flown without a pilot'. This has been interpreted by many, such as by ICAO,[16] EU[17] and scholars,[18] to mean aircraft whereby the pilot is not physically on board and is instead located in a remote site (RPAS). Therefore, the conclusion is made that UAS fall explicitly within the provision of the CC44.

This interpretation, however, appears to contradict the term 'pilotless' as this suggests that the aircraft operates without a pilot being located in or outside of the vehicle. If a literal interpretation is adopted, as prescribed under Article 31 of the Vienna Convention on the Law of Treaties (VCLT),[19] then Article 8 is to be interpreted as to include only autonomous aircraft, whereby there is a complete absence of a pilot on board and outside of the aircraft (unmanned drones). Additionally, a contextual interpretation provided for under Article 31 of the VCLT may also lead Article 8 to be more appropriately applied to unmanned drones. This is because unmanned drones, such as the V1, are the context in which the CC44 was born. This would, however, leave a lacuna in international law as RPAS would be excluded, and because of this, a teleological interpretation which would look at the object and purpose of the CC44 may be more appropriate.

---

[15] See UK CAA (2014a).

[16] See ICAO (2011). See also CC44 Annex 2 and 7.

[17] See European Commission (2014b), p. 1.

[18] See Masutti (2012), Chap. 2.

[19] Although the VCLT has only 41 signatory states, its provisions, including Article 31, are widely regarded as a codification of customary international law (Vienna Convention 1969).

It is unclear from only the text of the Article which interpretation prevails. However, this distinction appears to be only academic as the majority of entities put RPAS well within the scope of the CC44.

Environmental Protection

Within the area of traditional aviation, there is growing concern with environmental issues. Firstly, emissions from aircraft are a global concern, and it is a core element of the EU's aviation policy as it is currently trying to implement its emissions trading system (ETS).[20] The second environmental aviation issue is regarding noise, which is currently slowing down airport development around the world. The aviation industry is also taking emissions and noise seriously, as for example, they are having an influence in the production of aircraft as manufactures such as Airbus and Boeing are striving to create greener and quieter aircraft. Therefore, it seems logical that such considerations would flow into the UAS context. However, '[n]o reviews have been conducted for determination for the need for noise and emissions control with respect to UAS' (ICAO 2011, p. 36).

Nevertheless, this does not mean that there are not general environmental rules governing UAS. This is because 'it is generally accepted that the existing noise and emissions standards for manned aircraft should be applied to UAS ICAO' (ICAO 2011, p. 36). The appropriate rules are contained in Annex 16 to the CC44 titled 'Environmental Protection':

- Annex 16, Volume I—'Aircraft' (noise)
- Annex 16, Volume II—'Aircraft Engine Emissions'

Therefore, whilst environmental considerations for UAS have yet to become a major concern, the UAS that fall under the scope of the CC44's Annexes will have to conform to the standards set out under Annex 16 on emissions and noise. It may prove over time that these rules are as suitable for UAS as they are for manned aircraft. Alternatively, it may become more appropriate to construct new environmental rules for UAS.

Other Provisions of the CC44

If UAS fall under the scope of the CC44, then it also follows that the other applicable obligations set forth in the CC44 apply to UAS, for example, Article 20 which declares that:

> Every aircraft engaged in international air navigation shall bear its appropriate nationality and registration marks.

[20] See European Commission (2014c).

Thus, those UAS involved in international air navigation must be registered and bear certain marks that indicate nationality and registration, of which ICAO has recently made amendments to Annex 7 in order to facilitate this. This brief examination of Article 20 was to serve as an example to show that there are other applicable rules under the CC44 rather than to provide an exhaustive list of these rules.

EU Law

The EU Acquis

There is currently a lack of *lex specialis* for UAS within the EU *acquis* and the Commission has observed that there is currently a

> lack of a European regulatory framework encompassing civil and military unmanned aircraft [which] prevents the development of legally authorized unmanned aircraft operations. (European Commission 2009, p. 2)

Pursuant to Article 100(2) of the Treaty of the Functioning of the European Union (TFEU; ex-Article 80 TEC), the EU member states have divested some of their aviation competencies to the EU.[21] Consequently, the EU has constructed a body of European air law, which, as UAS fall under the scope of aviation, must be examined in order to assess whether it affects UAS.

Market Access Under Regulation (EC) No. 1008/2008

One of the foundational legal documents within EU air law is Regulation 1008/2008 (2008) which regulates

> the licensing of Community air carriers, the right of Community air carriers to operate intra-Community air services and the pricing of intra-Community air services. (Regulation 1008/2008 2008, Article 1)

Therefore, the ability to operate an aircraft within the EU is pivotal to the satisfaction of this Regulation's articles, and it is likely to have an important impact on UAS.

There are two important terms within the Regulation that will have a determination on whether it applies to UAS, and these must be examined. Firstly, is the term 'Community air carrier' as defined under Article 2 of Regulation 1008/2008.

[21] 'The European Parliament and the Council, acting in accordance with the ordinary legislative procedure, may lay down appropriate provisions for sea and air transport. They shall act after consulting the Economic and Social Committee and the Committee of the Regions' (TFEU 2008, Article 100(2)).

> [A]n air carrier with a valid operating licence granted by a competent licensing authority in accordance with Chapter II.

Chapter II, Article 3(3)(a) of Regulation 1008/2008 declares that air services carried out by ultralight power-driven aircraft are not required to hold a valid operating licence. Although the weight is not quantified, it is usually regarded that light UAS fall under the term 'ultralight power-driven aircraft'. Should this assumption be correct, then a significant amount of EU UAS operators are not required to apply for and hold an operating licence under this EU Regulation. Secondly, in order for Regulation 1008/2008 to apply, the UAS activity must be an 'air service' as defined under Article 2:

> A flight or a series of flights carrying passengers, cargo and/or mail for remuneration and/or hire.

A factual analysis of current UAS activities shows that they do not carry passengers. This may change with technological advancements as autopilots will become more advanced, and there may come a time where passenger transportation services by aircraft no longer require a pilot to be on board. Furthermore, the carriage of cargo and mail by UAS are also currently very limited. Therefore, the majority of UAS activities will not be 'air services'.

A fundamental component of EU air law—Regulation 1008/2008—does not apply to many UAS activities. As the majority of EU UAS will fall outside its scope, it will therefore be down to the national authorities of the 28 EU member states to grant the specific permission.

Insurance

The EU has regulated aviation insurance under Regulation 785/2004 (2004). The Regulation's objective is specified under Article 1(1):

> The objective of this Regulation is to establish minimum insurance requirements for air carriers and aircraft operators in respect of passengers, baggage, cargo and third parties.

UAS fall under the scope of Regulation 785/2004. However, the Regulation is vastly limited within the context of UAS as it does not apply to:

- State aircraft as referred to in Article 3(b) of the CC44
- Model aircraft with a maximum takeoff mass (MTOM) of less than 20 kg
- Aircraft, including gliders, with an MTOM of less than 500 kg, and micro lights, which are used for non-commercial purposes or are used for local flight instruction which does not entail the crossing of international borders
- Other types of aircraft listed in this Regulation (Regulation 785/2004 (2004), Article 2 (2)(g))

It would appear that the majority of EU UAS activities would be excluded under the existing rules on aircraft insurance.

The Single European Sky

A pertinent piece of EU legislation for the present subject is that on air traffic management (ATM) as laid down in the Single European Sky (SES) regime.[22] This regime promotes the implementation of a 'common transport policy' (Regulation 549/2004 (2004), Preamble (1)).[23] In this context, the question could be asked if and to what extent UAS fall under the common transport policy in cases where those aircraft do not transport? Transport by air could be defined, as noted above, as the carriage of persons, in most cases passengers, their baggage, whether checked in or not, and cargo by an air transport undertaking.

On the other hand, the SES regime is designed to meet the requirements of all airspace users, meaning 'operators of aircraft operated as general air traffic' (Regulation 549/2004 (2004), Article 2(8). General air traffic is defined as

> all movements of civil aircraft, as well as all movements of state aircraft (including military, customs and police aircraft) when these movements are carried out in conformity with the procedures of ICAO. (Regulation 549/2004 (2004), Article 2 (26)

Consequently, the SES regime tries to implement Article 3(a) of the CC44, as it should, as confirmed by the statement that the SES regime is 'without prejudice to the rights and duties of Member States under…the Chicago Convention' (Regulation 549/2004 (2004), Article 1(3).

The SES regime is not as consequent as it could be as it only excludes 'military operations and training (Regulation 549/2004 (2004), Article 1(2)' from its scope. Therefore, the question is whether the operation of aircraft, including UAS which are used for policy, customs and other typically public service purposes, is or is not subject to the provisions of the SES regime.

Within this context, SERA, as analysed above in the section 'The Notion of Aircraft', will also need to be considered.

Air Service Agreements Concluded by the EU and Its Member States

A state may allow a foreign aircraft to enter its sovereign airspace through prior expressed permission which is usually given through negotiations between states. Within the EU, this role may be taken over by the Commission which, in exceptional cases, such as the EU—US Agreement on air transport, negotiates for the EU as a whole rather than a collection of 28 separate states, or in close cooperation with EU states, to reach agreements on international air transport. UAS operators may be able to utilise these agreements in order to enter the airspace of a foreign state. The

---

[22] See Regulation 1070/2009 (2009).

[23] See TFEU (2008), Article 100.

US–EU Open Skies Agreement will be explored here as it is one of the most liberal air service agreements.

On 30 March 2008, the first phase of the EU–US Open Skies Air Transport Agreement (2007) entered into force which replaced the bilateral air service agreement (BASA) regime between the individual EU states and the US (Association of European Airlines 2008). Consequently, the Open Skies Agreement, if it provides designation for UAS, would open up the US and EU market. Article 3(1)(a) and (b) of the agreement permits 1st and 2nd freedom (transit) rights, that is, the rights for operations to pass through the airspace of the airspaces of the other contracting states and make technical stops on their territories.

> Each Party grants to the other Party the following rights for the conduct of international air transportation by the airlines of the other Party: (a) the right to fly across its territory without landing; (b) the right to make stops in its territory for non-traffic purposes.

Article 3(1)(c)(ii) of the agreement allows for traffic rights, that is, the right for operators to carry traffic, including passengers, their baggage and cargo between agreed points located in the territories of the States party to the agreement.

> [F]rom points behind the Member States via the Member States and intermediate points to any point or points in the United States and beyond; for all-cargo service, between the United States and any point or points; and, for combination services, between any point or points in the United States and any point or points in any member of the European Common Aviation Area.

UAS activities are, however, unlikely to fall within the scope of this agreement. This is because, if an EU or US UAS operator wishes to exercise transit and/or traffic rights, then this activity must conform with Article 3. Article 3 of the agreement makes it clear that transit and traffic rights will only be granted to 'air transportation' which is defined as

> carriage by aircraft of passengers, baggage, cargo, and mail, separately or in combination, held out to the public for remuneration or hire. (US—EU Agreement 2007, Article 3)

It follows that UAS do not currently facilitate such a service, and, subsequently, they do not conduct air transportation and so cannot execute such rights. There is an absence of prior state consent, even within the context of a liberal agreement on air transport. Consequently, consent will have to be sought either on an ad hoc basis or through a new special agreement between the concerned states.

Initiatives from the European Commission

The First Steps

The Commission has observed that there is currently a lack of a European regulatory framework for UAS. Consequently, it has drawn the conclusion that this is significant as the current EU legal regime does not promote the growth of civil UAS as it does not permit the industry to build pertinent business plans and to develop new products adapted to its clients (European Commission 2009, p. 8). The Commission has since begun to assess its role in the support of this emerging sector. It is taking prudent steps before it establishes legally binding rules as it believes it is

> necessary to fully understand the potential European industry baseline, the potentialities and benefits offered by UAS to the European citizens, and the existing obstacles to the market emergence. (European Commission 2009, p. 3)

For this to be achieved, the Commission has opened up a communication with interested parties whereby it has invited all of the 'stakeholders to build together a policy framework for the development of a competitive drones market as well as rules that will tackle all citizens' concerns' (European Commission 2014b, p. 3). Therefore, in light of the lack of EU regulation, the Commission aims to construct a policy framework that will concern civil and commercial operations, in line with EU competence, to which it has currently constructed a *six action* proposal (European Commission 2014b). This will now be assessed as this will undoubtedly shape the future regulation of UAS within Europe.

Common Certification Processes and Standards

Safety is a paramount objective for EU aviation policy, so Action 1 of the Commission's proposal has recommended a common certification process and standards for UAS. A common certification process and standards for UAS is deemed necessary as

> in many States, the grant of an aerial work license to an UAS operator is almost impossible, as no appropriate framework for certification of the unmanned aircraft systems exist. (European Commission 2009, p. 8)

Therefore, this is required to ensure that UAS 'will have an equivalent level of safety in comparison to regular, manned, aviation' (European Commission 2014a, p. 5).

The Commission has anticipated the significance of small UAS within the EU as the 'UAS sector below 150 kg is composed of aerial vehicles of very different types, capabilities, size and weight'; therefore, regulation should encompass this wide scope (European Commission 2009, p. 8). Therefore, 'in this respect, the restricted scope of EASA competence to unmanned aircraft above 150 kg on the basis of traditional airworthiness considerations is an arbitrary cut off point and should be reconsidered' (European Commission 2014a, p. 5).

Support Market Development and European Industries

The Commission has proposed to support and encourage the young growing civil UAS market by promoting the development of UAS

> applications and the related technologies, stimulate user-driven innovation, and foster the creation of cross sectorial industrial value chains, appropriate support infrastructures and clusters. (European Commission 2014b, p. 4)

Therefore, the Commission has proposed two actions. Firstly, Action 2 notes that the Commission will, within reasonable limits, ensure that certain research and development needs are taken into account. Secondly, Action 6 states that the Commission will integrate UAS into its other initiatives, such as Horizon 2020 and COSME. Actions 2 and 6, whilst potentially having beneficial consequences as they can enhance the growth of the market and lower costs for the consumer, will also increase the need for appropriate regulation. This is because market facilitation may increase the number of participants, so increasing strain on the licensing system, as well as exposing more people to potential violations of the law. Therefore, the rules and institutions should be in place to cope with an expanding market.

Tough Controls to Ensure Security

Action 3 of the Commission's proposal is there to ensure that security aspects are dealt with accordingly. UAS are not immune from unlawful actions, for example, they can be used as a weapon, suffer from jamming or be subjected to hacking (European Commission 2014a, p. 7). The EU is not competent to deal with all security issues, such as military aviation, as for example EASA's

> Regulation *shall not apply* when products, parts, appliances, personnel and organisations referred to in paragraph 1 are engaged in military, customs, police, or similar services. The Member States shall undertake to ensure that such services have due regard as far as practicable to the objectives of this Regulation. (Regulation 216/2008 (2008), Article 1)

Therefore, the regulation of military UAS will not be undertaken by the EU as this would be an ultra vires act and will instead be up to each member state.

UAS are also not immune from cyber security risks. Firstly, UAS may be used by the operator or those authorised by the operator in a manner that produces risks to cyber security. Secondly, they may be accessed by unauthorised external sources which will compromise security. Due to the potential abuse of UAS, cyber security is an important factor to consider within this emerging area. This is because acts may occur through the utilisation of cyber technologies, and access to the UAS may produce cyber security risks. Such acts may include jamming[24] and spoofing.[25] Thus, to stop or at least mitigate unwanted security violations, rules on security are required.

---

[24] Jamming can be defined as the act of 'transmitting a high-power electronic signal that causes the bit error' (Zielinski et al. 1996, p. 25). See Infosec Institute (2013).

[25] Spoofing can be defined as the act of 'taking over a…system by appearing as an authorised user' (Zielinski et al. 1996, p. 25). See Aerospace Engineering and Engineering Mechanics (2014).

Protect Citizens' Fundamental Rights

The Commission's Action 4 has a human rights ethos as it sets out to protect citizens' fundamental rights, including the respect for the right to private and family life, and the protection of personal data.[26] This is a significant point as one of the main uses of civil UAS is to remotely record information, such as in remote sensing.[27] Directive 95/46/EC (1995) and the Framework Decision 2008/977/JHA (2008) have been cited under Action 4 as already existing EU rules that could deal with the protection of citizens' fundamental rights within the scope of UAS.[28] The Commission has decided that

> privacy situation would need continuous monitoring by the competent authorities including the national data protection supervisory authorities (European Commission 2014a, p. 8)

in order to ensure that the data is recorded for legitimate grounds.

The implementation of these rules is one thing, but the enforcement within the area of UAS is another. For example, the UK Civil Aviation Authority (CAA) has dealt with several cases regarding the illegal recording of information. In an unreported case, Mr Clift received a caution for selling photographs obtained from a UAS without an appropriate licence (UK CAA 2014b). This case came to light only once attempts to sell the photos were made. Therefore, in the absence of such a public display, such violations may go unreported, so more would need to be done at a national and EU level in order to ensure compliance.[29]

Liability/Insurance

Under Action 5 of the proposal,[30] the Commission has addressed third-party liability and insurance.

> The Commission will assess the current liability regime and third-party insurance requirement. It will, subject to the impact assessment, take the appropriate initiatives to ensure that adequate regulatory provisions are in place.

The Commission, whilst striving for high levels of safety for EU aviation, is aware that accidents are an inherent element of aviation; therefore, victims should be ap-

---

[26] See Finn et al. (2014).

[27] 'Amongst the wide range of potential civil RPAS applications a number may involve collection of personal data and raise ethical, privacy or data protection concerns, in particular in the area of surveillance, monitoring, mapping or video recording' (European Commission 2014a, p. 7).

[28] See Roma (2014), p. 28.

[29] This could be in the form of a light-touch approach such as including warnings on UAS packaging and producing easy-to-understand user guides. Alternatively, a harder-touch approach could be adopted such as implementing an effective enforcement body that actively looks for infringements and enforces the rules.

[30] See Steer Davies Gleave (2014).

propriately compensated for any injury or death. Therefore, 'there is a need for a liability insurance regime that ensures that at fault parties can meet their financial obligations' (European Commission 2014a, p. 8). The current third-party insurance regime within the EU was constructed with manned aircraft in mind, and because of this, Regulation 785/2004 (2004), which has been noted above in the section 'Insurance', has limitations for the application of UAS.

Addressing Safety Concerns Under European Law

European Aviation Safety Agency

As demonstrated by the Commission's action proposals, safety is a key element of EU aviation. Consequently, there are already well-established safety rules within the EU, in which EASA plays a crucial role, and these may be applicable to UAS.

EASA's competencies are set out in Regulation 216/2008 (2008) (the Regulation) which defines its role in establishing and maintaining a high uniform level of civil aviation safety in Europe.[31] The Commission is currently in the process of revising the Regulation, with EASA making the technical proposals, so that it better suits UAS.[32] The Commission's desire to regulate UAS is likely to also provoke a new Regulation focused on UAS in the EU which will involve EASA in the area of technical proposals. However, these are still in the preparatory stages, and their content is not yet known, so it is not possible to discuss these any further. However, when the Regulation as it presently stands is applied to UAS, it has certain limitations which could have an impact on the regulation of UAS in the EU, and these must be considered when drafting the new rules.[33]

Annex II(i) of the Regulation declares that 'unmanned aircraft with an operating mass of no more than 150 kg' are exempt. This exemption creates a clear division in the competencies of EASA whereby aircraft below 150 kg are outside the scope of EASA's competencies and are for national legislators to regulate. Thus, there is a fragmented EU legal system which is in contradiction to the Commission's goal of comprising a comprehensive set of rules governing UAS across Europe. Hence, the current weight restrictions should be rethought. EASA, in its *Concept of Operations*

[31] 'Drafting aviation safety legislation and providing technical advice to the European Commission and to the Member States; Inspections and training to ensure uniform implementation of European aviation safety legislation in all Member States; Airworthiness and environmental type-certification of aeronautical products, parts and appliances; Approval of aircraft design organisations world-wide and of production and maintenance organisations outside the EU; Coordination of the European Community SAFA (Safety Assessment of Foreign Aircraft) programme; [and] Coordination of safety programmes, data collection, analysis and research to improve aviation safety' (EASA 2014d).

[32] See EASA (2014a). See also, European Commission (2014a).

[33] See EASA (2009).

for Drones paper, has however stated that it will move away from the weight-based approach and will categorise UAS as open, specific or certified based upon the risk that the operations pose (EASA 2015).

The focus on weight comes from traditional aviation activities, and this may be redundant for UAS. This is because a 10 kg UAS is capable of surveying the ground below in the same way as a 500 kg Predator (Military Factory 2013). Secondly, the amount of damage that an aircraft can cause does not necessarily relate to weight. For example, in the *Smethwick Chinese Lantern Fire,* a Chinese lantern, a lighter-than-air object, caused a fire involving 100,000 t of plastic-recycling material which produced a 6000-ft smoke plume and £6 million worth of damage on 30 June 2013.[34] A UAS laden with fuel also has the potential energy to cause devastating fires. Finally, a UAS is capable of operating in both the segregated and non-segregated airspace. Therefore, they pose a potential problem to other aviation activities. An object as small as a bird or volcanic ash can cause an aircraft to crash, so a small UAS that enters an aircraft's engine could also have the same effect. There are also risks of collisions as larger UAS often operate within the D airspace which is shared by commercial passenger aircraft. Therefore, this has led some to draw sole attention away from weight and incorporate population density and activity into the equation. It has also been suggested that the focus should be on energy. The purpose behind these additions is to provide a more practicable and proportionate solution to the categorisation of UAS. It is still unclear if these alternative methods will be adopted or to what extent.

Article 1(2) of the Regulation declares that those UAS 'engaged in military, customs, police or similar services' fall outside the scope of EASA's competencies. Unlike traditional aircraft, UAS do not necessarily need to be modified in order to move between state and civil use. This is because UAS, for example, fitted with surveillance equipment can equally be used for monitoring state borders as well as filming a family event. In addition to this, the term 'similar services' is vague and can potentially lead to a wide interpretation. The last-mentioned provision, as with manned aircraft, will result in a portion of UAS not having to comply with EASA's rules.

UAS '[s]pecifically designed or modified for research, experimental or scientific purposes, and likely to be produced in very limited numbers', pursuant to Annex II(b) of the Regulation, also fall outside the scope of EASA's competencies. Many UAS will be designed and utilised specifically for these purposes, and this may lead to them being exempt.

Annex II(c) of the Regulation declares that

> aircraft of which at least 51 % is built by an amateur, or a non-profit making association of amateurs, for their own purposes and without any commercial objective

are exempt. In many cases, people are buying individual standard parts and putting them together to create a unique UAS. This would, therefore, put them within this exemption.

---

[34] See BBC (2013).

Finally, pursuant to Annex II(d) of the Regulation, those aircraft that have been used in the service of military forces are exempt. The history of UAS places the majority of the technology well within the military context, and because of this, it is possible that certain models or technologies within the civil market will be exempt. However, this is limited as those which are 'of a type for which a design standard has been adopted by the Agency' will fall under the scope of the Regulation.

There are other limitations to EASA's competences; however, these are not applicable to UAS (EASA 2009). Part 21 is currently applicable to the certification of UAS within its competencies (EASA 2009, in section 'Addressing Safety Concerns Under European Law').[35]

National Safety Rules

As concisely shown above, there are numerous international and European rules regulating aviation safety, but they are not applicable to UAS in all circumstances. However, the gaps are being filled as in recent years several EU member states, including Belgium, Czech Republic, Denmark, France, Germany, Italy, Lithuania, Sweden and the UK, have adopted legislation to regulate the operations of UAS (European Commission 2014b, p. 2). Each national law varies in content, and it is beyond the scope of this work to examine them all. The UK and Germany will serve as illustrative examples.

Under national laws, the UK, for example, has a well-developed body of law governing the operation of military and all other, that is, civil aircraft. A civil aircraft registered in the UK which is exempted from the above EASA Regulation must have a certificate of airworthiness and a permit to fly issued by the UK CAA pursuant to the terms of the Air Navigation Order and Regulations (CAP393 2014). Special rules apply to small UAS, that is, aircraft weighing less than 20 kg, and even less stringent rules to very small UAS, that is, craft weighing less than 7 kg (CAP393 2014, Article 166). A differentiated regime applies small surveillance UAS which must meet more stringent operational constraints (CAP393 2014, Article 167).

Whilst some governments, such as the UK's, have been active in constructing its UAS regulations to cover all weights, this is not the case elsewhere. For example, the German UAS rules leave a lacuna of unregulated territory for UAS between 25–150 kg (Luftverkehrs-Ordnung 1963, Articles 15(a), 16 & 16(a)). The German national rules regulate UAS below 25 kg and leave those above this demarcation to be regulated by EASA. However, as noted above, EASA is restricted to aircraft above 150 kg. Consequently, such operations between 20–150 kg are prohibited.

Therefore, this brief analysis of national rules shows that attempts are being made at a state level, but there is by no means a harmonised approach.

[35] See EASA (2014c).

Liability for Damage Caused by UAS

Liability for those at fault has had a strong presence in air law for decades, both for second and third parties. There are international conventions and national rules regulating this area, and these may be applicable to UAS.

Second-Party Liability

Under international law, the Montreal Convention (1999) has become the most notable Treaty for second-party liability.[36] The Montreal Convention is only intended to unify 'Certain Rules for International Carriage'; therefore, other rules may be applicable. Additionally, the applicability of the Montreal Convention is only relevant to carriage between signatory states, so, for example, a service between Thailand and the Netherlands would not fall under its scope as Thailand is not a signatory state. Finally, the Montreal Convention belongs to a long history of second-party liability regimes; thus, there is a patchwork among the international community. From this patchwork, the Montreal Convention will be the focus as it is the most recent.

The Montreal Convention will only be applicable to UAS if it satisfies the prescribed criteria. Firstly, the vehicle must be an *aircraft* (Montreal Convention 1999, Article 1). This has already been established above in the section 'The Notion of Aircraft'. Secondly, it

> applies to all international carriage of persons, baggage or cargo performed by aircraft for reward. It applies equally to gratuitous carriage by aircraft performed by an air transport undertaking. (Montreal Convention 1999, Article 1)

Therefore, it is limited to the scope of *international* carriage, which, as noted in the section 'Application to International Aviation Only', is currently very limited. Additionally, it is evident that the current situation does not involve the *carriage* of persons, as UAS aircraft are totally unmanned by pilots, crew and passengers. However, as noted above, there is a growing emergence of cargo transportation, which is demonstrated by Amazon.com's recent licensing request to the US FAA (CNN 2014).[37] Therefore, the relevance of second-party liability is likely to increase.

If the Montreal Convention becomes applicable to UAS, then Articles 17 and 18 which set forth the requirements for the applicability of liability under the convention will then need to be considered in order to establish liability.[38]

---

[36] See Regulation 889/2002 (2002).

[37] Amazon.com is not alone as other companies, such as DHL, have plans to conduct a portion of their delivery service using UAS. See The Verge (2014).

[38] 'The carrier is liable for damage sustained in case of death or bodily injury of a passenger upon condition only that the accident which caused the death or injury took place on board the aircraft or in the course of any of the operations of embarking or disembarking' (Montreal Convention 1999, Article 17).

Third-Party Liability—Damage on the Ground

The Rome Convention (1952) provides the international rules for the compensation of damage sustained by third parties on the territory of a contracting state. Article 1(1) states that

> Any person who suffers damage on the surface shall, upon proof only that the damage was caused by an aircraft in flight or by any person or thing falling therefrom, be entitled to compensation as provided by this Convention. Nevertheless there shall be no right to compensation if the damage is not a direct consequence of the incident giving rise thereto, or if the damage results from the mere fact of passage of the aircraft through the airspace in conformity with existing air traffic regulations. (Rome Convention 1952, Article 1(1))

There are many limitations to this convention. The damage must be caused by an aircraft in another signatory state of the Rome Convention. The convention does not apply to damage inflicted to aircraft flying in air space or to persons or goods on board such aircraft. There is the requirement of an international element which, as noted in the previous section, is not the case for many UAS activities.

There are currently only 42 member states to the 1952 Rome Convention, and this number does not include many important aviation states, such as the US and China. Accordingly, there is by no means a universal application of the rules for the compensation of damage sustained by third parties. Finally, there are problems with the identification of the operator of the aircraft. This is because Article 2(1) declares that the 'liability for compensation contemplated by Article 1 of this Convention shall attach to the operator of the aircraft (Rome Convention 1952, Article 2(1))' and an 'operator'

> shall mean the person who was making use of the aircraft at the time the damage was caused, provided that if control of the navigation of the aircraft was retained by the person from whom the right to make use of the aircraft was derived, whether directly or indirectly, that person shall be considered the operator. (Rome Convention 1952, Article 2(2))

In the case of private recreational UAS, the person making use of the aircraft may involve several people during the course of a day. In the case of registration, identification is easier (Rome Convention 1952, Article 2(3)); however, most states do not have the facilities to accommodate a UAS registration. Many UAS aircraft will then go unregistered. Therefore, the person making use of the aircraft may be difficult to locate.

Due to the limited scope of this convention and the potential difficulties in applying it to UAS, reference to national law, in principle tort law and product liability, may be sought.

Criminal Offences Committed in Relation to UAS

There are a number of international criminal aviation treaties that attempt to criminalise certain acts, and some of these may be applicable for UAS, and because of this, they must be identified and analysed. In addition to this, national law may yet again also be relevant. This point will be further illustrated in the section 'CAA v Robert Knowles in the UK' discussing a recent case in the UK.

Criminal Acts Committed on Board Aircraft

The Tokyo Convention (1963) was an international reaction to cases, such as US v Cordova (1950) and R. v Martin (1956) whereby the cases exposed that aircraft during international flights were often in an 'oases of lawlessness, where no law was applicable' (Cheng 1987, p. 25). The reason behind this was that the application of a state's criminal law had not been extended to aircraft hosting their nationality. Therefore, Article 1(1) of the Tokyo Convention declares:

> This Convention shall apply in respect of:
> (a) offences against penal law;
> (b) acts which, whether or not they are offences, may or do jeopardize the safety of the aircraft or of persons or property therein or which jeopardize good order and discipline on board.

The scope of this provision is markedly limited within the context of UAS as Article 2 declares that the 'convention shall apply in respect of offences committed or acts done by a person on board any aircraft'. Therefore, the offence must be committed by a person *on board the aircraft,* which is not possible with current UAS technology. This limitation is not exclusive to the Tokyo Convention as the Hague Convention on Hijacking (1971)[39] suffers from the same limitation as its aim is the criminalisation of acts by persons who are on board an aircraft in flight when they hijack, attempt or help to hijack that aircraft (Cheng 1987, p. 25). Therefore, the application to UAS at the moment is none.

Montreal Convention 1971 on Sabotage

Whilst the above two conventions aim at criminalising certain acts on board of an aircraft, the Sabotage Convention (1971) criminalises acts against the aircraft. Therefore, under Article 1, if a person attacks or sabotages a UAS or its air navigation facilities, it will be deemed a criminal offence.[40] Within the civil UAS sector,

[39] 'Any person who on board an aircraft in flight: (a) unlawfully, by force or threat thereof, or by any other form of intimidation, seizes, or exercises control of, that aircraft, or attempts to perform any such act, or (b) is an accomplice of a person who performs or attempts to perform any such act commits an offence (hereinafter referred to as 'the offence')' (Hague Convention 1971, Article 1).

[40] '1. Any person commits an offence if he unlawfully and intentionally: (b) destroys an aircraft in service or causes damage to such an aircraft which renders it incapable of flight or which is likely to endanger its safety in flight; or (c) places or causes to be placed on an aircraft in service, by any means whatsoever, a device or substance which is likely to destroy that aircraft, or to cause damage to it which renders it incapable of flight, or to cause damage to it which is likely to endan-

sabotage is currently virtually unreported. The possibilities of sabotage were highlighted by Assistant Professor Todd Humphreys of Cockrell School of Engineering, whereby he demonstrated to the US Department of Homeland Security the ability to gain control of another user's UAS (Aerospace Engineering and Engineering Mechanics 2014). There have been cases within a military context; for example, on 4 December 2012, a cyber-warfare unit of the Iranian Government captured a US Lockheed Martin RQ-170 UAS in Iran near the Iran–Afghanistan border (Infosec Institute 2013).

The Sabotage Convention, like with the Tokyo and Hijacking Conventions, is limited to only international flights. However, if a person attacks or sabotages a UAS or its air navigation facilities, it will be deemed a criminal offence even without them being on board the UAS. Consequently, due to this and UAS potential vulnerability to sabotage, the Sabotage Convention is much more relevant than the Tokyo and Hague Conventions.

Beijing Convention 2010 Modernising Criminal Aviation Treaties

The most recent instalment in international aviation criminal law results out of ICAO's conference which was held in Beijing between 30 August and 10 September 2010, whereby 80 states adopted two international air law instruments for the suppression of unlawful acts relating to civil aviation. These two instruments are the Convention on the Suppression of Unlawful Acts Relating to International Civil Aviation (Beijing Convention 2010) and the Protocol Supplementary to the Convention for the Suppression of Unlawful Seizure of Aircraft (Beijing Protocol 2010).

These are relevant instruments for UAS as they criminalise the act of using civil aircraft as a weapon; the use of dangerous materials to attack aircraft or targets on the ground; the transport of biological, chemical and nuclear weapons and their related material; and threats against civil aviation. In addition to this, those leading attacks against aircraft and airports will have no safe haven from criminal prosecution. Finally, the convention also implicitly addresses the threat of cyberattacks on aviation. Thus, many uses of UAS or unwanted acts against aUAS, such as jamming and spoofing, will be criminalised under the Beijing rules.

Whilst they have a wide scope and may prove to be applicable to UAS activities, they are currently not in force, and it does not appear that this will change anytime soon.[41]

ger its safety in flight; or (d) destroys or damages air navigation facilities or interferes with their operation, if any such act is likely to endanger the safety of aircraft in flight; or (e) communicates information which he knows to be false, thereby endangering the safety of an aircraft in flight' (Sabotage Convention 1971, Article 1).

[41] See Piera and Gill (2014).

CAA v Robert Knowles in the UK

Each national criminal law relevant to UAS varies in content, and it is beyond the scope of this work to examine them all; therefore, a case study from the UK will be explored in order to highlight the applicability of national criminal law for UAS in Europe.

In 2014, the UK saw its first ever successful prosecution for the dangerous and illegal flying of a UAS in the unreported case of the *UK CAA* v *Robert Knowles*.[42] BAE Systems is a public limited defence company which has UK Government contracts.[43] On 25 August 2013, a BAE Systems' employee recovered a UAS from the waters located near to its nuclear submarine testing facility in Barrow-in-Furness. The UAS was then surrendered to the police. The UAS was a £2000 delta wing which had a wingspan of 1.35 m and weighed 1.86 kg (Drone Makers 2014), and attached to the aircraft was a camera which was then analysed by the police which then linked it to Mr Robert Knowles.[44]

Knowles was then charged with two offences. The relevant UK law is contained within CAP393 Air Navigation: The Order and the Regulations 2009.[45] There is a weight distinction made within UK law, and because Knowles' UAS weighed less than 20 kg, it was classified as a 'small unmanned aircraft' pursuant to Article 255 which is defined as 'any unmanned aircraft, other than a balloon or a kite, having a mass of not more than 20 kg without its fuel but including any articles or equipment installed in or attached to the aircraft at the commencement of its flight' (CAP393 2014, Article 255).

The analysis of the footage 'revealed that during its flight it had skimmed over the busy Jubilee Bridge over [the] Walney Chanel' (UK CAA 2014b). Knowles' first charge pertained to the flying within 50 m of a structure pursuant to Article 167(2)(c) of the Air Navigation Order.

> The person in charge of a small unmanned surveillance aircraft must not fly the aircraft… within 50 metres of any vessel, vehicle or structure which is not under the control of the person in charge of the aircraft. (CAP393 2014, Article 167)

The flight path and the location of the crash also took the UAS within close proximity of several vessels. This would result in further breaches of Article 167(2)(c). In addition, the offences could have been more as the bridge is used by pedestrians which could have resulted in a violation of Article 167(2)(d) if any persons were

42 See Scott (2015).

43 See BAE Systems (2014).

44 See YouTube (2014).

45 The overriding article within the legislation is Article 138, which covers the subject of endangerment and applies to all aviation activity at all times: 'A person must not recklessly or negligently cause or permit an aircraft to endanger any person or property' (CAP393 2014, Article 138).

within 50 m. Furthermore, the UAS landed near vessels, and if they were inhabited at the time would amount to a breach of Article 167(3).[46]

In addition to this, the footage revealed that it 'had also flown through restricted airspace around the nuclear submarine facility before it inadvertently landed in the water' (UK CAA 2014b). Therefore, the second charge was for flying the UAS in restricted airspace over a nuclear installation pursuant to Air Navigation Regulations (2007), Regulation 3(2).[47]

The breaches of UK law by Knowles were deemed by the UK CAA to be of 'sufficient seriousness' to warrant prosecution.[48] Consequently, the case was brought by the UK CAA to Furness and District Magistrates' Court and presented in front of District Judge Gerald Chalk.[49] On 1 April, 2014, Knowles was found guilty and fined £800 plus costs of £3500.[50] This may prove to be the first of many cases, and it also highlights the potential necessity for national laws in the absence of comprehensive international rules.[51]

Conclusion

While the technical capabilities of UAS are increasing, and their civil, military and strategic uses are being enhanced, the regulation of their operation is still in a phase of development. Several bodies are closely researching the subject and formulating rules or proposals for rules. Among them are ICAO as the worldwide regulator, EUROCONTROL as the European organisation for the safety and efficiency of ATM in Europe and the EU.

For the time being, national regulations must fill gaps which international and European regulations have not yet filled. As is shown in this work, bodies are undertaking efforts to contribute to a more comprehensive body of rules, taking into account the special characteristics of UAS.

As always in aviation, safety takes a prominent role in rule-making. The distinction between civil and military rules importantly affect the level of regulation, that is, international, European or national, and the competencies of the concerned bod-

[46] 'Subject to paragraph (4), during take-off or landing, a small unmanned surveillance aircraft must not be flown within 30 metres of any person' (CAP393 2014, Article 167(3)).

[47] 'Subject to regulations 4–13, no aircraft is to fly over a nuclear installation to which this regulation applies below the height above mean sea level specified in Column 4 of the Second Schedule opposite its name' (Air Navigation Regulation 2007, Regulation 3(2)).

[48] See UK CAA (2009).

[49] Justices of the peace are locally sourced volunteers with no required formal legal qualifications, and these usually preside over cases at the magistrates' court. However, in more complex or sensitive cases, district judges, which are legally qualified, paid and full-time professionals, may hear the case.

[50] Most cases held at a magistrates' court are brought to court by the Crown Prosecution Service (CPS), but due to the content of the case, the UK CAA prosecuted.

[51] See BBC (2014).

ies. Next to safety, security protection of rights of individuals, management of the operations, liability and insurance must be addressed.

All in all, operation of UAS presents a complex and challenging subject for further research. The Institute of Criminology of the University of Ljubljana should be congratulated for tackling it in a multifaceted approach.[52]

References

Aerospace Engineering and Engineering Mechanics. (2014). Todd Humphreys' research team demonstrates first successful GPS spoofing of UAV. www.ae.utexas.edu/news/features/todd-humphreys-research-team-demonstrates-first-successful-gps-spoofing-of-uav. Accessed 15 Dec 2014.

Air Navigation Regulations. (2007). The air navigation (Restriction of Flying) (Nuclear Installations) regulations.

Amendment 43. (2014). Adoption of Amendment 43 to Annex 2 AN 13/1.1-12/19, 10th April, 2014.

Amnesty International. (2013). 'Will I Be Next?' US drone strikes in Pakistan. *ASA 33/013/2013.*

Association of European Airlines. (2008). Open skies: The EU–US air transport agreement. www.files.aea.be/News/News020408.pdf. Accessed 15 Dec 2014.

BAE Systems. (2014). Home. www.baesystems.com/home. Accessed 15 Nov 2014.

BBC. (2013). Smethwick fire: Chinese lantern 'Caused Largest Blaze'. www.bbc.com/news/uk-england-birmingham-23123549. Accessed 15 Dec 2014.

BBC. (2014). Man held over drone flight at Manchester City stadium. www.bbc.co.uk/news/uk-england-29689360. Accessed 15 Dec 2014.

Beijing Convention. (2010). Convention on the suppression of unlawful acts relating to international civil aviation, ICAO Doc. 9960. (Not in Force).

Beijing Protocol. (2010). Protocol supplementary to the convention for the suppression of unlawful seizure of aircraft, Doc 9959. (Not in Force).

CAP393. (2014). CAP393 air navigation: The order and the regulations.

Cheng, B. (1987). The Conventions of Tokyo, the Hague, Montreal and a New Instrument Concerning Unlawful Violence at International Airports. *Proceedings of the Conference on Aviation Security, Organised by the International Institute of Air and Space Law*, January 1987, Peace Palace, The Hague.

Chicago Convention or CC44. (1944). Convention on international civil aviation, 61 Stat. 1180, 15 U.N.T.S. 295.

CNN. (2014). Amazon to FAA: Let us fly our drones. www.money.cnn.com/2014/07/11/technology/innovationnation/amazon-drone-faa/. Accessed 15 Dec 2014.

Decision 2008/977/JHA. (2008). Council Framework Decision 2008/977/JHA on the protection of personal data processed in the framework of police and judicial cooperation in criminal matters.

Degarmo, M. (2004). Issues concerning integration of unmanned aerial vehicles in civil airspace. *MP 04W0000323.*

Directive 2009/48/EC. (2009). Directive 2009/48/EC of the European Parliament and of the Council of 18th June, 2009 on the safety of toys.

Directive 95/46/EC. (1995). European Parliament and Council Directive 95/46/EC on the protection of individuals with regard to the processing of personal data and on the free movement of such data.

Drone Makers. (2014). First UK prosecution for dangerous usage of a drone. www.dronemakers.com/uk-prosecution-dangerous-usage-drone/. Accessed 15 Dec 2014.

[52] See Institute of Criminology of the University of Ljubljana (2014).

EASA. (2009). Policy statement airworthiness certification of unmanned aircraft systems (UAS). *E.Y013-01*.

EASA. (2014a). European commission policy initiative on aviation safety and a possible revision of regulation (EC) No 216/2008. *NPA 2014-12*.

EASA. (2014b). Notice of Proposed Amendment 2014-09: Transposition of Amendment 43 to Annex 2 to the Chicago Convention on remotely piloted aircraft systems (RPAS) into Common Rules of the Air. *NPA 2014-09*.

EASA. (2014c). Part 21. www.easa.europa.eu/regulations. Accessed 15 Dec 2014.

EASA. (2014d). What we do. www.easa.europa.eu/the-agency/easa-explained/what-we-do. Accessed 15 Dec 2014.

EASA. (2015). Concept of operations for drones: A risk based approach to regulation of unmanned aircraft. www.easa.europa.eu/system/files/dfu/204696_EASA_concept_drone_brochure_web.pdf. Accessed 11 July 2015.

EUROCONTROL. (2001). Definition of state aircraft. www.eurocontrol.int/sites/default/files/content/documents/single-sky/cm/library/cmac-definition-of-state-aircraft-200107.pdf. Accessed 15 Dec 2014.

EUROCONTROL. (2014). Requirements for state aircraft. www.eurocontrol.int/articles/requirements-state-aircraft.

European Commission. (2009). Hearing on light unmanned aircraft systems (UAS). *TREN F2/LT/GF/GCD*.

European Commission. (2014a). Communication from the commission to the European Parliament and the Council, a new era for aviation opening the aviation market to the civil use of remotely piloted aircraft systems in a safe and sustainable manner. *COM. 207 Final*.

European Commission. (2014b). Remotely piloted aviation systems (RPAS): Frequently asked questions. Memo, Brussels, 8th April, 2014.

European Commission. (2014c). The EU Emissions Trading System (EU ETS). www.ec.europa.eu/clima/policies/ets/index_en.htm. Accessed 15 Dec 2014.

Finn, R., et al. (2014). Privacy, data protection and ethical risks in civil RPAS operations. *D3.3: Final report for the European Commission*.

Hague Convention. (1971). Convention for the suppression of unlawful seizure of aircraft 860 U.N.T.S. 105 1971.

ICAO. (2011). Unmanned aircraft systems (UAS). *Cir 328 AN/190*.

ICAO. (2012). Manual on remotely piloted aircraft systems (RPAS). *First Edition (draft v1.0)*, 2nd April, 2012.

ICAO. (2014). UAS documents. www.icao.int/Meetings/UAS/Pages/UAS_Documents.aspx. Accessed 15 Dec 2014.

Infosec Institute. (2013). Hacking drones … overview of the main threats. www.resources.infosecinstitute.com/hacking-drones-overview-of-the-main-threats/. Accessed 15 Dec 2014.

Institute of Criminology of the University of Ljubljana. (2014). Spy in the sky: Regulatory issues of drones and unmanned aerial systems. www.inst-krim.si/index.php?id=193&L=0. Accessed 15 Dec 2014.

International Red Cross. (2014). The Geneva Conventions of 1949 and their Additional Protocols. www.icrc.org/eng/war-and-law/treaties-customary-law/geneva-conventions/index.jsp. Accessed 15 Dec 2014.

Lazarski, A. (2002). Legal implications of the uninhabited combat aerial vehicle. *Aerospace Power Journal, 16*(2), 74–83.

Luftverkehrs-Ordnung. (1963). Luftverkehrs-Ordnung, LuftVO, BGBl. I S. 652.

Masutti, A. (2012). The regulation of unmanned air system use in common airspace. *Journal of Law, Information and Science, 21*(2), 212.

Mendes de Leon, P. (2013). A spy in the sky. *Presentation given at the Institute of Criminology of the Faculty of Law of the University of Ljubljana*, Ljubljana, 24th May 2013.

Milde, M. (2008). *International law and ICAO*. Eleven International Publishing.

Military Factory. (2013). The MQ-1 predator unmanned aerial vehicle has found success as an unarmed reconnaissance platform and as a missile-armed hunter. www.militaryfactory.com/aircraft/detail.asp?aircraft_id=46. Accessed 15 Dec 2014.

Montreal Convention. (1999). Montreal convention for the unification of certain rules for international carriage by air 2242 U.N.T.S. 309; S. Treaty Doc. No. 106–45.

Paris Convention. (1919). Convention relating to the regulation of aerial navigation, 11 L.N.T.S. 173.

Peterson, M. (2006). The UAV and the current and future regulatory construct for integration into the national airspace system. *Journal of Air Law & Commerce, 71,* 521.

Petras, C. (2010). The law of air mobility—the international legal principles behind the U.S. mobility air forces' mission. *Air Force Law Review, 66*(1), 1–78.

Piera, A., & Gill, M. (2014). Will the new ICAO–Beijing instruments build a Chinese wall for international aviation security? *Vanderbilt Journal of Transnational Law, 47*(1), 145–237.

Regulation 889/2002. (2002). Regulation (EC) No 889/2002 of the European Parliament and of the council of 13th May, 2002 amending Council Regulation (EC) No 2027/97 on air carrier liability in the event of accidents.

Regulation 785/2004. (2004). Regulation (EC) 785/2004 on insurance requirements for air carriers and aircraft operators.

Regulation 549/2004. (2004). Regulation (EC) No. 549/2004 of the European Parliament and of the council of 10th March, 2004 laying down the framework for the creation of the single European sky.

Regulation 1008/2008. (2008). Regulation (EC) No 1008/2008 of the European Parliament and of the Council of 24th September, 2008 on common rules for the operation of air services in the Community (Recast).

Regulation 216/2008. (2008). Regulation 216/2008 on common rules in the field of civil aviation and establishing a European Aviation Safety Agency, and repealing Council Directive 91/670/EEC, Regulation (EC) No 1592/2002 and Directive 2004/36/EC, Amended by Commission Regulation (EC) No 690/2009 of 30th July, 2009; Regulation (EC) No 1108/2009 of the European Parliament and of the Council of 21st October, 2009; and Commission Regulation (EU) No 6/2013 of 8th January, 2013.

Regulation 1070/2009. (2009). EU Regulation 1070/2009, intended to improve the performance and sustainability of the European aviation system, and amending EC Regulations 549/2004 laying done the framework for the creation of the European Single Sky, EC Regulation 550/2004 on the provision of air services in the Single European Sky, Regulation 551/2004 on the organisation and use of the airspace in the Single European Sky and Regulation on the interoperability of the European Air Traffic Management network.

Regulation 923/2012. (2012). Regulation (EU) No 923/2012 of 26th September, 2012 laying down the common rules of the air and operational provisions regarding services and procedures in air navigation and amending Implementing Regulation (EU) No 1035/2011 and Regulations (EC) No 1265/2007, (EC) No 1794/2006, (EC) No 730/2006, (EC) No 1033/2006 and (EU) No 255/2010.

Reinhardt, D. (2007). The vertical limit of state sovereignty. *Journal of Air Law and Commerce, 72,* 65.

Roma, A. (2014). Remotely piloted aircraft systems. *Aviation & Space Journal, 13,* 22–34.

Rome Convention. (1952). Convention on damage caused by foreign aircraft to third parties on the surface, ICAO Doc. 7364/310 UNTS 182.

R v Martin. (1956). R v Martin 2 Q.B. 272 [1956].

Sabotage Convention. (1971). Convention for the suppression of unlawful acts against the safety of civil aviation, 974 UNTS 177; 24 UST 564; 10 ILM 1151.

Scott, B. (2015). The first UK conviction for the illegal use of an unmanned aircraft and how it can help improve regulations within the EU. *Aviation & Space Journal, 14,* 11–22.

Steer Davies Gleave. (2014). Study on the third-party liability and insurance requirements of remotely piloted aircraft systems (RPAS). Final Report 22603201 SI2.661592.

TFEU. (2008). Consolidated version of the treaty on the functioning of the European Union, O.J. C 115/47.

The Guardian. (2013). Drone strikes by US may violate international law says UN. www.theguardian.com/world/2013/oct/18/drone-strikes-us-violate-law-un. Accessed 15 Dec 2014.

The Verge. (2014). DHL drone will make deliveries to German island starting Friday. www.theverge.com/2014/9/24/6838443/dhl-drone-making-deliveries-to-german-island-juist. Accessed 15 Dec 2014.

Tokyo Convention. (1963). Convention on offences and certain other acts committed on board aircraft, ICAO Doc. 8364/704 UNTS 220/[1970] ATS 14/2 ILM 1042 (1963)/ 58 Am. J. Int'l L. 566.

UK CAA. (2009). Conduct of prosecutions by the CAA as a prosecutor. www.caa.co.uk/docs/503/CAAProsecutionsPolicyV2_16092009.pdf. Accessed 15 Dec 2014.

UK CAA. (2014a). EASA—air operations. www.caa.co.uk/default.aspx/default.aspx?catid=620&pagetype=90&pageid=11051. Accessed 15 Dec 2014.

UK CAA. (2014b). First conviction for illegal use of an unmanned aircraft. www.caa.co.uk/application.aspx?catid=14&pagetype=65&appid=7&mode=detail&nid=234. Accessed 15 Dec 2014.

US v Cordova. (1950). United States v Cordova 89 F. Supp. 298 (E.D.N.Y., 1950).

US FAA. (2014). Unmanned aircraft (UAS): Questions and answers. www.faa.gov/about/initiatives/uas/uas_faq/#Qn3. Accessed 15 Dec 2014.

US–EU Agreement. (2007). US–EU air transport agreement, 04/30/07.

Vienna Convention. (1969). Vienna convention on the law of treaties, UN Doc. A/Conf.39/27; 1155 UNTS 331; 8 ILM 679.

YouTube. (2014). Walney Channel, Robert Knowles prosecution flight. www.youtube.com/watch?v=hriyv8WRHg4. Accessed 15 Dec 2014.

Zielinski, R., et al. (1996). Star trek exploiting the final frontier: Counterspace operations. Air Force, 2025 SF298.

Pablo Mendes de Leon is a professor of air and space law and director of the International Institute of Air and Space Law of Leiden University, which is one of the leading international scientific research and teaching institutes in the world, specialised in legal and policy issues regarding aviation and space activities. The Faculty of Law of Leiden University founded the institute in 1986. In addition to his duties as director of the International Institute of Air and Space Law of Leiden University, Pablo Mendes de Leon maintains a vast range of memberships in organisations that work to combine law and practice of aviation law and policy. For instance, he is president of the European Air Law Association, visiting professor at various universities outside the Netherlands, a judge at the District Court of Haarlem, membre titulaire de l'académie de l'air et de l'espace, Toulouse, France, member of the Dutch Aviation Accident Board, a board member of the magazines Air and Space Law, Journal of Air Law and Commerce and the Italian ANIA Insurance Newsletter, and the director of the international Aviation Law and Policy Series with Kluwer Law International. He is the author of a large number of publications on topical issues regarding aviation law and policy.

Benjamyn Ian Scott studied for an LLB in law with European legal studies at the University of Kent and the University of Amsterdam and graduated in 2011 with honours. He then went on to study for an LLM in international commercial law at the University of Kent where he was awarded a distinction. During this time, he became interested in aviation and space law and consequently wrote his thesis on the demarcation of air and space. In 2013, he went on to study for the LLM (Adv.) in Air and Space Law at Leiden University where he graduated cum laude. While at Leiden, he conducted extensive research into unmanned aircraft and was able to work on this topic during a traineeship at the European Aviation Safety Agency. Benjamyn also has a strong research interest in suborbital vehicles. He has published an article on suborbital passenger liability in the April 2014 *Aviation ... Space Journal*. He is also due to publish an article in the spring 2015 edition of *Issues in Aviation Law and Policy* on international suborbital passenger transportation. Finally, Benjamyn is currently developing a website, www.rpassage.com, and writing about 'roadable aircraft'.

Part V
Domain-Specific Uses of Drones

Chapter 10
Droning on About Journalism: Remotely Piloted Aircraft and Newsgathering

David Goldberg

Introduction

> We are not using it as a drone. That is completely the wrong terminology to use to describe it. We see it as a flying camera (Hannen 2014).

Newer technologies generate activity for policy-makers and legislators as well as techno-panics for citizens, activists and academics (International Council on Human Rights Policy 2011):

> A moral panic occurs when a segment of society believes that the behavior or moral choices of others within that society poses a significant risk to the society as a whole. By extension, a 'techno-panic' is simply a moral panic that centers around societal fears about a specific contemporary technology (or technological activity) instead of merely the content flowing over that technology or medium (Technology Liberation Front 2014).

Thus, 'air law' became established on 23 April 1784, when:

> a French police directive was issued aimed directly and exclusively at the balloons of the Montgolfier Brothers: in order to protect the population, flights were not to take place without prior authorization (Giemulla and Weber 2011).

The response to the developing drones market is no exception to this sociological rule. In addition, to account for some of the anxiety so evident in certain quarters, it is suggested that one particular reason is because of the way images of drones are presented. These make them resemble scary flying creatures from the dinosaur era or tiny insects, thus playing into fear of dinosaurs (ornithoscelidaphobia) or fear of insects (entomophobia).

D. Goldberg (✉)
Glasgow, Scotland
e-mail: davgoldberg@gmail.com

A. Završnik (ed.), *Drones and Unmanned Aerial Systems*,
DOI 10.1007/978-3-319-23760-2_10

In this chapter, the term 'remotely piloted aircraft' (RPA) will be substituted for the word 'drone'.[1] It is technically precise and corresponds to contemporary usage (International Civil Aviation Organization 2011; UK Civil Aviation Authority 2012). It also pays homage to a little-known historical fact: present day *remotely* piloted aircraft owe their technology to the brainchild of Nikola Tesla, who patented the 'teleautomaton'. He presented his invention at Madison Square Garden's first electrical exhibition in September 1898:

> On a rainy September day in 1898 Nikola Tesla presented at Madison Square Garden's first Electrical Exhibition a new invention that he called a 'teleautomaton'. The invention was the first ever radio controlled device in the form of a miniature boat. He had two devices one that could be remote controlled above water and another that had a hidden loop antenna and could be controlled under water (Sachs 2010).[2]

Further, the term foregrounds two crucial aspects of this type of vehicle. First, it is often imagined that RPAs are unmanned because there is no human being on board. But the RPA is not literally 'unmanned' as there is a human operator (the pilot-in-command) operating and (hopefully) controlling/monitoring it ideally in conjunction with an observer too. Actually, the RPA is usually part of a system:

> …the unmanned aircraft (UA) [sic] and all of the associated support equipment, control station, data links, telemetry, communications and navigation equipment, etc., necessary to operate the unmanned aircraft (Federal Aviation Administration 2014c).

Second, from a regulatory and legal perspective, the remotely piloted vehicle is an *aircraft.* This point will be amplified later.

However, in order not to become entangled in what Mark Corcoran calls a 'definition dogfight', the phrase 'drone journalism' is acceptable (and corresponds to normal usage) when referring to the use of RPAs by mainstream media enterprises and journalists as well as citizen journalists in the pursuit of journalism (Corcoran 2014). Unlike military applications, most drone journalism uses a *remotely controlled quod or octo helicopter* with an on-board camera attached (Goldberg et al. 2013).[3]

There are already many civilian, that is, nonmilitary and non-state, applications for RPAs and these are set to increase. The future will be limited only by human imagination (Increasing Human Potential 2014a). Amongst such uses are, inter

---

[1] Widely so-called, the origin of the word 'drones'is less well known. It seems that, in 1931, '… the British developed the Fairey "Queen" radio-controlled target from the Fairey IIIF floatplane, building a batch of three, and in 1935 followed up this experiment by producing larger numbers of another RC target, the "DH.82B Queen Bee", derived from the de Havilland Tiger Moth biplane trainer. Through some convoluted path, the name of "Queen Bee" is said to have led to the use of the term "drone" for remote-controlled aircraft' (Goebel).

[2] See also Weber (2010), Tesla Memorial Society of New York (n.d.) and Nikola Tesla Museum (n.d.). For the patent, method of, and apparatus for controlling mechanism of moving vessels or vehicles, see Tesla (1898).

[3] See also, BBC News (2013). A tweak to the phrase 'drone journalism' is "sensor journalism": see Pitt (2014).

alia: carrying cargo (widely tipped to become a major application—see Insinna 2014; Flying Donkey Challenge 2012–2014); film, television and advertisement production;[4] wildfire detection and management (Campbell 2014); monitoring threats to vulnerable animals and birds (e.g., Shadowview Foundation n.d.)—as well as drone-assisted hunting; pollution monitoring; property selling by local estate agents; event security; traffic and road accident monitoring; disaster relief (e.g., Humanitarian UAV Network n.d.); search and rescue; post-natural disaster services; first responder assistance—for example, fire services, fisheries management, pipeline monitoring and oil and gas security, meteorology—storm tracking and tornado and hurricane research; airplane safety checking; remote aerial mapping (e.g., Drone Adventures n.d.); book and other consumer products delivery (e.g., Amazon's PrimeAir); distribution of medicinal products; tree-mapping (BBC News 2014b); transmission line inspection; infrastructure security monitoring, for example, ports (The National 2014); and sports analytics (Increasing Human Potential 2014a; McClatchy 2013). Some believe that in addition to cargo, the biggest sector will be 'precision agriculture', for example, crop inspection and management as well as surveying.[5] There will be quirky uses too, for example, the Archdiocese of Washington purchased a 'hubcap size' RPA which was deployed to:

> …videotape crowds participating in a procession marking the canonizations of popes John Paul II and John the 23rd (now called St. John Paul II and St. John the 23rd). The video images captured by the drone-mounted camera were used in a YouTube mashup the archdiocese made, mixing soaring classical music and scenes from an earlier indoor Mass (Increasing Human Potential 2014b).[6]

---

[4] Colloquially, 'drone cinematography,' see http://www.rollingstone.com/movies/news/drones-and-the-future-of-movies-20131028. See also, http://www.npr.org/blogs/alltechconsidered/2014/05/16/312487924/are-filmmakers-using-drones-illegally-looks-like-it; and http://www.iol.co.za/capetimes/film-industry-concerned-about-camera-drone-ban-1.1689174#.U3YcqPldWSp. Speaking for the US film industry, Howard Gantman, Motion Picture Association of America (MPAA) spokesperson, has stated: 'What we are looking for is line-of-sight things [sic] that can be utilized in innovative ways…These could be used much more safely than going up a tree and much more cheaply than renting a helicopter'; and now, see Federal Aviation Administration (2014a).

As Timothy Ravich (2014) notes, in the context of movies, 'In any event, while this is a step in the right direction—rewarding credentialed, safety-conscience "drone" users—this seems out of sorts with the First Amendment guarantee of free speech. Isn't a movie a form of expression (yes, the Supreme Court of the United States said in 1952)? Will the FAA exempt certain movie studios and not others?' Ravich is referring to *Joseph Burstyn, Inc. v. Wilson* (1952).

[5] 'You can take a simple UAV and repurpose imagery for a farmer's field for cents on the dollar compared to using traditional aircraft. The biggest potential for Unmanned Aerial Vehicles is aerial images and data acquisition'(Green 2013). See also the US Senate Committee on Commerce, Science, and Transportation (2014); Federal Aviation Administration (2014a); Association for Unmanned Vehicle Systems International's (AUVSI's) Unmanned Systems Mission Critical (2014); and O'Connell (2014).

[6] Potentially problematic was the use of the RPA within a prohibited radius of Reagan National Airport—the 'Flight Restricted Zone (FRZ), a roughly 10-nautical-mile area centered around Reagan National Airport in Virginia' (Miller 2014).

However, the only civil commercial application dealt with in this chapter is using RPAs to gather information for journalism (Frontline Club London 2013; Collins 2014)—as noted, supra, colloquially called 'drone journalism'[7]—by media companies, journalists and/or any citizen journalist engaged in newsgathering (Biederman Institute Conference 2014; Osterreicher 2014).

The key claim of the chapter is that this application engages a fundamental human right, namely, the general right to freedom of expression and, specifically, the right to receive ideas and information (as well as the component rights needed to make it a reality). Thus, the use of RPAs by media companies and/or citizen journalists raises unique concerns and issues because such use engages (elements of) the right to freedom of expression. Actually, the threshold right is arguably the right to access the communications technology that an RPA is in such situations (Herr 2013). Any *general* restriction(s) on using RPAs, which, a fortiori embrace newsgathering, would amount to a prima facie infringement of this right. Only if, *in casu, extremely strong compelling overriding* considerations defending and promoting another protected interest(s), for example, someone's purported right to respect for their private and family life or home, would the right to use an RPA in that specific, fact-limited context be trumped. At the very least, in the absence of carrying out an explicit exercise balancing the competing interests involved, any restriction would be challengeable as procedurally flawed.

Before concluding this introduction, the point already made, namely, that from a regulatory and legal perspective, the RPA is an *aircraft,* will be amplified. Understanding that RPAs are 'aircraft' produces a novel situation.[8] It is a truism that media organisations and media lawyers currently engage with a range of media regulators and laws. For example, there are laws protecting the reputation of subjects of media reporting, states jealously guard their national security by punishing disclosure of official secrets, the revenue of the media is affected by advertising rules, broadcasting is subject to being granted a national regulator's licence, and community standards are upheld by laws criminalising obscene speech, etc. But one regulator and legal regime that tends to be far removed from the consciousness of media companies, citizen journalists and their advisers, is the national aviation regulator and air law and rules, whether national, regional, or international. This is set to change. The safety dividend, small size, portability and low cost of RPAs will mean that the media companies etc. will themselves own and operate the vehicles directly. Such users will not only be more aware of the rules and regulations governing such activities but they will *have* to be so because they will be directly legally liable when things go wrong—as they inevitably will do (Merlin 2013). Conversely, national aviation regulators will have to understand that their decisions—in this context, who, if anyone, to permit to fly and under what conditions—might be met by a challenge using human rights law and jurisprudence concerning freedom of expression. In a nutshell, when media companies and citizen journalists (and content producers)

[7] See Drone Journalism Lab and Professional Society of Drone Journalists.

[8] An RPA is an 'aircraft' because it falls within the ICAO definition of an aircraft: 'Any machine that can derive support in the atmosphere from the reactions of the air other than the reactions of the air against the Earth's surface' (UK Civil Aviation Authority 2012).

deploy RPAs, the new tool in the newsgathering toolbox in particular for the professional media lawyer becomes—aviation regulation and law.

RPAs and Drone Journalism: Twenty-First-Century Photojournalism

As noted in the introduction, this chapter is concerned with one and only one specific RPA application, namely, the pursuit of newsworthy information whether by mainstream news media or citizen journalists using RPAs as *one* tool in the journalistic toolbox. The focus is on what is known as 'responsible journalism', i.e., what the Australian Law Reform Commission describes as 'journalistic activities [offering] significant public benefit'. This is to be distinguished from activities which are not journalistic in nature, where the public interest in a matter is trivial, or where the matter is merely of interest to the public or for the purposes of gossip (Australian Law Reform Commission 2014).[9]

Thus, in this chapter, 'Journalism' is understood to mean public interest reporting and publishing and all that it entails and not anything and everything that appears in a publication, whether perceived as a 'newspaper' or not.

Becoming ever more cheap, light and portable, the desire to deploy RPAs with HD cameras and other sensors aboard is set to increase over time.[10] This is nothing but the contemporary manifestation of the urge that has animated photojournalism since its earliest days: to deploy the (relatively) latest technologies when perceived benefits to telling the story warrant doing so.[11] For example, in 1906:

> the devastation of San Francisco after the…earthquake and fire was captured by George R. Lawrence, using a camera attached to a string of kites high above the city. His specially designed large-format camera had a curved film plate to provide panoramic images, which remain some of the largest aerial exposures ever taken. The camera, which was large and

[9] 'Historically, "responsible journalism" was developed as a defence to defamation in *Reynolds v. Times Newspapers Ltd.* Despite being crafted in the context of defamation, several of the matters listed by Nicholls LJ are relevant in the context of surveillance. For example, the seriousness of the conduct being investigated by a journalist, the likely strength of the individual under surveillance as a source of information, the likely nature of the information obtained, and the urgency of the matter may be relevant considerations (Australian Law Reform Commission 2014, Para 14.65).

[10] 'Parrot has added a new model to its line of airborne drones—the Bebop…the lightweight Bebop includes some snappy features, such as built-in image stabilization, a 14-megapixel camera with a fish-eye lens, and support for the Oculus Rift virtual reality headset' (Mello 2014). See also Stone (2014).

[11] 'The first known aerial photograph was taken in 1858 by French photographer and balloonist, Gaspar Felix Tournachon, known as "Nadar". In 1855 he had patented the idea of using aerial photographs in mapmaking and surveying, but it took him 3 years of experimenting before he successfully produced the very first aerial photograph' (PAPA International 2014). PAPA has a webpage about RPAs, asking members to respect local laws and regulations: http://www.papainternational.org/uas.asp. In 2011, The Museum of London held an exhibition of street photography dating from 1860 (see Urban75 2011).

extremely heavy, took as many as 17 kites to lift it 2000 ft into the air. Lawrence also used ladders and high towers to capture lower level 'aerial' photographs (PAPA International 2014).

'Photojournalism' as such:

> …emerged as a distinctive form of photography in the late 1920s and early 1930s. The term denoted picture making that was spontaneous, topical and rapid. This was facilitated by the introduction of small, hand-held cameras such as the Ermanox and the Leica, which enabled photographers to record fast-moving events and catch their subjects unawares (Victoria and Albert Museum 2014; Collins, various dates).

Drone journalism is simply twenty-first-century photojournalism—deploying a newish platform for aerial photography for the purpose of newsgathering. Another illustration in the development of *aerial* photojournalism is the invention of the 'newscopter'. In 1958, American John Silva, the 'father of helicopter journalism':

> …converted a small helicopter into the first airborne virtual television studio. The KTLA 'Telecopter,' as it was called by the Los Angeles station where Mr. Silva was the chief engineer, became the basic tool of live television traffic reporting, disaster coverage and that most famous glued-to-the-tube moment in the modern era of celebrity-gawking, the 1994 broadcast of O. J. Simpson leading a motorcade of pursuers on Los Angeles freeways after his former wife and a friend of hers were killed (Pool 2012).

There were numerous technical challenges to overcome:

> For one, the standard camera equipment literally weighed a ton. Keeping the project closely guarded, lest competing stations get wind of it, Silva designed and machined lighter equipment using aluminum parts to bring the weight below the 368-pound FAA limit. He also added a shock absorbing stabilization system and a helical antenna that extended below the body of the helicopter (Eschoff 2013).

In conformity with the sociological rule described in the introduction, social reactions to LA area copters were analogous to the current reaction to RPAs:

> Lots of homeowners and a few orchestra conductors (who've walked off the Hollywood Bowl stage in protest) are tired of the noisy company of tourist, paparazzo, news, and police helicopters, their jet engines roaring and blades thwacking the night air… The imagery of a circling bird is appropriate. When a police helicopter works a crime scene, it follows a tight orbit, generally at three or four hundred feet for better observation, aided at night by million-candlepower searchlights. The lights, the engine noise, and the staccato of rotor blades biting into the air can feel menacing to anyone on the ground, law-abiding or not (Waldie 2013).[12]

2011 was a seminal year. It was reported that the US Federal Aviation Authority was 'looking into' a NewsCorp publication called *The Daily* (now defunct) because it had used a small RPA to monitor storm damage in Alabama and flooding in North Dakota (Hill 2011). In November, Matt Waite launched the *Drone Journalism Lab* at the College of Journalism and Mass Communications, University of Nebraska-Lincoln '…as part of a broad digital journalism and innovation strategy' and has since received a $50,000 grant from the Knight Foundation (Drone Journalism Lab (n.d.); Waite 2013). In another 2011 development, the Professional Society of

[12] The ACLU does think there is a difference between helicopters and drones (see Stanley 2013).

Drone Journalists was created by Matthew Schroyer;[13] it now boasts more than 300 members in 35 countries. Since then, in many countries globally, the use of RPAs for newsgathering has been reported as follows (Levine 2014):

- Russia

Russians went to Bolotnaya Square in Moscow to protest election fraud; AirPano sent up a remote-controlled hexacopter to take pictures showing the scale of the crowd (Nurlybayeva 2012).

- Italy

The Milan leg of the tour Tsunami Beppe Grillo was picked up by over 100 m high, and *CBS* commissioned an Italian company to film the wreck of the *Costa Concordia* with a drone (Caputo 2013).

- Poland

A civilian-operated RC copter filmed riots in Warsaw (Reed 2011).

- Argentina

One group of citizen journalists took matters into their own hands by flying a drone over the crowds to show the large scale of a rally (Al Jazeera 2012).

- Ukraine

Drone covers the chaos in Ukraine (Ariens 2014).
Dramatic drone footage captures battle for central Kiev square (Globally 2014).

- Balkans

Aerial drone footage shows scale of flooding (The Telegraph 2014).

- Turkey

Police clash at Taksim Gezi Park (Drone Journalist 2014; ABC News 2014b).

- Philipines

Ten days after Super Typhoon Haiyan ripped through the Philippines, CNN used an aerial drone to get a bird's eye view of what was left standing in Tacloban (Penhaul 2013).[14]

- Pakistan

Geo News, Pakistan's most popular news channel, is in the process of experimenting with 2 DJI Phantoms, plans for which call for assisting in gathering footage for the network (Usmani 2014).

---

[13] See the Code of Ethics for Drone Journalists on the PSDJ website.

[14] CNN actually has a webpage that requests people to submit their drone images saying, 'We'll feature the best on CNN.com/Tech' (http://ireport.cnn.com/topics/1098143). Accessed 22 Sept 2014.

In Latin America, RPAs have been deployed for drone journalism to quite some extent:

> …El Salvador newspaper La Prensa Gráfica became one of the nation's first outlets to gather news with drones after purchasing three unmanned aerial vehicles, a pattern that other news media in Latin America are following, according to news website GlobalPost. The Salvadoran outlet uses its drones primarily to shoot aerial video or photographs of big crowds gathered for events, long traffic jams, or even simply natural and artificial landmarks around the nation's capital of San Salvador…. This is one of a growing roster of Latin American nations, including Brazil, Mexico and Peru, where news outlets are deploying the aircraft in reporting. Lima's major daily El Comercio dispatched a drone to cover a massive downtown fire in December, and Mexico's Grupo Reforma flew one over student protests in Mexico City. Vladimir Lara, chief of photography at El Salvador's La Prensa Grafica, said pictures taken from the sky garner up to 70 percent more views than traditional photographs on the newspaper's website (Stark 2014).[15]

However, ironically, on one occasion, a gossip news company *denied* it was planning to use RPAs for its business:

> TMZ is NOT getting in the DRONE business…we don't have a drone…we don't want a drone…we never applied for a drone…despite a bogus report to the contrary (TMZ 2012).

Finally, it is also worth pointing out that not all arguments made for accelerating the regulatory conditions favouring the legal deployment of RPAs for drone journalism are predicated on freedom of expression grounds. A wholly different argument concerns enhancing the safety of and protection for human lives. This consideration is urged by news broadcasting services' correspondents and citizen journalists who see significant benefits in being able to witness and report *indirectly* on happenings in potentially hazardous trouble spots, without risking, possibly unnecessarily or recklessly, the life and limb of the reporter:

> Small drones offer considerable advantages for news staff deployed on high risk assignments such as wars, civil unrest and natural disasters…Drone technology offers great potential for news-gatherers…but there are some important qualifications. Journalism is about people and personal contact and UAVs should not be seen as an easy substitute for the journalist or news team on the ground. The drone is a camera platform, a tool to be incorporated among all the other news-gathering technology and professional skills a journalist uses on hazardous assignments… (Corcoran 2014).

In conclusion, what primarily differentiates drone journalism from any other application is that it engages the right to freedom of expression and specific elements thereof. This is not about justifying 'speaker's right(s)', but rather, the right(s) of the reader/viewer/audience to receive video data/information. The use also engages the component right, namely, to access and use any communications technology.

[15] For that region generally see Sánchez (2014)..

Facilitating Maximum Feasible Use of RPAs for Drone Journalism

This section of the chapter sets out three scenarios and corresponding arguments that drone journalists should make to facilitate the maximum feasible use of RPAs for drone journalism. However, any such use is fully acknowledged to be without prejudice to the duty of the aviation regulator to order the use of the national airspace in the paramount interest of safety, the key normative and operational challenge being to integrate the use of RPAs in the domestic airspace (Dolan and Thomson 2013; Committee on Autonomy Research for Civil Aviation 2014; ABC News 2014c; Rosen 2014). Actually, at least a degree of safety may be achieved with a technological fix. In the context of avoiding certain air spaces, for example, around airports, one manufacturer's RPAs can be programmed with a so-called 'geo-fence', software that provides global positioning satellite (GPS) guidance (no fly zone technology).[16]

> The drones will be blocked from operating near 350 airports around the world by creating an electronic 'geo-fence' around airports to reduce the risk of collision between drones and manned aircraft.

With that caveat, the three scenarios are as follows.

i) Scenario One

There is either a general prohibition on using RPAs for 'commercial' or 'business' purposes deemed to include drone journalism or no norm(s) positively facilitating such use.

ii) Scenario Two

Using RPAs for drone journalism is legal either generally or on an ad hoc basis—but is opposed by the claim that to do so shall, or is likely to, seriously infringe civil liberties/human rights—most frequently stated to be someone's 'right to privacy'.[17]

iii) Scenario Three

A specific problem with drone journalism using, for example, a nano- or micro-RPA is alleged to be that the subjects of investigation might not realise that they are being surveilled, or being surveilled in an RPA-specific manner, because of the smallness of the RPAs and/or other technical capacities, for example, silence, mobility and endurance.

---

[16] 'DJI's No Fly Zone system creates a curious technological and sovereignty precedent. The initiative will effectively give a Chinese company indirect control over the movement of unmanned aircraft in Australian airspace—and in the skies of dozens of other nations. While DJI says its initiative is solely motivated by safety, there are concerns that drone flying restrictions could be easily exploited for political censorship' (ABC News 2014a).

[17] This phrase conveniently ignores the legal fact that there is no 'right to privacy' simpliciter in the European Convention on Human Rights (or in the US Constitution).

With regard to each scenario, a range of pro-deployment arguments can be identified, as follows.

Re Scenario One There is either a general prohibition on using RPAs for 'commercial' or 'business' purposes deemed to include drone journalism or no norm(s) positively facilitating such use.

Just because drone journalism is neither a military nor state nor recreational use of an RPA, it is simply wrong to claim it is a commercial or 'business' use of an RPA. It is a conceptual confusion to conflate them. Deploying an RPA for the purpose of drone journalism prima facie engages the (human) right to freedom of expression and, in particular, the right to pursue activities precedent to facilitating people's right to receive ideas and information. As has been said:

> …a complete ban misunderstands journalism as a purely commercial activity rather than a constitutionally-protected right to gather and disseminate news, covered in the First Amendment…This overly broad policy [sic], implemented through a patchwork of regulatory and policy statements and an ad hoc cease-and-desist enforcement process, has an impermissible chilling effect on the First Amendment newsgathering rights of journalists (Cruz 2014).

A general ban would constitute a prima facie infringement of the user's right to freedom of expression under, for example, Article 10 of the European Convention on Human Rights or an equivalent regime, for example, the US Constitution's First Amendment. The most systematic argument that a ban on using RPAs for newsgathering as part of a general ban on 'commercial' uses is an impermissible infringement of the right to expression (in that jurisdiction, the First Amendment) has been presented by 16 American news organisations in an *amici* brief submitted to the US National Transportation Safety Board (May 2014). Regarding the FAA's general ban on 'business' uses by RPAs, it states:

> The FAA's position is untenable as it rests on a fundamental misunderstanding about journalism. News gathering is not a 'business purpose': It is a First Amendment right. Indeed, contrary to the FAA's complete shutdown of an entirely new means to gather the news, the remainder of the federal government, in legislation, regulation and adjudication, has recognized that, in the eyes of the law, journalism is not like other businesses. The government in a myriad of measures has long accommodated the bedrock First Amendment principle that 'without some protection for seeking out the news, freedom of the press could be eviscerated (*Branzburg v. Hayes*, 408 U.S. 665, 702, 1972)'.[18]

Earlier, the US Congressional Research Service (CRS) published a report, 'Integration of Drones into Domestic Airspace: Selected Legal Issues', which contains a section entitled 'First Amendment and Newsgathering Activities'. Having considered other interests (e.g., privacy), the CRS counsels, giving proper regard to:

> …the public's countervailing concern in securing the free flow of information that inevitably feeds the 'free trade of ideas' (Dolan and Thomson 2013).

Further, even if the regulator permits deployment of RPAs for drone journalism on an ad hoc, case-by-case decision-making process authorising (or otherwise) X to fly an RPA, a question arises: Has the authority explicitly conducted the necessary balancing exercise weighing freedom of expression considerations, for example,

[18] See, United States of America National Transportation Safety Board (2013).

is any ban 'necessary in a democratic society' in coming to its determination? Has there been consistency between the agency's decisions? Is there an independent appeal process from an agency decision? For example, the FAA has announced it is considering requests from

> …seven aerial photo and video production companies [that] have asked for regulatory exemptions that would allow the film and television industry to use unmanned aircraft systems (UAS) with FAA approval for the first time (Federal Aviation Administration 2014a).[19]

Commenting, Timothy Ravich notes:

> In any event, while this is a step in the right direction—rewarding credentialed, safety-conscience 'drone' users—this seems out of sorts with the First Amendment guarantee of free speech. Isn't a movie a form of expression (yes, the Supreme Court of the United States said in 1952)? *Will the FAA exempt certain movie studios and not others?* [emphasis added] (Ravich 2014)[20]

The key claim of the chapter requires understanding that, as the CRS report notes, what is protected is not only *forms* of speech or content, but also:

> …conduct that is 'necessary for, or integrally tied to, acts of expression'…other conduct that is not expressive in itself, but is 'necessary to accord full meaning and substance to those guarantees'.

In like manner, the European Court of Human Rights has been supportive of activities which underpin not only publication but also newsgathering activities required to obtain material on which to base subsequent publication; a fortiori it is argued here that this support would extend to the use of RPAs to do this work (see Voorhoof 2014b). Key has been the court's jurisprudence regarding protecting the identity of journalists' sources and the protection from unjustified search and seizures of journalistic material.

The 'leading case', decided by the Grand Chamber of the Court, is *Sanoma Uitgevers B.V. v. the Netherlands* (European Court of Human Rights 2010). The case concerned photographs to be used for an article on illegal car racing, which a Dutch magazine publishing company was compelled to hand over to police investigating another crime, despite the journalist's strong objections to being forced to divulge material capable of identifying confidential sources. The Court found that the interference with the applicant company's freedom of expression had not been 'prescribed by law', there having been no procedure with adequate legal safeguards available to the applicant company to enable an independent assessment as to whether the interest of the criminal investigation overrode the public interest in the protection of journalistic sources. The Court reaffirmed its position in *Telegraaf Media Nederland Landelijke Media B.V. and Others v. the Netherlands* (European Court of Human Rights 2012). The Court stated:

> Freedom of expression constitutes one of the essential foundations of a democratic society and the safeguards to be afforded to the press are of particular importance. Whilst the press

[19] The FAA has also licenced BP and AeroVironment to fly an AeroVironment Puma AE for aerial surveys in Alaska—the first time the FAA has authorized a commercial UAS operation over land (Federal Aviation Administration 2014b) rized a commercial UAS operation o.

[20] Ravich is referring to *Joseph Burstyn, Inc. v. Wilson*, 343 U.S. 495 (1952).

> must not overstep the bounds set, not only does the press have the task of imparting such information and ideas: the public also has a right to receive them. Were it otherwise, the press would be unable to play its vital role of 'public watchdog' (*Observer and Guardian v. the United Kingdom*, 26 November 1991, § 59, Series A no. 216) The right of journalists to protect their sources is part of the freedom to 'receive and impart information and ideas without interference by public authorities' protected by Article ten of the Convention and serves as one of its important safeguards. It is a cornerstone of freedom of the press, without which sources may be deterred from assisting the press in informing the public on matters of public interest. As a result the vital public-watchdog role of the press may be undermined and the ability of the press to provide accurate and reliable information to the public may be adversely affected.
>
> The Court has always subjected the safeguards for respect of freedom of expression in cases under Article ten of the Convention to special scrutiny. Having regard to the importance of the protection of journalistic sources for press freedom in a democratic society, *an interference cannot be compatible with Article ten of the Convention unless it is justified by an overriding requirement in the public interest* [emphasis added].

One last argument: RPAs deployed with payloads such as electro-optic cameras to relay video data to journalists or media organisations should be understood as indirectly 'communications technologies'.[21] The RPA is simply the 'donkey'/platform facilitating the carriage of a camera (or other sensors—see Mesich 2014), which records and/or relays images to a ground receiver/station. Robin Elizabeth Herr has argued, in another situation:

> Human rights scrutiny is a necessary component of any effort to ensure that communication technology can be effectively adopted and used.

Herr has proposed

> a model to solve…*access* problems based on the case law of the European Court of Human Rights…. Khurshid Mustafa and Tarzibachi v. Sweden supports the adoption of that [sic] technology without unjustified restriction by the state or private individuals…no matter what type of communication technology is used, there exists a general right of access to all forms of information [emphasis added] (Herr 2013).

Re Scenario Two: Privacy Rights (Do Not) Trump Using RPAs for drone journalism is legal either generally or on an ad hoc basis—but is opposed by the claim that to do so is, or is likely to, seriously infringe civil liberties/human rights—most frequently stated to be someone's 'right to privacy'.

Even if RPAs are legally permitted to fly for the purpose of drone journalism—whether on an ad hoc, case-by-case basis, or in virtue of a general, sectoral permission—opposition, robust concerns and challenges are expressed in the name of civil liberties, most usually the so-called 'right to privacy'. The paradigmatic tone is rather of the 'what if' variety, for example,

> …the next privacy scandal in waiting is the story of drones. Not military drones, but increasingly widespread use of drones for agriculture, disaster areas and emergencies, archaeology, forestry and property management, among others.

[21] RPAs are part of what has been called ENG—'electronic news gathering'. See McGrath (n.d.).

> Drones are banned in London and can't be used below a certain height in residential areas. But how many uses could there be for a small, silent, fast, remote-controlled drone? How long before the first sunbathing politician is snapped on holiday? If the public is banned from a venue, or refused access to private land, or if a property is under siege from journalists, how long before a drone is used for high-quality aerial video? (Kiss 2014)

Many states within the USA have considered bills and resolutions regarding RPAs. Most of them are privacy protection measures that would restrict the use of drones and set limits on the collection and storage of data (See Domestic Drone Information Center n.d.; Williams 2014; and Johnson 2014). Yet, at best, the interests asserted in these anti-RPAs laws are simply *competing* or *conflicting* interests. Conflicting values or interests are just that—conflicting. The UK House of Lords (now Supreme Court) identified the correct approach when rights compete, for example, the right to gather information to facilitate the public's right to receive information on the one hand and a right to respect for another right, *in casu,* someone's private and family life on the other. First, neither article has *as such* precedence over the other. Secondly, where the values under the two articles are in conflict, an intense focus on the comparative importance of the specific rights being claimed in the individual case is necessary. Thirdly, the justifications for interfering with or restricting each right must be taken into account. Finally, the proportionality test must be applied to each (Re S (FC) (a Child) 2004).

Thus, in any given fact, pattern, or situation, there may well be, for example, an infringement of X's privacy—but, *in casu,* it is *justifiable* because the appropriate balancing exercise would give preeminence to the right to freedom of expression. Simply asserting that some activity is a 'breach of privacy' is not in per se a conclusive, knock-down argument, but the beginning of a complex exercise weighing competing interests. The words of the UK Information Commissioner's Draft CCTV Code of Practice are instructive:

> The use of conventional cameras (not CCTV) by the news media or for artistic purposes such as for film making are not covered by this code as an exemption within the DPA applies to such activities that are carried out for journalistic, artistic and literary purposes (Information Commissioner's Office 2014).

What needs to be foregrounded in the dispute between pro-privacy restrictionists and pro-drone journalism RPA deployers is the relevance and status of the exemption or defence for activities in pursuit of newsgathering. The matter is well set out by the Australian Law Reform Commission (ALRC). The issue is whether, in the circumstances, even if there is an infringement of someone's right to respect for their privacy that is *defensible* or comes within the scope of an exemption. The ALRC states:

> …some legitimate uses of surveillance devices by journalists may place journalists at risk of committing an offence under existing surveillance device laws. Responsible journalism is an important public interest and should be protected. Journalists and media organisations should not be placed at risk of committing a criminal offence in carrying out legitimate journalistic activities. *The ALRC has therefore proposed a 'responsible journalism' defence to surveillance device laws*. This defence should be confined to responsible journalism

> involving the investigation of matters of public concern and importance, such as the exposure of corruption [emphasis added] (ALRC 2014).[22]

In sum, there are a number of problems with the claim that RPA's deployment is inherently problematic because they intrude on civil liberties, are a threat to privacy and are 'spies in the sky'.[23] In the context of pursuing responsible journalism deploying RPAs, any right to respect for private and family life should give way to the public interest in the public's right to be informed. Further, no general privacy-protecting regulation can be useful as it will inevitably be overbroad and general, basing regulations on hypothetical or imaginary 'threats' or 'harms'. Indeed, much of the RPAs' discourse contesting their deployment is fuelled by the so-called precautionary principle. This approach simply ignores or discounts those who do not think RPAs do constitute (much of) a threat : one might ask, who is the 'we' who objects (there are significant socioeconomic discontinuities, for example, low rental communities prioritise the protection that low-flying RPAs can facilitate); the issue is not RPAs or even the nature/technical capacity of the payload, but only any intentional, systemic misuse of personal information/data constituting 'serious' invasions of privacy (incidental, inadvertent acquisition of personal data quickly disposed of cannot seriously be said to give rise to any intrusion of privacy concern). Finally, prioritising privacy is a relatively soft and easy concern (and hardly raises any issues at all not already covered in general and human rights law). More seriously, it loftily ignores the serious issues involved in deployment of RPAs: How to safely integrate them into the national unsegregated airspace, certification, airworthiness, pilot training, sense and avoid, spectrum allocation, command and control processes, security of data links, liability, third-party insurance, etc….

Re Scenario Three: Distinguish Between Snooping and Legitimate Subterfuge A specific problem with drone journalism using, for example, a nano- or micro-RPA is alleged to be that the subjects of investigation might not realise that they are being surveilled, or being surveilled in a RPA-specific manner, because of the smallness of the RPAs and/or other technical capacities, for example, silence, mobility and endurance.

The issue here is the concern that micro-RPAs may not be noticed by those who are being surveilled either because they are extremely small and/or are very quiet. However, even if the pro-privacy/anti-surveillance lobbies may have a (weak) argument in this situation, *the putative infringement may well engage an exercise of legitimate subterfuge.*

The classic statement is contained in a report published by the UK Press Complaints Commission (PCC). The report was an inquiry into interception/tapping of

---

[22] See also Australian House of Representatives, Roundtable on Drones and Privacy (2014). Canadian regulators etc are also exercised by this issue: see, Office of the Privacy Commissioner of Canada (2013); Cavoukian (2012); Gersher (2014); and Block (2013).

[23] See, for example, http://www.bing.com/videos/search?q=spies+in+the+sky+drones&qpvt=spies+in+the+sky+drones&FORM=VDRE.

phone messages at the (then) *News of the World*. Illegitimate 'snooping' was definitely ruled out as a journalistic practice:

> It is essential that the type of snooping revealed by the phone message tapping incidents at the News of the World is not repeated at any other newspaper or magazine. Such events threaten public confidence in the industry, despite the considerable change in culture and practice that has undoubtedly occurred over the last decade and a half, leading to greater accountability and respect by the press for the privacy of individuals (ALRC 2014).

But the PCC warned against 'overreaction', stating:

> it is similarly important that the industry guards against overreaction. There is a legitimate place for the use of subterfuge when there are grounds in the public interest to use it and it is not possible to obtain information through other means. It would not be in the broader public interest for journalists to restrain themselves unnecessarily from using undercover means because of a false assumption that it is never acceptable (ALRC 2014).

A concrete illustration is the PCC's adjudication in the matter of *Bell Pottinger Group v. The Independent*:

> A number of Bell Pottinger executives had been secretly recorded by journalists from the Bureau of Investigative Journalism (BIJ) who were posing as 'clients' seeking advice on a public relations strategy for the Uzbekistan government…the Commission noted that the newspaper's actions were a clear prima facie breach of Clause ten of the Code [sic] which states that 'the press must not seek to obtain or publish material acquired by using hidden cameras or clandestine listening devices'. The test was whether a sufficient public interest defence could be established (Press Complaints Commission (PCC) 2012).

The Commission:

> noted that the journalists had been investigating various claims that had been made about the activities of Bell Pottinger and other public relations firms, rather than as a means of confirming a specific hypothesis about Bell Pottinger in particular, but ruled that 'the means employed by the journalists had been appropriately tailored to explore the allegations made by confidential sources about the firm's activities, which raised issues of significant public interest'. It acknowledged the firm's position that no 'serious impropriety' had been exposed but decided that the public interest was served by subjecting the complainants' methods to 'wider scrutiny and comment, particularly at a time when the possibility of imposing greater regulation on the [lobbying] industry was being debated' (Press Complaints Commission (PCC) 2012).

The PCC, therefore, did not uphold Bell Pottinger's complaint. Thus, using RPAs for drone journalism even in a manner which raises the above concern should acknowledge that although there may be a prima facie infringement of any ban on using clandestine or subterfuge methods for acquiring information; nonetheless, *in casu,* the test is (per the PCC) whether the public interest in exploring suspicions or allegations entails that:

> the means employed by the journalists had been appropriately tailored to explore the allegations made by confidential sources about the firm's activities, which raised issues of significant public interest (Press Complaints Commission (PCC) 2012).

Subterfuge is only justifiable—but it should be understood *is* justifiable:

…only when there are grounds in the public interest for using it. Undercover investigative work has an honourable tradition and plays a vital role in exposing wrongdoing. It is part of an open society. But it risks being devalued if its use cannot be justified in the public interest (Press Complaints Commission (PCC) 2012).

Endnote

The desire and, as is asserted in this chapter, the right of drone journalists to use RPAs as part of their professional toolkit (just as 'regular' photojournalists use 'conventional' cameras, even with long-range telephoto lenses) is currently experiencing robust resistance and pushback by counter-lobbies, concerns, fears, anxieties, opposition—and downright hostility towards RPAs *in toto*.[24] As is well known, the resistance functions around three main axes: military and weaponised use of RPAs, in particular, when civilian deaths ensue (see, e.g., the UK All Party Parliamentary Group on Drones (APPG) n.d.; Drone Wars UK n.d.; and Fair et al. 2014); the perceived threat to individual privacy (see, e.g., the Australian House of Representatives Standing Committee on Social Policy and Legal Affairs, round table on drones and privacy—Commonwealth of Australia Parliamentary Debates 2014); and the technical capacity of the new generation of RPAs (actually it is not the RPA so much as the *payload*) to conduct wide-ranging and intense surveillance (see, e.g., the European Group on Ethics in Science and New Technologies to the European Commission 2014). Indeed, opposing the claimed 'anti-drone consensus' meets the riposte that efforts to counter RPAs deployment:

> are being impeded by those who mock the idea that domestic drones pose unique dangers (often the same people who mock concern over their usage on foreign soil). This dismissive posture is grounded not only in soft authoritarianism (a religious-type faith in the Goodness of US political leaders [sic] and state power generally) but also ignorance over current drone capabilities (Greenwald 2013).

Such concerns have led, at least in the USA, as noted above, to an anti-drone lawfare strategy, i.e., introducing and/or adopting state-level legislation to either curtail the use of RPAs or to delay their introduction. There are also Federal-level initiatives.[25] At the time of writing, no similar legislation is known to have been proposed or adopted anywhere else in the world. Indeed, in Australia, in May 2014, the Civil Aviation Safety Authority announced a consultation on the *relaxation* of the rules for RPAs under 2 kg., not least because it can see that there is no point having very strict but ultimately unenforceable rules (Civil Aviation Safety Authority 2014).

---

[24] Thierer (2013) argues that, 'If steps must be taken to address these concerns, education and empowerment-based solutions represent superior approaches to dealing with them compared to a precautionary principle approach, which would limit beneficial learning opportunities and retard technological progress.' See also Thierer (2012) and Cesca (2013).

[25] See for example: National Association of Criminal Defense Lawyers (2014) see also, Targeted Lethal Force Transparency Act2014. https://beta.congress.gov/bill/113th-congress/house-bill/4372.

And, in the USA, the FAA has decided to consider holding discussions with various sectors about licensing RPA use, where the usual grounds of opposition to using them, for example, intrusion into personal privacy, hardly apply, namely, precision agriculture, film-making, pipeline and power-line inspection and oil-and-gas flue stack inspection industries.

Against any carte blanche opposition is the claim made in this chapter that use of RPAs for drone journalism engages a significant human right (or constitutionally or legally protected value or interest) whether established through the prism of the European Convention on Human Rights or any functionally equivalent regime, for example, the US First Amendment.[26] As such, RPAs deployment cannot prima facie be generally prohibited. Of course, not every drone journalism use of RPAs will be *absolutely* protected any more than is the current use of technologies and techniques for such purposes (BBC News 2014a). It is sanguine to remember that in the UK, nearly 100 journalists have been arrested in connection with the so-called 'phone hacking scandal', computer misuse and…. Not all means justify the ends. What is argued though is that a general measure(s) precluding RPA use and a fortiori by the media would only be justifiable in specific circumstances when it was (as the ECtHR would put it) prescribed by law, for a legitimate aim, 'necessary in a democratic society'; and proportionate. The damage to the public's right to receive information and ideas that any such prohibition would entail means that such a restriction would be highly exceptional and could only be justified in the most extreme situations. Equally, where there is no general permission but only a case-by-case decision on the part of an aviation regulator to allow a drone journalism flight(s) to take place would be subject to strict scrutiny and possible judicial review in circumstances where the requisite weighing of the public's right to receive information had not been done explicitly.

Echoing the point made by Tom Hannen, supra, 'We see it as a flying camera'; this use of RPAs can also be simply understood as a development of photography in public places.[27] As such, what points can be deployed by media lawyers to counter anti-RPAs lawfare? In the UK, for example, according to Dr Michael Pritchard (Director-General, The Royal Photographic Society):

> …the law is very clear in that there is no restriction on photography in public places. The Terrorism Acts do not really affect this *general principle*…. Sometimes defining what is public and what is private space can be problematic and there are a few spaces such as the Royal parks and Trafalgar Square which have bylaws restricting photography although these should not affect normal amateur photography. There are some restrictions on photographing certain designated building e.g. military bases which predate the Terrorism Acts. Such buildings would be signed. It is not unlawful to photograph a police officer except where it is done to support the commissioning of terrorism (Pritchard 2013) [emphasis added].[28]

---

[26] Osterreicher (n.d.) has discussed whether there is a (US) constitutional right to use drones to collect aerial imagery.

[27] See, for example, *Photography Is Not a Crime: Be the Media*.

[28] Personal email to the author, 13 February 2013; see the Metropolitan Police website: http://content.met.police.uk/Site/photographyadvice. There are a few UK websites that cover this topic, for example, Digital Camera World (2012). See also Macpherson (2009).

In 2008, the UK government was asked to state 'what plans they have for reviewing the rules on street photography'. Replying for the government, Lord Bassam of Brighton stated:

> My Lords, the freedom of the press and media is one of the bedrocks of democracy in this country. Although police officers have the discretion to ask people not to take photographs for public safety or security reasons, *the taking of photographs in a public place is not subject to any rules or statute*. There are no legal restrictions on photography in a public place and no presumption of privacy for individuals in a public place. There are no current plans to review this policy (Lord Bassam 2008) [emphasis added].[29]

It is suggested here that, at least in states members of the Council of Europe, the core demands of freedom of expression in a democratic society and, in particular, the public's right to receive information, mean that the default position warrants the use of RPAs as 'flying cameras' in the pursuit of drone journalism and that any restriction(s) must be for a proper, legitimate aim pursued through a very narrowly and precisely crafted exception. Such is the fundamental role of a free and responsible press in a democratic society that any challenge not only to publishing information but also to *exercising the means to realise it* must necessarily overcome a very high threshold (see Scheinin n.d.).

In conclusion, using RPAs for newsgathering—'drone journalism'—raises unique normative concerns vis-à-vis other RPAs applications:

- It engages the public's right to receive information and ideas and hence is part of the right to freedom of expression.
- It engages the right to access communications technologies as well as the right to photograph in public places.[30]

---

[29] Of course, there is the offence of photography contrary to Section 1 of the Official Secrets Act 1911 and the whole matter of 'prohibited places', but the offences are for behaviour which is "for any purpose prejudicial to the safety or interests of the State." See also the Security Industry Authority Guidance (2011) provided to private security guards to ensure that it sufficiently reflects the right of the public to take photographs, British Transport Police (n.d.) With respect to dissenting judgements, see Voorhoof (2014a).

[30] One might wish that the right and its scope—even in the USA—were even clearer: 'The scope of the right to take photos or record video is still ambiguous, so tread carefully. The Supreme Court has never ruled that there is an affirmative right to take photos or video in public places. To be fair, they have never said there isn't such a right. But the exact scope of any such right has never been defined and federal courts are split on whether such a right exists at all. For example, in Glik v. Cunniffe, the First Circuit found that there was a First Amendment right to take video of police officers arresting someone. However, in Kelly v. Borough of Carlisle, the Third Circuit ruled that a police officer was entitled to qualified immunity on a First Amendment claim because there is no clearly established right to make a video of traffic officers on a traffic stop. Although there is some very good case law supporting such a right, and one may believe there should be such a right, the law is not perfectly clear. In some jurisdictions, those who take photos or videos, even in public places, run the risk of being arrested if they fail to comply with police orders or for violating state wiretap laws. Until the courts make clear there is an affirmative First Amendment right to record and what the scope of that right is, journalists should not be overconfident in their rights' (AEJMC 2014).

- Although users will have to be aware and take account of aviation law and the decisions of national, regional and international aviation regulators, for the first time, those regulators will have to take account of the concerns, rights and principles advanced by media lawyers and the right(s) bundled within the rubric of the right of freedom of expression. The mandate of the aviation regulator should not overstep measures in the interests of ordering the national airspace in the interests of safety, certification and airworthiness.

References

ABC News. (2014a). Chinese manufacturer programs Phantom drones with no-fly zones to protect Australian airports. http://www.abc.net.au/news/2014-04-14/chinese-made-drones-programmed-with-no-fly-zones/5388356. Accessed 22 Sept 2014.

ABC News. (2014b). Footage from drone flying over Taksim Square. http://www.abc.net.au/news2014-04-14/footage-from-drone-flying-over-taksim-square/5389438. Accessed 22 Sept 2014.

ABC News. (2014c). Perth airport near miss prompts ATSB drone safety warning. http://www.abc.net.au/news/2014-05-27/atsb-says-drones-pose-risk-to-aviation-following-near-collision/5480156. Accessed 22 Sept 2014.

AEJMC. (2014). Media law notes (Winter). http://aejmc.us/law/wp-content/uploads/sites/18/2013/10/Media-Law-Notes-Winter-2014.pdf. Accessed 24 Sept 2014.

Al Jazeera. (2012). Drone camera captures Argentina's #8N protests. http://stream.aljazeera.com/story/ 201211100124-0022394. Accessed 23 Sept 2014.

All Party Parliamentary Group on Drones (APPG). (n.d.) http://appgondrones.wordpress.com/. Accessed 12 Jan 2015.

ALRC. (2014). Serious invasions of privacy in the digital era (DP 80). http://www.alrc.gov.au/publications/serious-invasions-privacy-dp-80. Accessed 23 Sept 2014.

Amici News Media Brief in Support of Respondent Raphael Pirker (urging the engagement of the US First Amendment). (05 June 2014). http://www.hklaw.com/files/Uploads/Documents/CaseBriefs/MediaLaw/Drones.pdf. Accessed 22 Sept 2014.

Ariens, C. (2014). Drone covers the chaos in Ukraine. http://www.mediabistro.com/tvnewser/drone-covers-the-chaos-in-ukraine_b214311. Accessed 22 Sep 2014.

Australian House of Representatives. (2014). Roundtable on drones and privacy. http://www.aph.gov.au/parliamentary_business/committees/house_of_representatives_committees?url=spla/drones/index.htm. Accessed 23 Sept 2014.

Australian Law Reform Commission. (2014). Serious invasions of privacy in the digital era. http://www.alrc.gov.au/sites/default/files/pdfs/publications/final_report_123_whole_report.pdf. Accessed 30 Sept 2014.

AUVSI''s Unmanned Systems Mission Critical. (2014). http://issuu.com/auvsi/docs/mission_critical_agriculture_2014/1?e=2783066/7674504. Accessed 23 Sept 2014.

BBC News. (2013). 'Hexacopter' drone flying camera. https://www.youtube.com/watch?v=ZTWHP80hei0. Accessed 14 Sept 2014.

BBC News. (2014a). Phone hacking. http://www.bbc.co.uk/news/uk-14045952. Accessed 23 Sept 2014.

BBC News. (2014b). Tree-mapping drone start-up has sky-high ambitions. http://www.bbc.com/news/technology-27485418. Accessed 19 Sept 2014.

Biederman Institute Conference. (2014). Using drones in the news and entertainment industries: The legal and regulatory issues. http://www.swlaw.edu/academics/entertainmentlaw/instevents/droneconf2014/#schedule. Accessed 23 Sept 2014.

Block, G. (2013).Watching below: Dimensions of surveillance-by-UAVs in Canada. http://www.blockg.ca/portfolio-item/watching-below-dimensions-of-surveillance-by-uavs-in-canada. Accessed 23 Sept 2014.

British Transport Police. (n.d.). http://www.btp.police.uk/advice_and_info/travelling_safely/rail_enthusiasts.aspx. Accessed 24 Sept 2014.

Campbell, D. (2014). UAF unmanned aircraft joins Funny River firefighting efforts. http://uaf-cornerstone.net/acuasi_funnyriver/. Accessed 19 Sept 2014.

Caputo, L. (2013). Drone journalism, uno sguardo inedito sugli eventi. http://piazzadigitale.corriere.it/2013/03/04/drone-journalism-uno-sguardo-inedito-sugli-eventi/. Accessed 23 Sept 2014.

Cavoukian, A. (2012). Privacy and drones: Unmanned aerial vehicles. information & privacy commissioner Ontario, Canada. http://www.ipc.on.ca/images/Resources/pbd-drones.pdf. Accessed 23 Sept 2014.

Cesca, B. (2013). The most terrifying drone ever! run away! http://thedailybanter.com/2013/03/the-most-terrifying-drone-ever-run-away/. Accessed 23 Sept 2014.

Civil Aviation Safety Authority. (2014). NPRM 1309OS—remotely piloted aircraft systems. http://casa.gov.au/scripts/nc.dll?WCMS:STANDARD::pc=PC_102028. Accessed 23 Sept 2014.

Collins, K. (2014). Police, paps and privacy: The challenges of drone journalism. http://www.wired.co.uk/news/archive/2014-02/12/drone-journalism-legal-and-privacy. Accessed 24 Sept 2014.

Collins, R. (various dates). A brief history of photography and photojournalism. http://www.ndsu.edu/pubweb/~rcollins/242photojournalism/historyofphotography.html. Accessed 22 Sept 2014.

Committee on Autonomy Research for Civil Aviation. (2014). Autonomy Rresearch for Ccivil Aaviation: Toward a Nnew Eera of Fflight (US National Research Council). http://www.nap.edu/catalog.php? record_id=18815. Accessed 22 Sept 2014.

Commonwealth of Australia Parliamentary Debates. (2014). http://parlinfo.aph.gov.au/parlInfo/download/chamber/hansardr/19c32232-832c-45aa-8efa-783c9a3446a4/toc_pdf/House%20of%20Representatives_2014_03_17_2298_Official.pdf;fileType=application/pdf#search='chamber/hansardr/19c32232-832c-45aa-8efa-783c9a3446a4/0014. p. 1983. Accessed 23 Sept 2014.

Corcoran, M. (2014). Drone journalism: Newsgathering applications of Unmanned Aerial Vehicles (UAVs) in covering conflict, civil unrest and disaster. http://www.flinders.edu.au/ehl/fms/law_files/Drone%20Journalism%20During%20Conflict,%20Civil%20Unrest%20and%20Disasters%20March%201%202014.pdf. Accessed 11 Jan 2015.

Cruz, D. (2014). Journalism in the Americas. https://knightcenter.utexas.edu/blog/00-15587-us-media-defend-drone-journalism-federal-government-arguing-freedom-press. Accessed 22 Sept 2014.

Digital Camera World. (2012). http://www.digitalcameraworld.com/2012/04/14/photographers-rights-the-ultimate-guide/ and http://www.digitalcameraworld.com/2013/02/12/is-it-legal-to-take-pictures-of-buildings-photography-law-questions-answered-by-experts/. Accessed 23 Sept 2014.

Dolan, A. M., Thomson, R. M. (2013). Integration of drones into domestic airspace: Selected legal issues. http://www.fas.org/sgp/crs/natsec/R42940.pdf. Accessed 22 Sept 2014.

Domestic Drone Information Center (n.d.). http://www.nacdl.org/usmap/news/27169/39138/d/#PR

Drone Adventures (n.d.). http://droneadventures.org/. Accessed 19 Sept 2014.

Drone Journalism Lab (n.d.). http://www.dronejournalismlab.org/.

Drone Journalist. (2014). Drone footage: Police clash at Taksim Gezi Park. https://www.youtube.com/watch?v=MQcW9gijs_0. Accessed 22 Sept 2014.

Drone Wars UK (n.d.). http://dronewars.net/.

Eschoff, R. (2013). Telecopter inventor. https://alumni.stanford.edu/get/page/magazine/article/?article_id=60245. Accessed 22 Sept 2014.

European Court of Human Rights. (2010). Case of Sanoma Uitgevers B.V. v. the Netherlands. http://hudoc.echr.coe.int/sites/eng/pages/search.aspx?i=001-100448. Accessed 22 Sept 2014.

European Court of Human Rights. (2012). Case of telegraaf media nederland landelijke media B.V. and others v. the netherlands. http://hudoc.echr.coe.int/sites/eng/pages/search.aspx?i=001-114439. Accessed 22 Sept 2014.

European Group on Ethics in Science and New Technologies to the European Commission. (2014). Ethics of security and surveillance technologies. Opinion 28. http://ec.europa.eu/bepa/european-group-ethics/docs/publications/opinion_28_securityandsurveillancetechnologies.pdf. Accessed 23 Sept 2014.

Fair, C. C., Kaltenthaler, K., Miller, W. J. (2014). The foreign policy essay: What Pakistanis think about US drone strikes and why. http://www.lawfareblog.com/2014/06/the-foreign-policy-essay-what-pakistanis-think-about-u-s-drone-strikes-and-why/#more-35099. Accessed 23 Sept 2014.

Federal Aviation Administration (2014a). FAA to consider exemptions for commercial UAS movie and TV production. http://www.faa.gov/news/press_releases/news_story.cfm?newsId=16294. Accessed 19 Sept 2014.

Federal Aviation Administration (2014b). Surveys will check pipelines, infrastructure on Alaska north slope. http://www.faa.gov/news/press_releases/news_story.cfm?newsId=16354. Accessed 22 Sept 2014.

Federal Aviation Administration (2014c). Unmanned Aircraft Systems (UAS) frequently asked questions. http://www.faa.gov/uas/faq/. Accessed 11 Jan 2015.

Flying Donkey Challenge (2012–2014). http://www.flyingdonkey.org/. Accessed 19 Sept 2014.

Frontline Club London. (2013). Drone journalism: The future of newsgathering? https://www.youtube.com/watch?v=gu_0fh2IJms. Accessed 7 Jan 2015

Gersher, S. (2014). Canada's domestic regulatory framework for RPAS: A call for public deliberation, in Jjournal of Uunmanned Vvehicle Ssystems 2(1). http://www.nrcresearchpress.com/doi/abs/10.1139/juvs-2013-0016#.U5RtpfldXTo. Accessed 23 Sept 2014.

Giemulla E., Weber, L. (2011). International and EU aviation law: Selected issues. kluwer.

Globally, C. (2014). Ukraine dramatic drone footage captures battle for central Kiev square. https://www.youtube.com/watch?v=jR6N3Lo1Qs0. Accessed 22 Sept 2014.

Goebel, G. (n.d.). UAV universe. https://sites.google.com/site/uavuni/1920s-1930s. Accessed 11 Jan 2015.

Goldberg, D., Corcoran, M., Picard, R. (2013). Remotely piloted aircraft and journalism. https://reutersinstitute.politics.ox.ac.uk/sites/default/files/Remotely%20Piloted%20Aircraft%20and%20Journalism.pdf. Accessed 23 Sept 2014.

Green, M. (2013). http://www.thedailybeast.com/articles/2013/03/26/unmanned-drones-may-have-their-greatest-impact-on-agriculture.html. Accessed 23 Sept 2014.

Greenwald, G. (2013). Domestic drones and their unique dangers. http://www.guardian.co.uk/commentisfree/2013/mar/29/domestic-drones-unique-dangers. Accessed 23 Sept 2014.

Hannen, T. (2014). Global video unit, BBC world service. http://www.wired.co.uk/news/archive/2014-02/12/bbc-drone-journalism/viewgallery/ytKr2kdtZOFW0. Accessed 23 Sept 2014.

Herr, R. E. (2013). Can human rights law support access to communication technology? *Information and Communications Technology Law 22*(1),1–13. http://www.tandfonline.com/doi/abs/10.1080/13600834.2013.774517?journalCode=cict20. Accessed 22 Sept 2014.

Hill, K. (2011). FAA looks into news corp's daily drone, raising questions bout who gets to fly drones in the US. http://www.forbes.com/sites/kashmirhill/2011/08/02/faa-looks-into-news-corps-daily-drone-raising-questions-about-who-gets-to-fly-drones-in-the-u-s/. Accessed 22 Sept 2014.

Humanitarian UAV Network (n.d.). http://uaviators.org/. Accessed 19 Sept 2014.

Increasing Human Potential (2014a). http://increasinghumanpotential.org/espn-eyes-in-the-sports-sky-uas-are-coming-to-a-sports-field-near-you-and-not-just-for-pretty-video/. Accessed 19 Sept 2014.

Increasing Human Potential (2014b). Washington archdiocese takes to the heavens, with a UAS. http://increasinghumanpotential.org/washington-archdiocese-takes-to-the-heavens-with-a-uas/. Accessed 7 Jan 2015.

Information Commissioner's Office. (2014). CCTV code of practice: Draft for consultation. http://ico.org.uk/about_us/consultations/our_consultations. Accessed 22 Sept 2014.

Insinna, V. (2014). Sikorsky enters unmanned cargo aircraft market. National defense industrial association business and technology magazine. http://www.nationaldefensemagazine.org/blog/Lists/Posts/Post.aspx?ID=1505

International Civil Aviation Organization. (2011). Unmanned Aircraft Systems (UAS). http://www.icao.int/Meetings/UAS/Documents/Circular%20328_en.pdf. Accessed 16 Sept 2014.

International Council on Human Rights Policy. (2011). Navigating the dataverse. http://www.ichrp.org/files/reports/64/132_report_en.pdf. Accessed 16 Sept 2014.

Johnson, S. (2014). Balancing privacy, jobs in the domestic drone debate. http://www.usatoday.com/story/news/nation/2014/04/11/stateline-privacy-jobs-drones/7590409/. Accessed 22 Sept 2014.

Joseph Burstyn, Inc. v. Wilson, 343 U.S. 495. (1952). http://caselaw.lp.findlaw.com/scripts/getcase.pl?navby=CASE&court=US&vol=343&page=495. Accessed 23 Sept 2014.

Kiss, J. (2014). Worried about your privacy? Wait till the drones start stalking you. http://www.theguardian.com/technology/2014/feb/09/privacy-concerns-google-streetview-facebook-drones. Accessed 22 Sept 2014.

Levine, B. (2014). Flocks of airborne camera drones will change journalism—& spying. http://venturebeat.com/2014/05/23/flocks-of-airborne-camera-drones-will-change-journalism-spying. Accessed 22 Sept 2014.

Lord Bassam. (2008). Photography: Public places. http://www.publications.parliament.uk/pa/ld200708/ldhansrd/text/80716-0001.htm. Accessed 23 Sept 2014.

Macpherson, L. (2009). The UK photographers rights guide. http://www.sirimo.co.uk/wp-content/uploads/2009/05/ukphotographersrights-v2.pdf. Accessed 23 Sept 2014.

McClatchy, D. C. (2013). Industry looks to use drones for commercial purposes. http://www.mcclatchydc.com/2013/04/29/189893/industrylooks-to-use-drones-for.html#storylink=cpy. Accessed 19 Sept 2014.

McGrath, J. (n.d.) What is electronic news gathering? http://people.howstuffworks.com/electronic-news-gathering.htm. Accessed 22 Sept 2014.

Mello, J. P. (2014). Parrot uncages new drone. http://www.technewsworld.com/story/80435.html Accessed 19 Sept 2014.

Merlin, P. W. (2013). Crash course: Lessons learned from accidents involving remotely piloted and autonomous aircraft (NASA). http://www.nasa.gov/pdf/732725main_crash_course-ebook.pdf. Accessed 19 Sept 2014.

Mesich, A. (2014). How sensor reporting helps journalists find data where none exist. http://ijnet.org/blog/how-sensor-reporting-helps-journalists-find-data-where-none-exist. Accessed 22 Sept 2014.

Miller, Z. (2014). What you need to know about flying drones in Washington DC. http://time.com/88772/drones-washington-dc/. Accessed 23 Sept 2014.

National Association of Criminal Defense Lawyers. (2014). Domestic drone information center. http://www.nacdl.org/usmap/news/27169/39138/d/#FED.

Nikola Tesla Museum (n.d.). http://www.tesla-museum.org. Accessed 23 Sept 2014.

Nurlybayeva, N. (2012). The story behind those Russian drone protest images. http://ijnet.org/stories/story-behind-those-russian-drone-protest-images. Accessed 22 Sept 2014.

O'Connell, J. (2014). Idaho drone project studies potato stress. http://www.capitalpress.com/Idaho/20140527/idaho-drone-project-studies-potato-stress. Accessed 23 Sept 2014.

Office of the Privacy Commissioner of Canada. (2013). Drones in Canada. http://www.priv.gc.ca/information/research-recherche/2013/drones_201303_e.pdf. Accessed 23 Sept 2014.

Osterreicher, M. (n.d.). A constitutional right to use drones to collect aerial imagery? http://droneu.org/mickey-osterreicher. Accessed 23 Sept 2014.

Osterreicher, M. (2014). Charting the course for use of small unmanned aerial systems in newsgathering. http://www.auvsishow.org/auvsi2014/Custom/Handout/Speaker0_Session773_1.pdf. Accessed 23 Sept 2014.

PAPA International. (2014). History of aerial photography. http://professionalaerialphotographers.com/ontent.aspx?page_id=22&club_id=808138&module_id=158950. Accessed 22 Sept 2014.
Penhaul, K. (2013). CNN uses drone camera for aerial view of Tacloban. http://cnnpressroom.blogs.cnn.com/2013/11/18/cnn-uses-drone-camera-for-aerial-view-of-tacloban/. Accessed 22 Sept 2014.
Photography is not a crime: Be the Media. http://photographyisnotacrime.com/. Accessed 23 Sept 2014.
Pitt, F. (Ed.). (2014). Sensors and journalism. http://towcenter.org/wp-content/uploads/2014/05/Tow-Center-Sensors-and-Journalism.pdf. Accessed 14 Sept 2014.
Pool, B. (2012). John D. Silva dies at 92; introduced news helicopter. http://articles.latimes.com/2012/dec/07/local/la-me-john-silva-20121207. Accessed 22 Sept 2014.
Press Complaints Commission (PCC). (2012). Decision. http://pcc.org.uk/news/index.html?article=NzkxNQ==; Summary is available at http://www.pcc.org.uk/news/index.html?article=NzkxNg==.
Professional Society of Drone Journalists (n.d.). http://www.dronejournalism.org/.
Professional Society of Drone Journalists (n.d.). Code of ethics for drone journalists. http://www.dronejournalism.org/code-of-ethics. Accessed 30 Sept 2014.
Ravich, T. (2014). And the drone goes to... http://droninglawyer.com/2014/06/03/and-the-drone-goes-to/. Accessed 22 Sept 2014.
Re S (FC) (a Child). (2004). http://www.publications.parliament.uk/pa/ld200304/ldjudgmt/jd041028/inres-1.htm. Accessed 7 Jan 2015.
Reed, J. (2011). Civilian UAV films polish riots from above. http://defensetech.org/2011/11/16/video-civilian-uav-films-polish-riots-from-above/#ixzz2DhTJQx2D. Accessed 22 Sept 2014.
Rosen, A. (2014). A bizarre near-miss between a drone and a passenger plane is shrouded in mystery. http://www.businessinsider.com/the-mystery-of-a-near-miss-between-a-drone-and-a-passenger-plane-2014-5#ixzz33ZJ7bVwM. Accessed 22 Sept 2014.
Sachs, S. (2010). Teleautomaton. http://teleautomaton.com/post/1373803033/how-teslas-1898-patent-changed-the-world. Accessed 23 Sept 2014.
Sánchez, W. A. (2014). Drones in Latin America. http://www.coha.org/wp-content/uploads/2014/01/COHA_Sanchez_LATAM_Drones_Final_Jan122014.pdf. Accessed 22 Sept 2014.
Scheinin, M. (n.d.). Criteria for balancing. http://stream.hanaholmen.fi/media/speakingissilver/MartinScheinin.wmv. Accessed 23 Sept 2014.
Security Industry Authority. (2011). http://www.bsia.co.uk/web_images//publications/form_122.pdf. Accessed 24 Sept 2014.
Shadowview Foundation (n.d.). Unmanned ariel systems. http://www.shadowview.org/the-uas/. Accessed 19 Sept 2014.
Spies in the Sky. (2014). https://www.youtube.com/watch?v=YfY-vbstX9o. Accessed 7 January 2015.
Stanley, J. (2013). We already have police helicopters, So What's the Big Deal Over Drones? https://www.aclu.org/blog/technology-and-liberty-criminal-law-reform/we-already-have-police-helicopters-so-whats-big-deal. Accessed 22 Sept 2014.
Stark, J. (2014). News drones over El salvador. http://www.globalpost.com/dispatch/news/regions/americas/140502/news-drones-over-el-salvador. Accessed 22 Sept 2014.
Stone, B. (2014). DJI's drone is simple enough for anyone to use. http://www.businessweek.com/articles/2014-05-15/dji-innovations-drone-is-simple-enough-for-anyone-to-use. Accessed 19 Sept 2014.
Targeted Lethal Force Transparency Act. (2014). https://beta.congress.gov/bill/113th-congress/house-bill/4372. Accessed 23 Sept 2014.
Technology Liberation Front. (2014). Moral panics / techno-panics. http://techliberation.com/ongoing-series/ongoing-series-moral-panics-techno-panics/. Accessed 16 Sept 2014.
Tesla Memorial Society of New York (n.d.). http://www.teslasociety.com/. Accessed 23 Sept 2014.
Tesla, N. (1898). Patent for method of and apparatus for controlling mechanism of moving vessels or vehicles, 8 Nov 1898. http://www.google.com/patents/US613809. Accessed 7 Jan 2015.

The National. (2014). http://www.thenational.ae/business/industry-insights/shipping/eye-in-the-sky-abu-dhabis-ports-now-protected-by-drones. Accessed 19 Sept 2014.

The Telegraph. (2014). Aerial drone footage shows scale of Balkan floods. http://www.telegraph.co.uk/news/worldnews/europe/serbia/10839004/Aerial-drone-footage-shows-scale-of-Balkan-floods.html. Accessed 22 Sept 2014.

Thierer, A. (2012). Technopanics, threat inflation and the danger of an information technology precautionary principle. http://papers.ssrn.com/sol3/papers.cfm?abstract_id=2012494. Accessed 23 Sept 2014.

Thierer, A. (2013). Technopanics, threat inflation, and the danger of an information technology precautionary principle. http://techliberation.com/2013/02/14/technopanics-paper-published-in-minn-jour-of-law-science-tech/. Accessed 23 Sept 2014.

TMZ. (2012). We're not keeping up with the droneses. http://www.tmz.com/2012/11/27/tmz-drone-faa-bogus-report/. Accessed 22 Sept 2014.

UK Civil Aviation Authority. (2012). UK CAP 722, unmanned aircraft system operations in UK airspace—guidance. http://www.caa.co.uk/docs/33/cap722.pdf. Accessed 16 Sept 2014.

United States of America National Transportation Safety Board. (2013). Brief of news media amici in support of respondent Raphael Pirker (amici brief, urging the engagement of the US first amendment). http://www.hklaw.com/files/Uploads/Documents/CaseBriefs/MediaLaw/Drones.pdf. Accessed 22 Sept 2014.

United States Senate Committee on Commerce, Science and Transportation. (2014). The future of unmanned aviation in the US economy: Safety and privacy considerations (Statement of henio arcangeli). http://www.commerce.senate.gov/public/?a=Files.Serve&File_id=ba8490d0-e9fc-4c97-83cc-1ad211291bc3. Accessed 11 Jan 2015.

Urban75. (2011). http://www.urban75.org/blog/museum-of-london-london-street-photography/. Accessed 22 Sept 2014.

Usmani, B. (2014). How drones might save Pakistan's embattled news media. http://motherboard.vice.com/blog/how-drones-might-save-p. Accessed 22 Sept 2014.

Victoria and Albert Museum. (2014). Photojournalism. http://www.vam.ac.uk/content/articles/p/photojournalism/. Accessed 22 Sept 2014.

Voorhoof, D. (2014a). The European court of human rights Finland decision, Pentikäinen v. Finland, http://merlin.obs.coe.int/iris/2014/4/article2.en.html. Accessed 24 Sept 2014.

Voorhoof, D. (2014b). The right to freedom of expression and information under the European Human Rights system: Towards a more transparent democratic society. EUI RSCAS, 2014/12, Centre for Media Pluralism and Media Freedom (CMPF). http://cadmus.eui.eu/handle/1814/29871. Accessed 22 Sept 2014.

Waite, M. (2013). The drone age is here, and so are the lawyers. http://www.niemanlab.org/2013/12/the-drone-age-is-here-and-so-are-the-lawyers/. Accessed 22 Sept 2014.

Waldie, D. J. (2013). A short history of the intrusive helicopter. http://blogs.kcrw.com/whichwayla/2013/09/a-short-history-of-the-intrusive-helicopter. Accessed 22 Sept 2014.

Weber, A. (2010). Nikola Tesla: Father of unmanned vehicle technology. http://www.assemblymag.com/articles/87689-nikola-tesla-father-of-unmanned-vehicle-technology. Accessed 23 Sept 2014.

Williams, R. (2014). Current unmanned aircraft state law landscape. http://www.ncsl.org/research/civil-and-criminal-justice/current-uas-state-law-landscape.aspx.

David Goldberg is an academic lawyer; senior visiting fellow, Institute of Computer and Communications Law at the Centre for Commercial Law Studies, Queen Mary University of London; adjunct associate professor of law, Southwestern Law School, Los Angeles, USA; and directs deeJgee Research Consultancy. He is an occasional ad hoc expert/consultant for Organization for Security and Co-operation in Europe (OSCE) projects, works as a volunteer for the UK Campaign for Freedom of Information (since 1984), is an Access Info Europe International Advisory Board member and a fellow of the Royal Society of the Arts. He coauthored the ground-breaking report

Remotely Piloted Aircraft Systems and Journalism (Reuters Institute for the Study of Journalism, 2013) and co-organised the innovative conference Using Drones in the News and Entertainment Industries: The Legal and Regulatory Issues (Southwestern Law School, 2014). He is a legal and regulatory expert for Unmanned Experts and a member of the Royal Aeronautical Society's Unmanned Aircraft Systems Specialist Group. Goldberg gained a 'Ph.D. by Publication' (Glasgow Caledonian University, 2011).

Chapter 11
Drones, Resistance and Countersurveillance

Aleš Završnik

Introduction

In *Metamorphoses,* Ovid narrates that Medusa, the snake-haired and most deadly of the gorgons, had the power to turn to stone anyone who dared gaze into her eyes. By surprising her in her sleep, Perseus was able to cut off her head while looking at her reflection in the bronze shield given to him by Minerva. The hero, after having freed Andromeda and defeated Phineus thanks to the still intact petrifying power of Medusa, gave the head to Minerva, who used it to adorn her aegis, and then her shield, as a terrifying weapon to defeat the enemies of reason and knowledge, the virtues that she embodied.

If drones are Medusa's eyes petrifying the observed, either literally turning them into stone/ashes or metaphorically turning them into "dead" numbers emptied of life—numbers that begin to circulate and start to have a life on their own, as Haggerty and Ericson (2000) claimed for subjects in the information age and dubbed them "data doubles"—then one should not gaze at drones directly. If we stare at them directly, we may not see the whole "truth" of drones, which may be a part of something "larger" than flying machines. They may also be a part of the wider processes that are at work in the contemporary "all-seeing society" (Bauman 2011).

The other point is that drones' "negative" controlling power can be undermined or "hijacked" and redirected to the controllers. When put under the gaze, we may see that drone operators may have Orwellian ambitions, but they are far from achieving them. Operators might not be inspired by Orwell—or other twentieth-century dystopian novelists—but just want to fit into the current exhibitionist culture and take "dronies" (selfies with drones) of themselves. The motivations, aspirations and interests of those employing drones may be as varied as those using a kitchen knife.

The third point is that several contemporary uses of drones, such as for land surveying, mapping purposes and crop inspection, are very much removed from the

---

A. Završnik (✉)
Institute of Criminology at the Faculty of Law, Poljanski nasip 2,
SI-1000 Ljubljana, Slovenia
e-mail: ales.zavrsnik@pf.uni-lj.si

A. Završnik (ed.), *Drones and Unmanned Aerial Systems,*
DOI 10.1007/978-3-319-23760-2_11

classic intelligence, surveillance and reconnaissance (ISR) focused on people. They may have social implications for security and surveillance, but these may be more removed. While these are the more "neutral" uses of drones in terms of surveillance and personal data protection, there are other uses of drones that are clearly "positive" and even empowering and liberating. Drones are used for documenting genocide and empowering powerless groups, for instance in South Sudan, where drones are called the "new knights" and used because people are changing into apathetic robots and "maybe the robots can be the ones to change the seeds and roots of the remaining humanity" (Tomo Križnar in Vidrih 2014).

This chapter attempts to act as such a bronze shield decorated with the *Head of Medusa* when looking at drones. It shows how drones can be negotiated and resisted and even work to empower the observed and, what is more, to stand as the sublime *objet petit a*—a sublime empty object waiting to be filled with the gaze of the observer.

In the surveillance studies literature, there have been significant contributions on social sorting, the invasion of privacy, racial profiling, discrimination and other mechanisms of unequal treatment but to a lesser extent on the ways in which those who are subject to surveillance manage, negotiate and resist these processes. The other side and potential of surveillance—for individual autonomy and dignity, fairness and due process, community cooperation, empowerment and social equality—has been less explored, with some notable exceptions.

Lyon (2007) argued that surveillance studies are overly concerned with the analysis of omnipotent surveillance stemming from top to bottom, that is, from surveillance agencies to the surveilled individuals, and neglect the nature of contemporary surveillance, which is dynamic and not hierarchically structured. The new surveillance is forming "complex networks of power relations and resistances" (Green 1999). Tinic (2006) claimed individuals as objects of surveillance occupy a significant role in the execution of surveillance. The increasing power of the new surveillance not only increases the capacity of the authorities (watchers) to monitor the watched majority but also carries an emancipatory potential. Objects of surveillance can be seen as empowered agents not as passive and powerless parties (Ball 2003). They can be active entities reflecting on surveillance practices and trying to escape and avoid them or draw public attention to such illegitimate, discriminatory or otherwise unfair practices.

Resistance practices have been tackled from different theoretical perspectives. Mann (2002), for instance, deems that "sousveillance" occurs when objects of surveillance use oppressive tools against the oppressors themselves. Other authors denote the trend as "inverse surveillance" or "co-veillance" where surveillance and countersurveillance coexist. According to Monahan (2006), "countersurveillance" as a form of tactical interference often leads surveillance technologies to work against themselves in order to redress institutional power asymmetries. Koskela (2009) speaks of the "hijacking" of surveillance, that is, when surveilled persons resist and reverse their gaze. Yar (2003) showed how the surveilled objects at times respond emotionally to surveillance technologies, such as surveillance cameras, and find ways to avoid or confuse their gaze.

The chapter focuses on struggles against drones and the co-optation of drones to empower ends—as an "antidote to studies that stress merely the potential of new surveillance technologies, and thus produce a sense of powerlessness or pessimism" (Lyon 2007, p. 160). It is the "dialectic of control" (Lyon 2007, p. 161) that the chapter traces in the domain of resistance to and co-optation of drones. The dialectic takes into account three features of surveillance (Lyon 2007): surveillance as an ambiguous phenomenon—at the same time both socially "negative" and "positive"; surveillance as subtle and complex—perceived differently by different people and even differently by the same people at different times;[1] and surveillance as technologically imperfect—not immune to failure or error.

Can drones be designed, employed and regulated in a way so as to contribute to predominantly socially beneficial ends? Are they doomed to fail ab initio due to their birth in the realm of warfare and "targeted" surgical strike killings based on "behavioural patterns" associated with terrorist activities ("signature strikes")?

The problem of weaponization and surveillance that first comes to mind when imagining drones is primarily a product of "conditioned" human thought. When it comes to drones, we think of weaponization of drones and drone surveillance because we know that militaries co-opt such technologies on a regular basis. However, every technology can be bent into a weapon. This is definitely true, and that is why it is not the most important question to ask, especially since the answer is clearly positive. So these are, I believe, equally important questions to ask:

- What are the non-military applications of this technology for "socially constructive" uses?
- Why do we stick to technology labelling, instead of labelling the decision-makers: If extraterritorial use of drones for targeted killings outside the USA defines the whole framing of drones, why do we not explicitly say it is a violation of international law that is at stake? Perhaps because labelling the technology as "bad" is more politically correct and thus less "dangerous".
- How should "undesired" droning be resisted: Should droning be resisted as a whole—"as such"—as the corporate and state alliance to integrate drones into the civil airspace in order to develop a "promising industry" is doomed to serve "socially" disruptive ends? Or, on the contrary, is it theoretically conceivable and practically possible to counter undesired droning by designing out privacy violations or prohibiting them—who will decide this and under what criteria attuned to whose interests?
- How is drone surveillance then understood and responded to by those subjected to its gaze? Can they and do they use drones to monitor peers and/or organizations (governments and corporations)?

The chapter first presents "the other face" of surveillance: empowering surveillance, amusing surveillance, and resisting and countering surveillance practices.

[1] Closed-circuit television (CCTV), for instance, can make people feel safer in dark streets, or, on the contrary, can make them feel more anxious as they then assume the area must be "dangerous" (see Leman-Langlois 2008).

After presenting the existing studies, the chapter looks into possibilities of resistance to drones and with drones, i.e. denial, subversion and distortion of drone surveillance systems and using drones to empower citizens in relation to state and corporate power.

The "Other Face" of Surveillance

Foucault (1975) claimed that supervisory power generates resistance by itself.[2] Surveillance is a relational concept involving power dynamics (Lyon 2007) and both constrains and empowers individuals—it is caring and controlling at the same time. Surveillance may be inherent to every social condition, as it occurs everywhere where two people meet (Pečar 1991).

This "Janus-faced" nature of surveillance debated in the literature has been complemented by studies into recent changes in surveillance occurring either in the passage from the ancién regime to modernity or from modernity to postmodernity.

Surveillance is a part of modern organizations' appetite for ever more refined personal information, contends Rule (1980). The pressure towards the institutional collection of information on private citizens, and towards the use of such information to shape institutional action towards the latter, is not the result of "technology" but of the modern welfare-state and business systems (Rule 1980). Mass surveillance represents a distinctive social innovation of the twentieth century. "Whenever institutions face the task of dealing precisely and correctly with each one of millions of individuals, in light of the exact risks, opportunities, and deserts associated with each one, systems of mass surveillance are apt to arise" (Rule 2009, p. 11). Surveillance is thus not imposed unilaterally on unwilling members of the public. The public demands surveillance practices, for instance, to track down illegal aliens, to deny welfare payments to those not deserving them or to catch criminals. There is a public demand—often coupled with punitive aspirations—and an inherent institutional drive to collect, analyse and act upon the growing amount of personal information.

For other authors, contemporary surveillance practices are substantially different from previous types of surveillance. It is "new" surveillance (Marx 2002; Lyon 2007), which is (1) technologically enhanced, (2) oriented towards the body, (3) embedded in the mundane routines of everyday life and (4) universal—no one can meaningfully avoid its gaze. Surveillance has also intensified in the past 30 years and become central to the new order of global capitalism (Deleuze 1995). This acceleration has many sources, including the fact that surveillance systems compensate for the weakening of face-to-face social relationships in which the mechanisms of social integration are increasingly removed and abstract (Lyon 2003). Surveillance has increased and functions as a "glue" that keeps the society together and

[2] See Badalič in this book, showing how the power of military drones used in Pakistan triggers resistance.

builds trust throughout a society of strangers (see Bennett 2008, p. 11; Lyon 2003). It is an empowering and caring social foundation—the "other face" of surveillance.

Those carrying out surveillance are also individuals, as surveillance is not only about powerful organizations—states and corporations—controlling helpless subjects. The relationships between watchers and the watched go in both directions: organizations (public and private) keep tabs on individuals and vice versa. Bennett's (2008) four-cell "heuristic framework" or typology of surveillance includes (1) situations wherein individuals monitor other individuals ("peer-monitoring") and organizations ("sousveillance") and (2) situations wherein organizations monitor individuals ("surveillance") and other organizations ("oversight"). Lateral surveillance (Andrejevic 2003) and "sousveillance" (Mann 2002) may (or not, as Andrejevic shows) amount to resistance to "controlling" surveillance practices and seek to balance the surveillance powers in a society, decentralize observation and potentially achieve a state of "equiveillance" (Mann 2002). Surveillance technologies are used to oppose surveillance. Such resistance to surveillance takes many forms, includes many actors and employs several strategies (Bennett 2008). Thus, surveillance may well be empowering,[3] and, furthermore, it can also function as entertainment.[4] Surveillance may not only be positively protective but may be a comical, playful and amusing practice as well (Marx 1996).

In theorizing the surveillance/privacy dichotomy, authors have contended, it is necessary to avoid the simplistic dualisms—good/bad or helpful/harmful surveillance: it is counterproductive to claim that surveillance is "good" or "bad". Sewell and Barker (2001) propose that we should primarily try to understand surveillance by distinguishing the panopticon as an institutional arrangement (e.g. factory, prison, school) from "panopticism" as a cultural disposition which deems that the gathering of ever greater amounts of information will reveal the truth about the social world (Sewell and Barker 2001).

Similar attempts to transcend this dualism are related to the question of whether we are currently facing the demise of the public or the private sphere. Contrary to the conventional wisdom with regard to the demise of privacy in the age of ubiquitous computing and social media, Bauman (2011) claims that the private has colonized the once public space. While the ancient Greek *agora* was a physical space where heads of families were invited to discuss and decide on questions of common interest, and thus the function of the *agora* was in the reciprocal translation between the language of "private" family interests and the language of "public" interests, today we are witnessing the demise of this democratic process. The *agora* was expected to translate public affairs into individuals' rights and duties, contends Bauman, and any democracy can be measured according to the success of this translation. But today, the private interests of the minority are translated into public questions. The interests of the majority remain private "affairs". In this sense, we can thus speak of the demise of the "public" in contemporary society, as the existing imbalance only reflects the interests of an increasingly smaller minority of the world's population.

---

[3] See also the special edition of *Surveillance & Society*, 2010.

[4] See also the special edition of *Surveillance & Society*, 2005.

The rise of the "society of confessions", claims Bauman, promised the ultimate victory of privacy. However, privacy colonized the public sphere—but this did not come without a price as regards individual autonomy. The deregulation of the state and "individualization"—that is, where members of society are expected to find individual solutions to public problems—mislead many to think that a "biographic policy" will lead to the triumph of human rights. What the public is left with are "wardrobe communities": communities whose members are granted their privacy, where individuals gather, for instance, for sporting or cultural events, but when the events are over and they pick up their coats, the community is disbanded.

In order to transcend this dualism, Sheller and Urry (2003) contend, "public" and "private" are used by social scientists in often contradictory ways, as seen in Bauman's claims above. The world is neither private nor public: A more dynamic conceptualization of the fluidities and mobilities that have hybridized the public and private are needed, they claim (Sheller and Urry 2003). This hybridization of the public and private is extensive and occurs in complex and fluid ways, for example, as exemplified by automobility or information and communications technology (ICT) networks (Sheller and Urry 2003). If the former shows how "private zones of domesticity are reproduced on the road" and "the sensing of the public world is impoverished" when we speed along in moving private capsules (Sheller and Urry 2003, p. 331), the latter has "implanted zones of publicity into the once-private interior spaces of the self and home" (Sheller and Urry 2003, p. 332). Transportation and information technologies converge and furthermore refold what is public into what is private.

Surveillance and privacy are then notions that include changing power relationships, and this is highly relevant for resistance to surveillance and countersurveillance. Even more so because even privacy itself can be understood as a form of social control: "Privacy is not simply a way to extricate individuals from social control, as it is itself a form of social control that emerges from a society's norms. It is not an external restraint on society, but is in fact an internal dimension of society" (Solove 2007, p. 763).

The question is then how to confront and resist surveillance, by drones in particular, in surveillance regimes where it is difficult to distinguish "helpful" from "harmful" surveillance, on the one hand, and "public" from "private", on the other. If surveillance is growing not as a tree but as creeping weeds (Deleuze 1995), if it is flexible and mobile "rhizomatic surveillance" (Haggerty and Ericson 2000), a meaningful resistance can only be "liquid"—to paraphrase "liquid surveillance" (Bauman and Lyon 2013).

Due to such mobile, fluid surveillance, which has even more than two faces and several "other" sides, there are challenges as to how to conceptualize the resistance to surveillance too: Are we speaking of resistance when drone activists employ drones—as for instance in the example of animal activists spying on factory farms with drones (Abell 2013)—or is resistance an act that fundamentally changes the game, for example, by "engineering-out" some uses of drones, for example, by geofencing? Now we will turn to theorizing resistance and countersurveillance, labelled and conceived very differently in the surveillance studies literature. Similar to surveillance and its "other side", theories of subverting surveillance tackle practices of "resisting surveillance" (Fernandez and Huey 2009), "countersurveillance" (e.g. Monahan 2006), "surveillance and empowerment" (Monahan et al. 2010), "hijacking" surveillance (Koskela 2006), "inverse" surveillance and "sousveillance" (e.g. Mann 2002).

Resistance to Drones

Those surveilled may negotiate, modify, evade, deny or take pleasure in surveillance practices (Coleman and McCahill 2011, p. 144). The scales and types of resistance are manifold and range from avoidance to fighting back by using countermeasures to reverse the gaze; from an individual attempting to resist surveillance to group resistance escalating even to a social movement; from radical/vigilant resistance, for example, by shooting down drones (as examples, see the cases in Slovenia (STA 2015) and the USA (Newman 2015)), to legal resistance that seeks, for instance, to change personal data protection law; from ad hoc actions to systematic pressure applied by organizations. Resistance ranges as to the degree of ambition, means and strategies used (Bennett 2008) and the abilities of actors to contest the reach of surveillance into their daily lives (Coleman and McCahill 2011). It may be directed against peers, a state or a corporate actor or—their amalgam: the "Surveillance-Industrial Complex" (Ball and Snider 2013).

What then is "resistance" when "everything from revolutions […] to hairstyles […] has been described as resistance" (Einwohner 2004, in Coleman and McCahill 2011, p. 145)? "Who" is resisting and "what" is being resisted?

Resistance interactions occur between three actors: individuals (members of "civil society") and organizations—corporate and state institutions. A certain power dynamic—surveillance and resistance—is in place amongst all of them. However, this chapter is not focused on the interplay between corporate and state surveillance, where—as we have seen from the history of whistleblowing in the early twenty-first century—the relation between the two is an important area for power struggles as well: Corporations have been usurping states' security/surveillance functions while states have been pushed into imbalanced public–private partnerships due to austerity measures (e.g. states have been outsourcing specific law enforcement tasks, whole police stations, and military and intelligence chores to private companies).[5] There are many power struggles between the two. Instead of focusing on this interplay, the chapter focuses on individuals—civil society—resisting against corporate and state powers ("vertical" surveillance) and ("horizontal") surveillance conducted by their peers. Such focus does not exclude those working within "surveillance authorities" or "corporate organizations" since these subjects can distort and modify

[5] See the notorious case of Booz Allen Hamilton and the "revolving door" effects of its vice chairman Mike McConnell or James Clapper, the director of the National Security Agency and a former executive at Booz Allen. See Riley 2013.

surveillance practices as well—in this sense, resistance can go beyond the "subject-agent relationship" (Martin et al. 2009; in Coleman and McCahill 2011, p. 146).

Resistance to surveillance by individuals can then be scaled as (1) *avoidance,* (2) *countering* surveillance—as a reaction to the existing surveillance stemming from organizations and a "jujitsu-like" reversing of the gaze and (3) *empowering* surveillance—using surveillance tools to achieve socially productive ends.

These divisions are not hermetically closed, but they offer a more nuanced designation of resistance intensity on a scale from "passive" to "active". In terms of depth, resistance can be divided in terms of "thin" resistance—located in localized settings—or "thick" resistance—challenging structural power (Raby 2005). Countering surveillance by means of drones can be used to reverse the gaze of sensors ("thin") or to change the game by redesigning and completely disabling unwanted sensing ("thick"). In doing so, the latter challenges the wider power relations, while the former is merely "contestation"—an activity of undermining the rules in a micro-setting (Aggleton and Whitty 1985). In more fundamental terms, resistance is an active behaviour of individuals and interest groups (civil society) conducted either as micro-resistance practices to defeat drones or as political resistance to contest drones per se (cf. Coleman and McCahill 2011, p. 147).

In defining who is resisting "what", drone resistance opposes personal data collection and transcends the narrow legalistic end. Resistance is focused on "drone logic" (see Chap. 2) as well: "a dispositive of ubiquitous, always-on, sensor-based monitoring for the purposes of automated data collection, processing, and response" (Coleman and McCahill 2011).

Reflecting the fluidity and dynamism of surveillance, resistance to drones is dynamic too. It faces several challenges—social, political and technological (Lyon 2007). The social challenge relates to understanding the intensified surveillance and its consequences, and also the role ordinary people play within the surveillance process (Lyon 2007, p. 373). The power of surveillance systems depends on the knowledge of the system its subjects have. It depends on their responses to the fact of being under surveillance. People reproduce surveillance regimes; they can enable—or disable—such regimes. They are the crucial strength or weakness of these systems. They can comply with surveillance—or not. They may admire the "endless uses" of drones—or not.[6] Those who are surveilled can be addressed in order to spark a change in the cultural perception of drones. The point I would like to make here is that an individual is a site of surveillance, and while being "a very unstable entity" (Nietzsche), the individual represents a place of uncertainty. Surveilled individuals are not merely subject *to* surveillance, but subjects *of* surveillance. They will approach situations with expectations, hopes, fears and caution. They are inventive in finding ways to beat control systems and to avoid observation by several "generic techniques of neutralization" (Marx 2003) that individuals, those within or outside, and organizations can employ in order to subvert, distort, block, mask or

[6] See Chap. 3 in this book about the "moral economy" related to the unconditioned admiration of the "endless uses" of drone technology.

counter surveillance systems. This social challenge is concurrently an opportunity for resistance.

The political challenge relates to the ethical choices of new practices. "Categories have politics", argues Suchman (in Lyon 2007, p. 375). Even algorithmic surveillance—commonly perceived as "objective" mathematics—reflects the desires, intentions and biases of those designing and implementing it (Barocas and Selbst 2015). Politics is inscribed in technology, and software/code is never neutral—also with regard to resistance politics. Those resisting surveillance can thus use technology to not only, for instance, communicate amongst themselves but also to write code to watch the watchers in order to increase their visibility and accountability.

The technological challenges relate to understanding the technological dimensions of identifying, locating, monitoring and tracking individuals and groups. The technical challenge of resistance includes engineering out the undesired and unintended effects of the technology. Several technological solutions can be used: software products designed to enhance privacy protection, such as privacy-enhancing technologies (PET), privacy-by-design (PbD) principles (e.g. "geofencing" resistance software as a way to protect privacy through the design of the navigation software),[7] surveillance/data protection impact assessment tools, privacy audits and privacy certification schemes (cf. Finn 2014, pp. 329–350).

In fact, resistance technologies have been developed in every surveillance domain, for instance, encryption tools to prevent the hacking of information transmitted over the Internet, such as the Pretty Good Privacy (PGP) standard for e-mail encryption; encrypted biometric access systems that allow the use of a fingerprint to authenticate an individual's identity but which do not retain the actual fingerprint; and microphones that listen to voices on public streets but cannot hear conversations between people. This is a view that tries to ensure that certain forms of surveillance are not only illegal but also technically impossible. The focus on technology can challenge the dominant legalistic way of thinking about privacy protection. "Resistance" software products were often developed in the face of considerable opposition from law enforcement and intelligence communities (Bennett 2008, p. 82). This struggle for privacy goes back to the "crypto wars"—over the right to use and distribute products that contain strong encryption technology—in the mid-1990s, which has not ceased ever since (Zetter 2015). Drones can then be designed to resist surveillance—and attempts will be made to reverse-engineer them into surveillance tools.

The provision of personal privacy has become an industry of its own, a marketplace where products sell their PbD. Tech companies compete against each other as to which will improve the protection of privacy the most.[8]

Technologies should not be understood then as artefacts that have an impact on society—technology and society are not separable (Winner 1986). "The code is the law", is the famous observation of Lessig (2004), meaning that computer code

[7] See Cavoukian (2012) for features of privacy by design (PbD) with regard to drones.

[8] See the Electronic Frontier Foundation (2015) annual report, which evaluates how companies protect user data from government control and censorship.

has regulatory properties. Technological artefacts contain social and political biases (Winner 1986), and therefore their design should be monitored and reviewed for such biases. The very structure and architecture of technology is linked to particular patterns of power. Technology may thus carry, intentionally or unintentionally, a "valance that may be pro-privacy, or pro-surveillance" (Bennett 2008, p. 83).

Technologies are therefore a part of the solution to a problem: If drones provoke socially disruptive outcomes/challenges, then technological (re)design may be one of the ways to dissolve them. Anonymity can be built into information systems. Privacy cannot be ensured only by regulations but also by "physics and mathematics" (Bennett 2008, p. 84).

In fact, in limiting every surveillance technology a range of regulatory rules should be used because any single regulatory tool is either ill-suited or simply incapable of carrying out regulatory tasks (Bennett and Raab 2006). Drones can also be regulated and limited in a number of other ways. According to Lessig's well-known regulatory scheme (2004), this can be achieved by means of changes to the following: (a) law (e.g. personal data protection laws and aviation laws), (b) economic policy (e.g. the taxation of surveillance technologies), (c) social norms (e.g. by "naming and shaming" drone usage) and (d) design (e.g. with the implementation of the principle PbD). Let us now apply this fourfold scheme to drones.

Legal limitations on drones are being set up in, for instance, Europe (see the Article 29 Data Protection Party 2015), the USA, Canada (see, e.g. Cavoukian 2012) and the UK (see Civil Aviation Authority 2015) in the field of aviation[9] and data protection law.[10] Concern for personal data protection and other violations of rights with recommendations for laws have been expressed by international groups (see The International Working Group on Data Protection in Telecommunications 2013), researchers (e.g. Finn et al. 2014; Bracken-Roche et al. 2014) and NGOs (e.g. Stanley and Crump 2011; Jones 2014; Hayes et al. 2014) working in the field of privacy issues and also with regard to international military drone operations (see, e.g. Amnesty International 2014). In terms of legal limitations, rules have been adopted, for instance to prohibit flying over and near airports and other "security" areas; the in-line-of-sight rule; the obligation to ensure crash-avoidance capabilities; the prohibition on some capabilities, such as fully autonomous drones; the requirement of a "probable cause" warrant in order for law enforcement authorities to use drones to collect information to use against someone in court; an explicit ban on the weaponization of drones (see the Campaign to Stop Killer Robots); special protections from aerial surveillance for farmers or ranchers; reporting requirements on law enforcement agencies' drone usage; the limitation of such usage with regard to serious crimes; the stipulation that information that is collected incidentally cannot be used in court; and the requirement to delete incidental data within 24 h of collection (Hayes et al. 2014). Drone operators could develop their own voluntary code of conduct in the licensing phase of drone operators. And finally, "administrative" protection could increase accountability of drone operators. In her speech at

[9] See Chap. 9 in this book.

[10] See Chap. 4 in this book.

a conference on remotely piloted aircraft systems (RPAS) in Riga, European commissioner Bulc proposed an electronic identity card for each drone so the operator can be identified (Nielsen 2015).

The economic policy on limiting drones may focus on taxing this "risky" technology or on obliging operators to financially contribute to a fund offering assistance to victims of privacy violations. The insurance obligations of drone operators and owners could additionally influence the economics of the drone industry.

Social norms define the level of social acceptance of drones. It has been claimed[11] that the drone industry is making a great effort to change the morality of drones and that this is problematic on its own. However, it is interesting that in every category of drone employment support is lower when a drone is deployed instead of piloted aircraft (Bracken-Roche et al. 2014).

The design of drones can follow PbD principles that can "engineer-out" violations of individual rights. The design of drones can and should take into account security risks by using the methods of cyber defence and "geofencing". Drone manufacturers are developing navigation software for the "geofencing" of drones (e.g. a new model of the commercially sold Parrot model cannot be operated in the vicinity of airports). Location coordinates can be inserted by anyone who would want drones to treat their land as off limits. Another model related to the design of technology is to develop ISO standards with integrated software for no-fly GPS locations if the drone is to be able to reach consumers in terms of consumer protection law. Individuals could also employ privacy-protecting self-help countermeasures in the form of drone-proof umbrellas or jackets.

Finally, there are the physical interferences that can be used to disable drones. Military and intelligence services are looking into how to defend themselves and thwart civilian attacks: (1) by hacking to seize control of the drone's flight, by hacking the aircraft's control signals and tricking it into believing it is somewhere else; (2) by jamming the drone's signal to send it off track or disrupt communication, to block the radio signals linking the drone to its controller, obfuscating the drone's flight-control and video-broadcasting systems; and (3) by physically disabling drones by knocking the drone out of the air, for example, with a bird/falcon (Atherton 2014), laser (Agence France-Press Beijing 2014), anti-drone drone (Atherton 2015) or even catching it with the use of a net (Dredge 2015)

Countersurveillance: Resisting with Drones

Despite the trend of increasing "horizontal", "lateral" or "peer" surveillance, on the one hand, and the ability of drones to "watch" the watchers and increase their visibility, transparency and accountability, on the other, the vast majority of surveillance literature has centred on the monitoring of individuals by state or corporate organizations. Surveillance technologies may be used to reverse the gaze and monitor

[11] See Chap. 3 in this book.

those in power. These may be legal, but they can also cross the line of criminality. In the following section of this work, I will discuss civil uses of drones that change the power dynamic vis-à-vis peers and (corporate and governmental) organizations.

Peer-Monitoring, Provocation and Criminal Drones

While lateral surveillance used to be a significant marker of the autocratic regimes that maintained an extensive network of informers (Marx 1988), such surveillance has become a marker of contemporary "Western" societies as well, as voyeurism has reached unprecedented levels with the proliferation of camera phones—and personal do-it-yourself (DIY) drones with the rise of "dronies". As drones can carry a variety of payloads, drone operators may breach not only privacy and personal data protection laws (e.g. when drone users film others in nudist camps), in terms of security, individual users can also breach other legitimate interests of fellow citizens. Drones have carried and fired semi-automatic handguns strapped to their bodies (Unknown 2015). Taser drones can "taser" someone with an 80,000-volt stun gun (Knibbs 2014). These are all individual uses of drones expanding power of operators. Drones can shoot fire and Silly String, and illustrate graffiti on walls (Larson 2015). For instance, the graffiti artist Katsu attached a spray can to a quadrocopter and painted a large billboard in New York City—which was subsequently labelled as a "public vandalism drone" (Holland 2015). Wi-Fi-sniffing drones are able to detect unsecured Wi-Fi networks (Storm 2010). Criminals have also used drones to carry illegal cargo, for example, to smuggle drugs (Valencia and Martinez 2015) or to pick up a ransom (Poulsen 2015).

A drone carrying the flag of the non-existent "Great Albania" interrupted a soccer match between Serbian and Albanian teams. The provocation happened amidst the political tension that exists between the two states (K. S. 2015). The incident provoked a heated debate with regard to Albanian expansionism at the expense of the neighbouring Serbia. Similarly, a drone provocation occurred in Slovenia when a drone was used on the 70th anniversary celebration of the end of WWII in Ljubljana. The drone was carrying a ragged former Yugoslav flag and a banner with an equal sign between a red star—the emblem of the former socialist state—and a swastika, and buzzed over the participants. The drone crashed several minutes after take-off and injured one of the security guards at the event.

Some claim that the trend of peer-monitoring networks has amplified and replicated the power of government or corporate surveillance in a climate of perceived risk and savvy scepticism (Andrejevic 2005). Others claim that individuals' uses of drones, especially for lateral surveillance, may carry empowering potential. By focusing on webcams and "citycams", Koskela (2006) shows the other—empowering—side of surveillance in the domain of video monitoring. In contrast to surveillance cameras, she claims that interactive citycams, with detailed real-time videos used to promote spaces, have empowering roles (Koskela 2006, p. 169). Such image-making is based on strategies that aim to attract companies and investment, as well as inhabitants. Koskela claims that the immediate impact of citycams can be liberating or empowering for the monitored objects. However, even she admits that citycams may enable new forms of control. While she focuses on less "politically dangerous" examples, such as parental control over youth by means of citycams—this can cause youth to flee from central places in cities in order to escape the control of caring "helicopter parents"—there is a more substantive logic at work with citycam installations. Citycams expand the logic of the "free market", where everything, even cities, should compete on the market. There is a tendency at work here to submit all aspects of life to the prerogatives of the "free market" logic that subjugates entities such as cities and states to its rule. The logic is that even states have to compete for investments of supranational global capital by offering lower taxes as a sort of "special discount" for huge players on the capital market. There is thus a more fundamental control effect taking place behind the friendly citycams.

One can argue similarly with regard to Koskela's example of "home webcams". They indeed carry empowering potential—"webcams seem to be opening up radically new subjectivities. They describe how new technologies do not only serve surveillance but liberation, resistance, and 'escape'" (Koskela 2006, p. 173). "Webcams [...] can be interpreted as a form of confrontation, surveillance turned into spectacle—a form of resistance. Webcams clearly support active agency" (Koskela 2006, p. 175). People are not submitted to the passive role of "the observed" but instead play active roles with "surveillance" equipment. However, the underlying logic of the "all-seeing" or "sensor" society is replicated (cf. Andrejevic 2007). This produces a flood of images that the IT industry encourages.

Similar to webcams, DIY recreational drones amplify the contemporary "datasaur"—a beast hungry for digital data "emptied of life"—an idea that everything has to be digitized and stored for future decision-making purposes. Drone users are already inventing uses, such as escorting children to the school bus by drones (Kersey 2012)—a transition from "helicopter" to "drone parents".

As the examples of individual uses of drones show, drones can carry a variety of payloads—from cameras to Wi-Fi sniffers. Resistance with drones can then jeopardize several rights of peers and amplify other social processes at work in the given society. For instance, the use of drones can increase hatred between people of different nationalities (the Albanian flag case) or people with different political preferences and world views (the Slovenian flag case).

"Sousveillance" Drones

The other part of Bennett's (2008) "heuristic framework" of surveillance practices that positions individuals as monitoring others is the position wherein individuals use surveillance technologies to monitor or otherwise focus the gaze on organizations. This "sousveillance" perspective (Mann 2002) involves the recording of the activities of the observers by the observed—*sous*veillance, meaning monitored from "below" and reflecting the bottom-up nature of "viewing" (Mann et al. 2003).

While Mann's "wear-cam" innovations were designed for the purposes of monitoring the powerful by the weak, in practice they mostly resulted in the monitoring of the weak. The industrialization of wear-cams—as initiated, for example, by Google Glass—shows how wearable computing developed more into an "auto-surveillance" than a "sousveilance" tool. However, the idea behind Mann's practice of wearing cameras—the idea of watching the watchers—reflects the empowering potential of surveillance for those "below". It ranges from a situation when the individual is watching other *individuals* with more social power to situations when the individual is keeping tabs on *corporate* power and *state* organizations. Let us now turn to the latter two, which are more relevant for the drone domain.

Drone *corporate* sousveillance occurs, for instance, when drone activists keep tabs on polluters. Several environmental activist groups have been using drones: The Australian animal rights group Animal Liberation is using drones to keep tabs on industrial livestock operations (Bird 2013). The Sea Shepherd Conservation Society—a marine wildlife conservation organization—has been using drones to track and follow Japanese whaling fleets: "Drones cover hundreds of miles [...] and they have proven to be valuable assets for this campaign" (Shepherd 2011). Individual recreational users discovered evidence in the skies above Dallas in 2012 that a factory farm was discharging enormous amounts of pig blood into a creek (Mortimer 2012).

Drone sousveillance directed at *state* power was executed against Gaddafi in Tunisian protests (the Camera Project): Those "below" were trying to increase their power to resist the political regime with drones. While one might claim that this was a political struggle in which drones may not have played a central role, a project in Sudan shows how drones can directly save lives by increasing international community awareness of ongoing atrocities and document the evidence of international core crimes. In filming a third documentary in the Sudan region entitled "Drones above the roots of humanity", the activist Tomo Křižnar called drones the "new knights" because people are changing into apathetic robots: "[...] maybe robots can be the ones to change the seeds and roots of the remaining humanity" (Vidrih 2014). Drones may be, he claims, the only saviours of the indigenous people preserving the evidence of genocide—they are the "flying eyes and ears of God", documenting genocide in Sudan and Congo, where central governments are systematically killing the indigenous Nuba people. The use of drones from those below can also be understood as empowering drones. It not only documents evidence of mass atrocities but by increasing the global visibility of the conflict it supports the cause of the minority group.

Drones have been used to spy on the police. Protesters in Warsaw gained a drone's eye view of a phalanx of police in riot gear during a demonstration in 2011 (Ackerman 2011, cf. Sharkey and knuckey 2011). Such drone countermeasures are forming part of the sousveillance "cop-watching" activities that started by filming incidents of police brutality with mobile phones (cf. Sandra Bland case in the USA, Stern 2015).

Using drones for activism includes graffiti drones, such as the Icarus open-source spray-paint drone (Graffiti 2015). Drones carrying a paint payload entwine art and activist goals. Guerrilla activism also employs drones that carry video projectors to display images on buildings instead of putting images on a building physically (Redd 2015).

Sousveillance can be found in every surveillance domain—as noted before, supervisory power generates resistance by itself (Foucault 1975). However, the proliferation of big data, increasing computer power, and visualization techniques are leading to a new wave of computerized sousveillance programmes. Drones only form a part of this new wave of technologically enhanced sousveillance, which are illustrated herein with the following two examples.

In Slovenia, the "Supervizor" is an on-line application (http://supervizor.kpk-rs.si) that provides citizens with data on the business transactions of public-sector bodies and government spending (Commission for the Prevention of Corruption 2013). The application, awarded the "2013 United Nations Public Service Award", indicates the contracting parties, the largest recipients of funds, the related legal entities, the date and amount of the transactions and also the purpose of the money transfers. It enables the presentation of these data using graphs as well as printouts for specified periods of time and in other ways. The application uses and connects large amounts of government data in order to increase transparency and mitigate corruption risks and conflicts of interest. It is a special "sousveillance" system for citizens enabling them to monitor those "above", demonstrating how the accountability of those in power can be increased.

Another example of technologically enhanced sousveillance is the Slovenian platform named "Kdo vpliva?" (literally "Who influences?", www.kdovpliva.si), which shines a light on the connections between lobbyists, companies, politicians and state institutions through visualizations of three different kinds of networks: lobbying contacts, the network of transactions between the companies represented by the lobbyists and the public sector. The platform, a web application, shows the national lobbying landscape by allowing every Internet user to monitor the creation of new laws. By clicking the name of a deputy, it is possible to see the acts the deputy in question was most involved in and his or her statements in sessions of specific committees (Čebokli 2015).

To conclude, drones may be used to counter existing power relations. They are used to keep tabs on peers, to provoke them and even to commit crimes. They have been increasingly used to monitor state institutions and corporate misconduct as well. They fit in with the wider trend of countering surveillance by means of IT. However, they fail to address the underlying logic that feeds the "datasaur"—a tendency to submit everything to digitization, storing and acting upon large amounts of digital data.

Empowering Drones

In line with "Janus-faced" surveillance, the empowering aspects of surveillance can be located in every surveillance domain, including drones. Studies on video monitoring are a relevant body of research to start with. While tackling CCTV,

Ball (2003) argued that the objects of surveillance are empowered agents, not just passive and powerless parties. By being subjects of desire, CCTV operators can manipulate the operations of the system, for instance by looking at other people or actions than the expected "suspicious" ones. As objects of surveillance, individuals can avoid monitoring by taking alternative routes; for instance, the iSee art project's aim was to suggest to strollers via on-line maps and interactive application routes in Ljubljana and New York how to evade the gaze of CCTVs. A camera distribution project by B'Tselem empowered Palestinians living in high-conflict areas. In January 2007, B'Tselem launched a video advocacy project with the goal of providing Palestinians video cameras in order to bring the reality of Palestinian lives under occupation to the attention of the Israeli and international public.

In the domain of drones, the project called "Eyes and Ears of God: Video Surveillance of Sudan" by two Slovenian human rights activists, Tomo Křižnar and Klemen Mihelič, and the humanitarian organization H.O.P.E., shows how drones can be used to empower an entire indigenous population. Therein drones are being used for several purposes—documenting occupation, collecting evidence of crime and drawing worldwide public attention to attract humanitarian and developmental aid. As Křižnar reports, the Nuba people fought on the side of the southern Sudan Peoples' Liberation Army (SPLA) during the 20-year north–south civil war. Subsequently, when South Sudan had voted for independence, the people of South Kordofan, which remains in the North, faced brutal retaliation by the northern government. They are stranded in "northern" Sudan but are ethnically considered "black Africans", like the people in South Sudan. The Arab regime in Khartoum is led by the Arabic movement that teaches that black Africans belong to an "inferior race". In this political violence Křižnar is equipping the Nuba people with drones. Amidst several ethnic groups living in the area, for example, the Nuer and Dinka ethnic groups, the Arabs living in (north) Sudan, and the Russians and Chinese with economic interests in the area's oil, only the indigenous Nuba people do not oppose surveillance from drones. They do not have "anything to hide", as Křižnar assumes from the political analysis of the situation (Křižnar 2014).

Related to Sudan's "Eyes and Ears of God" project is the Satellite Sentinel Project—a high-tech project tracking troops and warning civilians of attacks. Dubbed the "Anti-genocide paparazzi" and supported by George Clooney and other celebrities, the project employs commercial DigitalGlobe satellites passing over Sudan and South Sudan and captures imagery of possible threats to civilians, detects bombed and razed villages and notes other evidence of pending mass violence. The Harvard Humanitarian Initiative analyses the imagery, open source data and information from sources on the ground to produce reports. The Enough Project contributes field reports, policy analysis and communication strategy and, together with *Not on Our Watch,* pressures policymakers by urging the public to act.

The two cases of civil society empowerment with drones and satellites show how technologies encapsulate ideology and values, and privilege one type of knowledge over others. Even mapping and supposedly "neutral" satellite imagery necessarily needs interpretation (Kurgan 2013). Satellite images—taken-for-granted artefacts supposedly "objectively" communicating reality—bear "political, military, and economic stakes that underwrite the creation and expansion of the database" (Kurgan 2013, p. 21). The visions of the planet Earth, Kurgan argues, "are not simply photographs taken by a person travelling in space with a camera. They are composites of massive quantities of remotely sensed data collected by satellite-borne sensors" (Kurgan 2013, p. 11). Thus, when we are looking at the default screen of an iPhone, we are in fact seeing a patchwork of satellite data, artificially assembled, and not an integrated vision of a particular person standing in a particular place or even floating in space (Kurgan 2013, p. 12). Its basis is remotely sensed data related to resolution, measurability and a reliable algorithm for translating between presentation and reality.

In the process of "returning from the battlefield", drones are following the pattern of the gradual transition of technology similar to the one taken by the Internet and satellite mapping—from state secret to commonplace everyday instruments.

This transfer cannot be the result of a singular force, such as one type of stakeholders—industry—advocating the use of drones and their transfer from the military to the civil domain. Rather, the transfer of drones should be observed in the context of the neo-liberal project of information technology development that has its origins in the late 1960s. Similar to the Internet—a developmental template model—a technology developed for military purposes gains public funding, but eventually, when it begins to be more remunerative, it is put on the market and slowly privatized through public–private partnership arrangements. Satellite images (cf. Kurgan 2013) and drones have been taking the same steps. Simultaneously, the "moral economy" gradually changes, and the technology is handed over to civil users—benevolent and malevolent users—including users that monitor those in power.

While analysing the civilian use of aerial and satellite imagery, Kurgan (2013) observes that the launch of Ikonos, the first satellite to make high-resolution image data publicly available, created the revolution because a new kind of deterrence was possible, whereby organizations and civilians could test, with meaningful certainty, the authority of official claims regarding, for example, the presence of nuclear facilities in other states (Kurgan 2013, p. 25). Thus, the satellite imagery that was used for military purposes and which only later entered the civil domain—only in August 1995—to show classified images to members of the UN Security Council as evidence of mass killings in Bosnia was transformed into a tool that provides an independent check on what the government is saying about mass graves and other wartime atrocities in the Balkans (p. 25). Similar to drones, "organizations may find it easier to respond quickly to sudden refugee movements, to document and publicize large-scale humanitarian atrocities, to monitor environmental degradation, or to manage international disputes before they escalate […]. But, there is no way to guarantee benevolent uses" (Kurgan 2013, p. 25). "Images are interpretations presented as facts. The interpretations made of images are anything but objective or self-evident" (Kurgan 2013).

Drones can be and are used for "benevolent" and "malevolent" purposes. In the hands of operators guided by humanitarian principles, drones too provide a potent new way of ensuring that the world witnesses threats to civilians" (cf. Van Rooyen, Director of the Harvard Humanitarian Initiative, from Kurgan 2013, p. 26).

The use of drones thus shows how surveillance is always a double-edged sword, more specifically, that there is a control–care continuum that can be traced in every surveillance regime, domain or technology. It is thus possible to acknowledge this Janus-faced surveillance in the drone domain. The "other" side of drones is manifold, ranging from escape-and-avoid techniques, and "finger-pointing" and "shaming-and-naming" strategies underlying its illegitimacy to subverting it to empower those "below" or lower on the social power scale.

Problems with Drone Resistance

Resistance to drone surveillance faces difficulties similar to resistance in other surveillance domains. There seems to be something inherent to the issues of privacy and surveillance that cannot rise to a higher level of mass consciousness—privacy violations tend to be hard to visualize (Bennett 2008). Compared to conventional crime, which leaves tangible consequences, privacy violations seem very abstract—there are no broken bodies, only broken souls.

In analysing resistance to surveillance, Clemet and Hurrell (2005, p. 15) note that advocacy organizations have great difficulty in mobilizing public support around anything other than dramatic public "scandals". Drones are different from other surveillance technologies in so far as they are—at least for the time being—prone to provoke public scandals. For instance, the above-mentioned case of a drone buzzing around a stadium during a football match between Serbian and Albanian teams would not have provoked such political tension and heated debate if the "offending message" of the Albanian spectators had been conveyed on the ground. Calo (2011) thus properly observes that drones are good "privacy catalysts": "The widespread domestic use of drones [...] may help restore our mental model of a privacy violation. They could be just the visceral jolt society needs to drag privacy law into the twenty-first century."

Drones are "suitable targets" for resistance to surveillance because we are used to dangerous, congested streets and vehicles dangerously passing by, but we are frightened of flying machines in the vicinity. The monitoring from the sky seems more inappropriate than other types of omnipresent monitoring, such as surveillance on the Internet. A public perception survey conducted in Canada (Bracken-Roche et al. 2014), for instance, demonstrates that in every area of awareness and support the use of drones rather than piloted aircraft changed public opinion about information collection. In nearly every category support is lower when a drone is deployed. So, for instance, if an organization starts using a drone for routine activities that the public had formerly considered acceptable, such would now be deemed unacceptable (Bracken-Roche et al. 2014, p. 39).

In this sense, drones are "suitable targets" for resistance. However, resistance to drones is not undifferentiated. The same survey shows that resistance is directed against individual uses of drones, for example, monitoring of people, but not in emergency situations, such as responding to disasters, finding missing persons and hostage situations (Bracken-Roche et al. 2014, p. 40). Resistance is also amplified towards specific operators, for example, private-industry use of drones is only supported when the drone's data collection is directed towards the oversight, maintenance and management of private property. In general, corporate uses of drones garner less support than governmental use. There is a distinct lack of support for drones used by "private investigators" (11 %), "industry and corporations" (12 %), "journalists and media" (12 %) and individual "hobbyists" (14 %), while "emergency responders" get significant support for the collection of data through the use of drones (65 %) (Bracken-Roche et al. 2014). Resistance to drones is very nuanced also according to a European study (Finn et al. 2014). The use of drones, particularly for the inspection of critical infrastructure, was viewed by respondents as the user-application combination that carries the fewest and least intense privacy, data protection and ethical risks, while uses of RPAS for visual surveillance by police and by private individuals were thought to carry the highest risk (Finn et al. 2014, p. 171). Public attitudes to drones are nuanced, and drone uses are not all targeted and resisted. On the contrary, the public supports empowering uses of drones, and these nuances should be the leading principle in the forthcoming regulation of drones.

Resistance to drones lacks another strength. It confirms Huey's insight that there is only a network of privacy advocates that remains fragmented with a relatively small and geographically biased core. Many anti-drone organizations are still focused on the USA, such as the "Network to Stop Drone Surveillance and Warfare" (NSDSW), "KnowDrones.com", "Global Drones Watch" and the "No Drones Network".

The problem with initiatives to resist drones is the extent of their reach—they do not tackle the obsession to collect and store massive amounts of data to digitize everything that may (or not) be relevant for decision-makers sometime in the future. Drones fit into the developing "sensor society", where they are not at all alone: wearable computers, "smart cities", "big data", "swarm intelligence", robotic revolution, etc., are all based on collecting, recording and storing huge amounts of data. Sensors are not only above us but also much closer—in the form of "wearable drones" (see flynixie.com) or even "bodily-drones" (Droneologist 2014). The issue at stake here is then in what Sewell and Barker (2001) termed *panopticism*—the cultural disposition that the gathering of ever greater amounts of information will reveal the truth about the social world. The social context of the proliferation of drones should then be understood as the new phase and acceleration of panopticism (cf. Chap. 2). For instance, drone "signature strikes" manifest this very clearly. They show how weapons can be used solely on the basis of huge amounts of data with which algorithms are supposedly able to make a "surgically precise" recognition of a "terrorist". The advent of "big drone data" then shows how the above-mentioned cultural disposition is being realized through the use of drones.

Today, resistance to technological development may be perceived as blasphemy. The "technological sublime" (Carey 2005) is the dream of a world perfected through technology. "Transcendence through technology" is the new project of the Global North states. Other big technological innovations, such as the contemporary advent of the "big data" revolution, carry—similar to drones—the same amount of fear and resentment, on the one hand, and enthusiasm and approval, on the other. The only meaningful resistance to drones is resistance through co-optation, but from resistance to drones to sousveillance and empowering drones is a huge leap that carries a huge risk too. The use of drones contributes to the reproduction of the existing knowledge/power relations. Only very limited uses of empowering drones may "save humanity, where humans have failed" (Křižnar 2014).

Acknowledgements The author wishes to express his thanks to Ajda Gorjanc (www.helivideo.si) and Dr. Danijel Skočaj (Computer Vision Laboratory, Faculty of Computer and Information Science, University of Ljubljana) for technological insights, and Laura Matjašec and Sara Železnik who assisted with research. Special thanks are due to Dean J. DeVos, an English language editor, for his patience and diligence during our collaboration.

References

Abell, D. (4 Sept 2013). Australian animal activists launch spy drone to monitor factory farms. *One Green Planet*. http://www.onegreenplanet.org/news/australian-animal-activists-launch-spy-drone-to-monitor-factory-farms/.

Ackerman, S. (17 Nov 2011). Occupy the skies! protesters could use spy drones. *Wired*. http://www.wired.com/2011/11/ows-drones/.

Agence France-Press Beijing. (3 Nov 2014). China unveils laser drone defence system. *Guardian*. http://www.theguardian.com/world/2014/nov/03/china-unveils-laser-drone-defence-system.

Aggleton, P. J. & Whitty, G. (1985). Rebels without a cause? socialization and subcultural style among the children of the new middle classes. *Sociology of Education, 58*(1), 60–72.

Amnesty International. (2014). *Left in the dark. Failures of accountability for civilian casualties caused by international military operations in Afghanistan*. London: Amnesty International.

Andrejevic, M. (2003). *Reality TV: The work of being watched*. New York: Rowman & Littlefield Publishers.

Andrejevic, M. (2005). The work of watching one another: Lateral surveillance, risk, and governance. *Surveillance & Society, 2*(4), 479–497.

Andrejevic, M. (2007). *iSpy: Surveillance and power in the interactive era*. Lawrence: University Press of Kansas.

Article 29 Data Protection Working Party. (16 June 2015). *Opinion 01/2015 on privacy and data protection issues relating to the utilisation of drones*. 01673/15/EN WP 231.

Atherton, K. D. (5 Dec 2014). Can birds be trained to bring down drones? *Popular Science*. http://www.popsci.com/can-birds-be-trained-attack-drones.

Atherton, K. D. (16 Jan 2015). Fight drones with drones: Rapere is an anti-drone interceptor. *Popular Science*. http://www.popsci.com/rapere-anti-drone-interceptor.

Ball, K., & Snider, L. (Eds.), (2013). *The surveillance-industrial complex: A political economy of surveillance*. London: Routledge.

Ball, K. S., & Webster, F. (Eds.). (2003). *The intensification of surveillance*. London, Sterling (Virginia): Pluto Press.

Barocas, S., & Selbst, A. D. (2015). Big data's disparate impact. *California Law Review. 104 (forthcoming 2016)*. http://ssrn.com/abstract=2477899.

Bauman, Z. (2011). *Collateral damage: Social inequalities in a global age*. Cambridge: Polity Press.

Bauman, Z., & Lyon, D. (2013). *Liquid surveillance: A conversation*. Cambridge: Polity.

Bennett, C. J. (2008). *The privacy advocates: Resisting the spread of surveillance*. Cambridge: MIT Press.

Bennett, C. J., & Raab, C. (2006). *The governance of privacy: Policy instruments in global perspective*. Cambridge: MIT Press.

Bird, S. (2 Oct 2013). Aussie activists deploy $17,000 drone to spy on factory farms. *Care2*. http://www.care2.com/causes/aussie-activists-deploy-17000-drone-to-spy-on-factory-farms.html.

Bracken-Roche, C., Lyon, D., Mansour, M. J., Molnar, A., Saulnier, A., & Thompson, S. (30 April 2014). *Surveillance drones: Privacy implications of the spread of unmanned aerial vehicles (UAVs) in Canada*. A report to the office of the privacy commissioner of Canada, under the 2013–2014 contributions program. http://www.sscqueens.org/sites/default/files/Surveillance_Drones_Report.pdf.

Calo, R. (2011). The drone as privacy catalyst. *Stanford Law Review Online, 64,* 29–33.

Carey, J. (2005). Historical pragmatism and the internet. *New Media & Society, 7*(4), 443–455.

Cavoukian, A. (2012). *Privacy and drones: Unmanned aerial vehicles*. Ontario: Information and Privacy Commissioner of Ontario, Canada. https://www.ipc.on.ca/images/Resources/pbd-drones.pdf.

Čebokli, A. (20 April 2015). Lobbying influence—unsupervised grey zone. *RTV SLO*. http://www.rtvslo.si/news-in-english/lobbying-influence-unsupervised-grey-zone/363285.

Civil Aviation Authority. (2015). *The dronecode guidance on flying drones and unmanned aircraft*. http://www.caa.co.uk/default.aspx??catid=1995&pagetype=90&pageid=17054.

Clement, A., & Hurrell, C. (2005). *Information/Communications rights as a new environmentalism?* CRACIN working Paper No. 3. https://tspace.library.utoronto.ca/bitstream/1807/32410/1/CRACIN%20Working%20Paper%20No%203.pdf.

Coleman, R., & McCahill, M. (2011). *Surveillance and crime*. London: Sage.

Commission for the Prevention of Corruption. (2013). About supervizor. https://www.kpk-rs.si/en/project-transparency/supervizor-73.

Deleuze, G. (1995). Postscript to control societies. In G. Deleuze (Ed.), *Negotiations: 1972–1990*. New York: Columbia University Press.

Dredge, S. (15 Jan 2015). Developers plan interceptor drone to hunt and crash other drones. *The Guardian*. http://www.theguardian.com/technology/2015/jan/15/drone-hunt-crash-rapere.

Droneologist. (27 March 2014). The future is here—with internal body drones. *The Droneologist*. http://www.thedroneologist.com/2014/03/the-future-is-here-with-internal-body-drones/.

Electronic Frontier Foundation. (2015). *In the electronic frontier foundation's annual "Who's got your back"*. https://www.eff.org/who-has-your-back-government-data-requests-2015.

Fernandez, L. A., & Huey, L. (2009). Is resistance futile? Thoughts on resisting surveillance. *Surveillance & Society, 6*(3), 198–202.

Finn, R. L., Wright, D., Jacques, L., & De Hert, P. (2014). *Study on privacy, data protection and ethical risks in civil remotely piloted aircraft systems operations. Final report*. Luxembourg: Publications Office of the European Union.

Foucault, M. (1975). *Surveiller et punir: naissance de la prison*. Paris: Gallimard.

Graffiti, D. (2015). *Tumblr*. http://dronegraffiti.tumblr.com.

Green, S. (1999). A plague on the panopticon: Surveillance and power in the global information society. *Information, Communication and Society, 2*, 26–44.

Haggerty, K. D., & Ericson, R. V. (2000). The surveillant assemblage. *British Journal of Sociology, 51*(4), 605–622. http://bigo.zgeist.org/students/readings/IPS2011/8/Haggerty%20ericson%20 2000.pdf.

Hayes, B., Jones, C., & Töpfer, E. (2014). *Eurodrones Inc*. http://www.statewatch.org/news/2014/feb/sw-tni-eurodrones-inc-feb-2014.pdf.

Holland, M. A. (30 April 2015). The age of drone vandalism begins with an epic NYC tag. *Wired*. http://www.wired.com/2015/04/age-drone-vandalism-begins-epic-nyc-tag/.

International Working Group on Data Protection in Telecommunications. (2013). *Working paper on privacy and aerial surveillance*. http://www.datenschutz-berlin.de/attachments/996/675.47.25.pdf?1385047665.

Jones, C. (2014). *Back from the battlefield. Domestic drones in the UK*. https://dronewarsuk.files.wordpress.com/2014/05/domestic-drones-in-the-uk-final.pdf.
K. S. (15 Oct 2015). Incident na tekmi Srbija-Albanija politična vroča tema: "Manjkala samo še bomba". *RTV SLO*. http://www.rtvslo.si/svet/incident-na-tekmi-srbija-albanija-politicna-vroca-tema-manjkala-samo-se-bomba/348805.
Kersey, B. (1 Dec 2012). Parent hacks quadcopter drone to follow his son to the school bus stop. *The Verge*. http://www.theverge.com/2012/12/1/3714654/quadcopter-drone-tracks-son.
Knibbs, K. (10 March 2014). Watch an intern get zapped by a taser drone at SXSW. *The Daily Dot*. http://www.dailydot.com/technology/drone-stuns-intern-taser-sxsw/.
Koskela, Hille (2006). The other side of surveillance. Webcams, power and agency. In D. Lyon (Ed.), *Theorizing Surveillance: The Panopticon and Beyond* (pp. 163–181). Cullompton: Willan.
Koskela, H. (2009). Hijacking surveillance? The new moral landscapes of amateur photographing. In: K. Franko Aas, H. Oppen Gundhus, H. Mork Lomell (Eds.), *Technologies of In Security: the surveillance of everyday life* (pp. 147–167). Abingdon, New York: Routledge-Cavendish.
Križnar, T. (2 Aug 2014). Z droni nad skrito vojno v Sudanu. *Delo*. http://www.delo.si/zgodbe/sobotnapriloga/z-droni-nad-skrito-vojno-v-sudanu.html.
Kurgan, L. (2013). *Close up at a distance: Mapping, technology, and politics*. New York: Zone books.
Larson, S. (15 March 2015). A fire-spewing drone lights up SXSW. *The Daily Dot*. http://www.dailydot.com/technology/fire-drone-sxsw/.
Leman-Langlois, S. (2008). The local impact of police videosurveillance on the social construction of security. In S. Leman-Langloisa (Ed.), *Technocrime. Technology, crime and social control* (pp. 27–45). Portland: Willan Publishing. (Cullompton [England]).
Lessig, L. (2004). *Free culture: How big media uses technology and the law to lock down culture and control creativity*. New York: Penguin Press.
Lyon, D. (2003). *Surveillance as social sorting: Privacy, risk, and digital discrimination*. New York: Routledge.
Lyon, D. (2007). *Surveillance studies: An overview*. Cambridge: Polity.
Mann, S. (2002). *Sousveillance, not just surveillance, in response to terrorism*. http://wearcam.org/metalandflesh.htm.
Mann, S., Nolan, J., & Wellman, B. (2003). Sousveillance: Inventing and using wearable computing devices for data collection in surveillance environments. *Surveillance & Society, 1*(3), 331–355. http://www.surveillance-and-society.org/articles1(3)/sousveillance.pdf.
Martin, A. K., Van Brakel, R. E., & Bernhard, D. J. (2009). Understanding resistance to digital surveillance: Towards a multi-disciplinary, multi-actor framework. *Surveillance & Society, 6*(3), 213–232.
Marx, G. T. (1988). *Undercover: Police surveillance in America*. Berkeley: University of California Press.
Marx, G. T. (1996) Electric eye in the sky: some reflections on the new surveillance and popular culture. In D. Lyon and E. Zureik (Eds.), *Computers, Surveillance and Privacy*. Minneapolis/London: University of Minnesota Press, 193–233.
Marx, G. T. (2002). What is new about the "new surveillance"? Classifying for change and continuity. *Surveillance & Society, 1*(1), 9–29. http://www.surveillance-and-society.org/articles1/whatsnew.pdf.
Marx, G. T. (2003). A tack in the shoe: Neutralizing and resisting the new surveillance. *Journal of Social Issues, 59*(2), 369–390. http://web.mit.edu/gtmarx/www/tack.html.
Monahan, T. (2006). Counter-surveillance as political intervention? *Social Semiotics, 16*(4), 515–534.
Monahan, T. Phillips, D. J. & Murakami Wood, D. (2010). Editorial: Surveillance and empowerment. *Surveillance & Society, 8*(2), 106–112.
Mortimer, G. (23 Jan 2012). Dallas meat packing plant investigated after drone images reveal pollution. *sUAS News*. http://www.suasnews.com/2012/01/11389/dallas-meat-packing-plant-investigated-after-drone-images-reveal-pollution/.

Newman, L. H. (29 June 2015). Maybe don't shoot down your neighbor's drone when it's not even over your property. *Slate*. http://www.slate.com/blogs/future_tense/2015/06/29/court_rules_that_neighbor_has_to_pay_for_drone_he_shot_down.html.

Nielsen, N. (23 July 2015). Civil drone almost collides with Lufthansa plane. *Euobserver*. https://euobserver.com/justice/129748.

Pečar, J. (1991). *Neformalno nadzorstvo: Kriminološki in sociološki pogledi*. Radovljica: Didakta.

Poulsen, K. (24 July 2015). Drones and spyware: The bizarre tale of a brutal kidnapping. *Wired*. http://www.wired.com/2015/07/mare-island/.

Raby, R. (2005). What is resistance? *Journal of Youth Studies, 8*(2), 151–171.

Redd, N. (30 June 2015). The intersection of drones, photos and art. [TV show]. http://live.huffingtonpost.com/r/segment/drones-and-the-future-of-photography/558c621e2b8c2a5a2c00040e.

Riley, C. (10 June 2013). Booz Allen Hamilton in spotlight over leak. *CNN Money*. http://money.cnn.com/2013/06/10/news/booz-allen-hamilton-leak/index.html.

Rule, J. B. (2009). The limits of privacy protection. In: B. J. Goold & D. Neyland (Eds.), *New Directions in Surveillance and Privacy* (pp. 3–17). Cullompton: Willan.

Rule, J., McAdam, D., Stearns, L., & Uglow, D. (1980). *The politics of privacy: Planning for personal data systems is powerful technologies*. New York: Elsevier.

Sewell, G., & Barker, J. R. (2001). Neither good, nor bad, but dangerous: Surveillance as ethical paradox. *Ethics and Information Technology, 3,* 183–196.

Sharkey, N., & Knuckey, S. (21 Dec 2011). Occupy wall street's 'occucopter'—who's watching whom? *The Guardian*. http://www.theguardian.com/commentisfree/cifamerica/2011/dec/21/occupy-wall-street-occucopter-tim-pool.

Sheller, M., & Urry, J. (2003). Mobile transformations of "public" and "private" life. *Theory, Culture and Society, 20*(3), 107–125. http://www.extrememediastudies.org/extreme_media/2_mobiles/pdf/capsules_shellerurry.pdf.

Shepherd, S. (24 Dec 2011). *Sea shepherd intercepts the Japanese whaling fleet with drones*. http://www.seashepherd.org/news-and-media/2011/12/24/sea-shepherd-intercepts-the-japanese-whaling-fleet-with-drones-1299.

Solove, D. J. (2007). 'I've got nothing to hide' and other misunderstandings of privacy. *San Diego Law Review, 44,* 745; GWU Law School Public Law Research Paper No. 289. SSRN: http://ssrn.com/abstract=998565.

STA. (25 June 2015). Na Bizeljskem 59-letnik streljal na daljinsko voden letalnik. *STA*. https://www.sta.si/2150015/na-bizeljskem-59-letnik-streljal-na-daljinsko-voden-letalnikn.

Stanley, J., & Crump, C. (2011). *Protecting privacy from aerial surveillance: Recommendations for government use of drone aircraft*. https://www.aclu.org/files/assets/protectingprivacyfromaerialsurveillance.pdf.

Stern, M. J. (23 July 2015). Did Sandra Bland have a right to film her arrest? *Slate*. http://www.slate.com/articles/news_and_politics/jurisprudence/2015/07/do_you_have_a_right_to_film_the_police_sandra_bland_was_arrested_and_told.html.

Storm, D. (18 Aug 2010). War-flying with a wi-fi-sniffing drone. *Computer World*. http://www.computerworld.com/article/2468870/endpoint-security/war-flying-with-a-wi-fi-sniffing-drone.html.

Tinic, S. (2006). (En)visioning the televisual audience: Revisiting questions of power in the age of interactive television. In: K. D. Haggerty, R. V. Ericson (Eds.), *The New Politics of Surveillance and Visibility* (pp. 308–326). Toronto: University of Toronto Press.

Unknown. (2015). Flying gun [Video file]. https://www.youtube.com/watch?v=xqHrTtvFFIs.

Valencia, N., & Martinez, M. (23 Jan 2015). Drone carrying drugs crashes south of U.S. border. *CNN*. http://edition.cnn.com/2015/01/22/world/drug-drone-crashes-us-mexico-border/.

Vidrih, P. (28 Dec 2014). Z letečimi očmi in ušesi boga zaščitimo korenine človeštva. *Primorske Novice*. http://www.primorske.si/Plus/Sobota/-Z-letecimi-ocmi-in-usesi-boga-zascitimo-korenine-.

Winner, L. (1986). *The whale and the reactor: A search for limits in an age of high technology*. Chicago: University of Chicago Press.

Yar, M. (2003). Panoptic power and the pathologisation of vision: Critical reflections on the foucauldian thesis. *Surveillance and Society, 1*(3), 254–271.
Zetter, K. (10 June 2015). Kaspersky finds new nation-state attack—in its own network. *Wired.* http://www.wired.com/2015/06/kaspersky-finds-new-nation-state-attack-network/.

Aleš Završnik is a Doctor of Law (LLD), assistant professor and a researcher at the Institute of Criminology at the Faculty of Law in Ljubljana. He was a postdoctoral Yggdrasil fellow at the Department of Criminology and Sociology of Law at the University of Oslo (2012) and was a postdoctoral fellow at the Max-Planck-Institute für ausländisches und internationals Strafrecht, Freiburg i. Br. (2009) and a fellow of the World Federation of Scientists, Geneva (2008). He collaborated in the European Cooperation in Science and Technology (COST) Action "Living in Surveillance Societies" and the COST Action "Cyberbullying". Aleš Završnik edited the book *Crime and Technology: How Computers Transform Surveillance and Privacy, Crime and Crime Control?* (2010) and published the book *Homo Criminalis: Images of a Criminal in High-Tech Risk Society* (2009). He has researched and published on theoretical criminology, criminal law and "law and technology", that is, cybercrime, cyber law and surveillance. He is a member of the editorial board of the *Masaryk University Journal of Law and Technology*.

Index

A. Završnik (ed.), *Drones and Unmanned Aerial Systems*,
DOI 10.1007/978-3-319-23760-2

Sudan, 244, 256, 258

MIX
Papier aus verantwortungsvollen Quellen
Paper from responsible sources
FSC® C105338

If you have any concerns about our products,
you can contact us on
ProductSafety@springernature.com

In case Publisher is established outside the EU,
the EU authorized representative is:
Springer Nature Customer Service Center GmbH
Europaplatz 3, 69115 Heidelberg, Germany

Printed by Libri Plureos GmbH
in Hamburg, Germany